Pitman Research Notes in Mathematics Series

Submission of proposals for consideration
Suggestions for publication, in the form of outlines and representative samples, are invited by the Editorial Board for assessment. Intending authors should approach one of the main editors or another member of the Editorial Board, citing the relevant AMS subject classifications. Alternatively, outlines may be sent directly to the publisher's offices. Refereeing is by members of the board and other mathematical authorities in the topic concerned, throughout the world.

Preparation of accepted manuscripts
On acceptance of a proposal, the publisher will supply full instructions for the preparation of manuscripts in a form suitable for direct photo-lithographic reproduction. Specially printed grid sheets can be provided and a contribution is offered by the publisher towards the cost of typing. Word processor output, subject to the publisher's approval, is also acceptable.

Illustrations should be prepared by the authors, ready for direct reproduction without further improvement. The use of hand-drawn symbols should be avoided wherever possible, in order to maintain maximum clarity of the text.

The publisher will be pleased to give any guidance necessary during the preparation of a typescript, and will be happy to answer any queries.

Important note
In order to avoid later retyping, intending authors are strongly urged not to begin final preparation of a typescript before receiving the publisher's guidelines. In this way it it hoped to preserve the uniform appearance of the series.

Longman Scientific & Technical
Longman House
Burnt Mill
Harlow, Essex, UK
(tel (0279) 426721)

Titles in this series

D F Griffiths and G A Watson (Editors)

University of Dundee

Numerical analysis 1991

Proceedings of the 14th Dundee
Conference, June 1991

Longman
Scientific &
Technical

Copublished in the United States with
John Wiley & Sons, Inc., New York

Longman Scientific & Technical
Longman Group UK Limited
Longman House, Burnt Mill, Harlow
Essex CM20 2JE, England
and Associated Companies throughout the world.

Copublished in the United States with
John Wiley & Sons Inc., 605 Third Avenue, New York, NY 10158

First published 1992

AMS Subject Classification: 65-06

ISSN 0269-3674

British Library Cataloguing in Publication Data

A catalogue record for this book is
available from the British Library

Library of Congress Cataloging-in-Publication Data

A catalogue record for this book is available

Printed and Bound in Great Britain
by Biddles Ltd, Guildford and King's Lynn

Contents

Preface

In June 1965, Professor A R Mitchell was the main organiser of a Symposium on the Solution of Differential Equations at the University of St Andrews. The event was judged a success, and was repeated in 1967. Later in 1967, Ron Mitchell came to the University of Dundee to become the first holder of the Chair of Numerical Analysis. Under his guidance, conferences on Numerical Analysis continued to be organised every second year in Dundee, and the series of biennial meetings is now the longest running such series in the world. Ron celebrated his 70th birthday on June 22nd 1991, and so it was very appropriate that the 1991 meeting should be a celebration of this. Many tributes were paid to Ron during the meeting, for in addition to his outstanding leadership of the numerical analysis group at Dundee, he has contributed greatly to numerical analysis, both at a national and an international level, and he has always been a source of great encouragement to young numerical analysts. He generates enormous respect for his technical skills, but also for his human qualities, and no picture of Ron would be complete without a mention of his personal magnetism and his marvellous sense of humour.

For the record, the 14th Dundee Biennial Conference on Numerical Analysis was held at the University of Dundee during the 4 days June 25-28, 1991. The meeting was attended by around 210 people from over 25 countries, with about half the participants coming from outside the UK. The technical programme consisted of 16 invited talks and 100 submitted talks, the latter being presented in 4 parallel sessions. The Conference Committee decided to invite an eminent numerical analyst to present a lecture in Ron's honour, and Professor G H Golub accepted an invitation to present the first A R Mitchell lecture. In addition to his outstanding contributions to numerical analysis, Gene Golub also has a long connection with Ron Mitchell and Dundee, and his talk "Matrices, moments and quadrature" opened the conference programme. This proceedings volume contains full versions of all the other invited papers; the titles of all contributed talks given at the meeting, together with the names and addresses of the presenters, are also listed here.

We would like to take this opportunity of thanking all the speakers, chairman and participants for their contributions. In particular we would like to record our appreciation of the excellent after dinner speech given by Professor J D Lambert. Jack Lambert was one of Ron's first PhD students, and has been a colleague of Ron in Dundee since 1967. His speech drew on many personal recollections and anecdotes (most of them, in characteristic fashion, vigorously denied by Ron as ever having happened). Jack has now retired from the University of Dundee, and we would like to take this opportunity to wish him well, and to thank him, in particular, for his years of service to numerical analysis.

We are grateful for help received from members of the Department of Mathematics and Computer Science before and during the Conference, in particular the secretarial assistance of Mrs S Fox. The Conference is also indebted to the University of Dundee for making available various University facilities throughout the week, and for the provision of a Reception for the participants in University House. We also gratefully acknowledge the financial support of the European Research Office of the US Army. Finally, we would like to thank the publishers of these Proceedings, Longman Scientific and Technical, for their co–operation during the pre–publication process.

D. F. Griffiths

G. A. Watson

November 1991

Invited Speakers

Å. Björck :   Department of Mathematics, Linköping University, S–581 83 Linköping, Sweden.

J. W. Demmel: Computer Science Division and Mathematics Department, University of California, Berkeley, CA 94720, USA.

J. J. Dongarra: Computer Science Department, University of Tennessee, Knoxville, TN 37996–1301, USA.

N. I. M. Gould :   Rutherford Appleton Laboratory, Numerical Analysis Group, Central Computing Department, Atlas Centre, Didcot OX11 0QX.

G. H. Golub: Department of Computer Science, Stanford University, Stanford CA 94305, USA.

J. A. Gregory: Department of Mathematics and Statistics, Brunel University, Uxbridge, Middlesex UB8 3PH.

A. Iserles: DAMTP, University of Cambridge, Silver Street, Cambridge CB3 9EW.

J. Lorenz: Department of Mathematics and Statistics, University of New Mexico, Albuquerque, NM 87131, USA.

K. W. Morton: Computing Laboratory, University of Oxford, 11 Keble Road, Oxford, OX1 3QD.

M. J. D. Powell: DAMTP, University of Cambridge, Silver Street, Cambridge, CB3 9EW.

J. M. Sanz–Serna: Departmento Matemática Aplicada y Computación, Universidad de Valladolid, Facultad de Ciencias, Valladolid, Spain.

H. Schwetlick: Sektion Mathematik, Martin Luther University, Postfach, D–(O)–Halle 4010, Germany.

I. H. Sloan: School of Mathematics, University of New South Wales, PO Box 1, Kensington, New South Wales 2033, Australia.

V. Thomée: Department of Mathematics, Chalmers University of Technology, S–402 20 Göteborg 5, Sweden.

L. N. Trefethen: Department of Computer Science, Cornell University, 405 Upson Hall, Ithaca, NY 14853, USA.

J. G. Verwer: Centre for Mathematics and Computer Science, Kruislaan 413, PO Box 4079, 1009 SJ Amsterdam, The Netherlands.

Å BJÖRCK
Pivoting and stability in the augmented system method

1   Introduction

Let $A \in \mathbf{R}^{m \times n}$, where $m \geq n$ and consider the following two problems

$$\min \|b - Ax\|_2^2 + 2c^T x. \tag{1.1}$$

$$\min \|b - y\|_2, \qquad A^T y = c, \tag{1.2}$$

The conditions for their solution is given by the symmetric indefinite (if $A \neq 0$) linear system involving A

$$\begin{pmatrix} I & A \\ A^T & 0 \end{pmatrix} \begin{pmatrix} y \\ x \end{pmatrix} = \begin{pmatrix} b \\ c \end{pmatrix}. \tag{1.3}$$

These equations are obtained by differentiating (1.1) to give $A^T(b - Ax) = c$, and setting y to be the least squares residual $y = r = b - Ax$. For (1.2) they are obtained by differentiating the Lagrangian $\|b - y\|_2^2 + 2x^T(A^T y - c)$, and equating to zero. Here x is the vector of Lagrange multipliers.

We call 1.3 the augmented system formulation of the two problems above. It covers two important special cases. Setting $c = 0$ in (1.1) gives the linear least squares problem. Setting $b = 0$ in (1.2) gives the problem of finding the minimum 2-norm solution of a linear underdetermined system, which occurs for example as a subproblem in optimization algorithms. The augmented system for the linear least squares problem was originally proposed by Bartels et al. [3].

The augmented system can be solved using symmetric Gaussian elimination giving an LDL^T factorization, where D is block-diagonal with 1×1 and 2×2 pivots. We remark that this approach is of interest mainly for *sparse problem*. If A is dense the diagonal blocks in the system matrix will in general fill in after a few elimination steps, and the method is not competitive with orthogonalization methods, see [8]. For sparse linear least squares problems the augmented system approach was first considered by Hachtel [20] and [21]. More recently Arioli et al [2] have developed a successful algorithm based on the Harwell program MA27. For applications to power system networks see Alvarado [1]. For an account of its use in optimization see [17].

The pivoting strategy used for general dense symmetric indefinite systems is that of Bunch and Kaufman [9], see also Fletcher [16]. These pivoting strategies have been extended to the sparse case by Duff et al. [13] for use in a multifrontal algorithm MA27 by Duff and Reid [15]. In [12] the pivoting strategy of MA27 has been modified to handle augmented systems in a more efficient way.

In section 2 we describe the Bunch-Kaufman pivoting strategy and its dependence on the scaling of the augmented system. In section 3 we derive a new formula for the optimal value of the scaling parameter based on minimizing an upper bound for the round-off error in the solution. The extension of these result to weighted least squares problems is discussed in section 4. In section 5 some relevant results on the inertia for augmented systems are reviewed, and in section 6 numerical results are presented which support the recommendations given in section 3 and 4.

2 Pivoting Strategy

The following partial pivoting strategy for computing the LDL^T factorization of a symmetric indefinite matrix PMP^T of order p, where P is a permutation matrix, is due to Bunch and Kaufman [9]. For simplicity of notations we restrict our attention to the first stage of the elimination. All later stages proceed similarly. First determine the off-diagonal element of largest magnitude in the first row $\lambda = |m_{1r}| = \max_{2 \leq j \leq p} |m_{1j}|$. If

$$|m_{11}| \geq \rho|m_{1r}|, \qquad \rho = (\sqrt{17}+1)/8 \approx 0.6404, \tag{2.1}$$

then m_{11} is used as a pivot. If not, determine the largest off-diagonal element in row r, $\sigma = \max_{1 \leq i \leq p} |m_{ri}|$, $i \neq r$. If $|m_{11}| \geq \rho\lambda^2/\sigma$, then again take m_{11} as pivot, else if $|m_{rr}| \geq \rho\sigma$, take m_{rr} as pivot. Otherwise take the 2×2 pivot

$$\begin{pmatrix} m_{11} & m_{1r} \\ m_{1r} & m_{rr} \end{pmatrix}.$$

Note that at most 2 rows need to be searched in each step, and at most p^2 comparisons are needed in all. Further it can be shown that if M is positive definite only 1×1 pivots will be chosen.

With the choice of ρ given in (2.1) the element growth is bounded by the factor

$$g_p = \frac{\max_{i,j,k} |m_{ij}^{(k)}|}{\max_{i,j} |m_{ij}|} \leq (1 + 1/\rho)^{p-1} < (2.57)^{p-1}, \tag{2.2}$$

where p is the dimension of the system. No practical example is known where significant element growth occur at every step. This bound can be compared to the bound 2^{p-1}, which holds for Gaussian elimination with partial pivoting.

A standard backward error analysis shows, see [19, Sec. 3.3.1] or [22] that the computed factors satisfy $\bar{L}\bar{D}\bar{L}^T = P(M + E)P^T$ where, using (2.2)

$$|e_{ij}| \leq 2(p-1)u \max_{i,j,k} |m_{ij}^{(k)}| \leq 2(p-1)ug_p \max_{i,j} |m_{ij}|. \tag{2.3}$$

Taking also rounding errors in the solution of $\bar{L}\bar{D}\bar{L}^T(Pz) = Pb$ into account it can be shown that the computed solution $\bar{z}$ satisfies $(M + E)\bar{z} = d$, where the norm of E satisfies a bound of the form

$$\|E\|_2 \leq \epsilon\|M\|_2, \qquad \epsilon = c(p)g_pu, \tag{2.4}$$

where $c(p)$ is a low degree polynomial. Hence, the method of Bunch-Kaufman is backward stable for *general* symmetric systems in the usual normwise sense, and the forward error in $\bar{z}$ can be bounded by

$$\|\bar{z} - z\|_2 \leq \frac{\epsilon\kappa_2(M)}{1 - \epsilon\kappa_2(M)}\|z\|_2. \tag{2.5}$$

Unfortunately the Bunch-Kaufman method is not strongly backward stable for the problems (1.1) and (1.2), since the perturbations introduced by round-off do not respect the structure of the augmented system (1.3). This can be made clear by introducing the scaled vector $\alpha^{-1}y$ the augmented system can be written

$$\begin{pmatrix} \alpha I & A \\ A^T & 0 \end{pmatrix} \begin{pmatrix} \alpha^{-1}y \\ x \end{pmatrix} = \begin{pmatrix} b \\ \alpha^{-1}c \end{pmatrix} \quad \Leftrightarrow \quad M_\alpha z_\alpha = d_\alpha. \tag{2.6}$$

Alternatively, a symmetric reordering of rows and columns in (2.6) gives

$$\begin{pmatrix} 0 & A^T \\ A & \alpha I \end{pmatrix} \begin{pmatrix} x \\ \alpha^{-1}y \end{pmatrix} = \begin{pmatrix} \alpha^{-1}c \\ b \end{pmatrix}. \tag{2.7}$$

In the following we assume that

$$0 \leq \alpha \leq \|A\|_2 = \sigma_1(A).$$

Then the element of maximum magnitude in M_α satisfies

$$\max_{i,j} |m_{ij}| \leq \sigma_1(A). \tag{2.8}$$

Using any of the pivoting strategies referred to above the choice of pivots will depend on the value of α, and also on the choice of formulation (2.6) or (2.7). However, it is important to note that for *a fixed pivot sequence* the numerical solution of (1.3) by Gaussian elimination is *independent* of the scaling parameter α, and α only affects the accuracy through the choice of pivots. (We assume that the scaling itself is done exactly.)

For sufficiently large values of α the Bunch-Kaufman strategy will choose the first m pivots from the diagonal $(1,1)$-block. Let us put $\alpha = 1$ and assume these pivots are chosen. The resulting reduced system equals the normal equations

$$- A^T A x = c - A^T b, \tag{2.9}$$

which as is well-known does not give a backward stable method. Using smaller values of α will introduce 2×2 pivots of the form

$$\begin{pmatrix} \alpha & a_{1r} \\ a_{1r} & 0 \end{pmatrix},$$

called *tile pivots* in [17], which may improve the stability. This raises the question of the *optimal choice of α for stability*.

It was shown by Golub [18] that the eigenvalues λ of M_α can be expressed in terms of the singular values σ_i, $i = 1, \ldots, n$ of A, see also Björck [4]. If $M_\alpha z = \lambda z$, $z = (x, y)^T \neq 0$, then

$$\alpha x + A y = \lambda x, \qquad A^T x = \lambda y,$$

or eliminating x, $\alpha \lambda y + A^T A y = \lambda^2 y$. Hence, if $y \neq 0$ then y is an eigenvector and $(\lambda^2 - \alpha\lambda)$ an eigenvalue of $A^T A$. On the other hand $y = 0$ implies that

$$A^T x = 0, \qquad \alpha x = \lambda x, \qquad x \neq 0.$$

3

It follows that the $m + n$ eigenvalues of M_α are

$$\lambda = \begin{cases} \frac{\alpha}{2} \pm \sqrt{\frac{\alpha^2}{4} + \sigma_i^2}, & i = 1, 2, \ldots, n; \\ \alpha. \end{cases} \tag{2.10}$$

If rank$(A) = r \leq n$, then the eigenvalue α has multiplicity $(m - r)$, and 0 is an eigenvalue of multiplicity $(n - r)$.

From (2.10) it is easily deduced that if $\sigma_n > 0$ then

$$\min_\alpha \kappa_2(M_\alpha) = \frac{1}{2} + \sqrt{\frac{1}{4} + 2(\frac{\sigma_1}{\sigma_n})^2} \approx \sqrt{2}\kappa_2(A) \ ,$$

is attained for

$$\alpha = \tilde{\alpha} = \frac{1}{\sqrt{2}}\sigma_n(A). \tag{2.11}$$

Because of the above result $\tilde{\alpha}$ (or σ_n) has been suggested as the optimal scaling factor in the augmented system method. Minimizing $\kappa_2(M_\alpha)$ will minimize the forward bound for the error in z_α, using (2.5)

$$\|\bar{z}_\alpha - z_\alpha\|_2 \leq \frac{\epsilon\kappa(M_\alpha)}{1 - \epsilon\kappa(M_\alpha)}\|z_\alpha\|_2, \qquad z_\alpha = \begin{pmatrix} \alpha^{-1}y \\ x \end{pmatrix}.$$

Hence α influences the norm in which the error is measured. In the next section we will slightly refine the error analysis and then instead minimize a bound for the error in $\bar{x}$.

3   Refined Error Analysis

By a standard analysis rounding errors in symmetric GE *with pivoting* are equivalent to a perturbation

$$\delta M_\alpha = \begin{pmatrix} E_1 & E_2 \\ E_2^T & E_4 \end{pmatrix}, \tag{3.1}$$

where using (2.8)

$$\|E_i\|_2 \leq c(m+n)g_{m+n}u \max\{\alpha, \|A\|_2\} \leq c(m+n)g_{m+n}u\sigma_1, \tag{3.2}$$

g_{m+n} is the growth factor, and u the machine precision.

If we perturb the matrix M_α by δM_α the corresponding perturbation δz_α satisfies the basic identity

$$\delta z_\alpha = -(I + M_\alpha^{-1}\delta M_\alpha)^{-1}M_\alpha^{-1}\delta M_\alpha z_\alpha.$$

Provided that $\rho(|M^{-1}||\delta M|) \leq \eta < 1$, it holds for all α that

$$|\delta z_\alpha| \leq |M_\alpha^{-1}||\delta M_\alpha||z_\alpha| + O(u^2). \tag{3.3}$$

where the inequalities are to be interpreted component-wise, see [6].

4

It is easily verified that the inverse of M_α is

$$M_\alpha^{-1} = \begin{pmatrix} \alpha^{-1} P_{\mathcal{N}(A^T)} & (A^\dagger)^T \\ A^\dagger & -\alpha(A^T A)^{-1} \end{pmatrix}, \tag{3.4}$$

where $P_{\mathcal{N}(A^T)} = I - AA^\dagger$ is the orthogonal projector onto the nullspace of A^T. From this we get the perturbation bound

$$\begin{pmatrix} \alpha^{-1} |\delta y| \\ |\delta x| \end{pmatrix} \leq \begin{pmatrix} \alpha^{-1} |P_{\mathcal{N}(A^T)}| & |A^\dagger|^T \\ |A^\dagger| & \alpha|(A^T A)^{-1}| \end{pmatrix} \begin{pmatrix} |E_1| & |E_2| \\ |E_2|^T & |E_4| \end{pmatrix} \begin{pmatrix} \alpha^{-1} |y| \\ |x| \end{pmatrix}. \tag{3.5}$$

where quantities of order $O(u^2)$ have been neglected. Taking norms, using (3.2) and

$$\|A^\dagger\|_2 + \alpha\|(A^T A)^{-1}\|_2 = \sigma_n^{-1}(1 + \alpha\sigma_n^{-1}).$$

we obtain the upper bounds for the round-off errors in the computed solution

$$\begin{pmatrix} \|\bar{y} - y\|_2 \\ \|\bar{x} - x\|_2 \end{pmatrix} \leq c(m + n)g_{m+n} u f(\alpha) \begin{pmatrix} \sigma_1(A) \\ \kappa_2(A) \end{pmatrix}, \tag{3.6}$$

where

$$f(\alpha) = (1 + \frac{\alpha}{\sigma_n})(\frac{1}{\alpha}\|y\|_2 + \|x\|_2) = \frac{1}{\alpha}\|y\|_2 + \frac{\alpha}{\sigma_n}\|x\|_2 + \frac{1}{\sigma_n}\|y\|_2 + \|x\|_2. \tag{3.7}$$

$f(\alpha)$ is minimized when the two first terms are equal, i.e., if $x \neq 0$

$$\alpha = \alpha_{opt} = \left(\frac{\sigma_n\|y\|_2}{\|x\|_2}\right)^{1/2}, \tag{3.8}$$

and

$$f_{min} = \left(1 + \frac{\alpha_{opt}}{\sigma_n}\right)^2 \|x\|_2 = \left(1 + \frac{\sigma_n}{\alpha_{opt}}\right)^2 \sigma_n^{-1}\|y\|_2. \tag{3.9}$$

Taking $\alpha = \sigma_n$ in (3.7) we get

$$f(\sigma_n) = 2\left(\frac{1}{\sigma_n}\|y\|_2 + \|x\|_2\right) \leq 2f_{min}. \tag{3.10}$$

Hence, although we may have $\alpha_{opt} \gg \sigma_n$, taking $\alpha = \sigma_n$ will at most double the upper bound. In particular when $y = 0$, i.e. $Ax = b$ (consistent), $\alpha_{opt} = 0$ and $f(\sigma_n) = 2\|x\|_2 = 2f_{min}$.

If $\alpha_{opt} \geq \sigma_1$ then it follows from (3.8) that $\sigma_1\|x\|_2 \leq (\sigma_n/\sigma_1)\|y\|_2$, and using (3.7)we get

$$f(\sigma_1) \leq \left(1 + \frac{\sigma_n}{\sigma_1}\right)^2 \frac{1}{\sigma_n}\|y\|_2 \leq 2^2 f_{min}.$$

Hence, taking $\alpha = \sigma_1$ can only increase the bound by a factor of 2^2. (Note that for the error analysis it was assumed that $\alpha \leq \sigma_1$, and so taking $\alpha > \sigma_1$ may increase the bound (3.6).)

To summarize, the error analysis in this section suggests the choice

$$\alpha = \begin{cases} \alpha_{opt}, & \sigma_n \leq \alpha_{opt} \leq \sigma_1; \\ \sigma_1, & \text{if } \sigma_1 < \alpha_{opt}; \\ \sigma_n, & \text{if } \alpha_{opt} < \sigma_n. \end{cases} \tag{3.11}$$

For sparse least squares problems taking $\alpha = \sigma_n$ will often give rise to unacceptable fill-in, see [2]. Therefore our result that often $\alpha_{opt} \gg \sigma_n$ has important implications. Note that α_{opt} depends on σ_n and $\|y\|_2$, both quantities which in general are not known *a priori*. Since M_α is symmetric its singular values are the absolute values of its eigenvalues. Assuming that rank $(A) = n$, it follows from (2.10) that

$$\sigma_n^2(A) = \sigma_{m+n}(M_\alpha)(\sigma_{m+n}(M_\alpha) + \alpha) \tag{3.12}$$

Using this relation and a condition estimator to estimate $\sigma_{m+n}(M_\alpha)$, it is possible to obtain an estimate of α_{opt} *a posteriori*.

An *acceptable-error stable* algorithm is defined to be one which gives a solution whose error size is never significantly worse than the error bound obtained from a tight perturbation analysis. Assume that we perturb the problem (1.3), such that

$$\|\delta A\|_2 \leq cu\|A\|_2.$$

Then using (3.5) with $E_1 = E_4 = 0$ and $\|E_2\|_2 = cu\|A\|_2$, we obtain tight upper bounds for the resulting perturbation in the solution

$$\begin{pmatrix} \|\delta y\|_2 \\ \|\delta x\|_2 \end{pmatrix} \leq cu(\|x\|_2 + \frac{1}{\sigma_n}\|y\|_2) \begin{pmatrix} \sigma_1(A) \\ \kappa_2(A) \end{pmatrix}. \tag{3.13}$$

Using

$$f_{min} \leq f(\sigma_n) = 2\left(\|x\|_2 + \frac{1}{\sigma_n}\|y\|_2\right). \tag{3.14}$$

in (3.6) shows that the error bounds obtained with α chosen by (3.11) are similar to those for a backward stable algorithm. Hence with this choice of α the augmented system method is acceptable-error stable. The analysis also shows that, as remarked by Duff and Reid [14], automatic scaling methods which aim to make all non-zero elements have modulus near unity should not be used on the whole of the augmented matrix.

4   Weighted Least Squares Problems

We now consider the more general problem

$$\begin{pmatrix} \alpha W & A \\ A^T & 0 \end{pmatrix} \begin{pmatrix} \alpha^{-1}y \\ x \end{pmatrix} = \begin{pmatrix} b \\ \alpha^{-1}c \end{pmatrix}, \tag{4.1}$$

where W is symmetric and positive semi-definite matrix. The system matrix in (4.1) is nonsingular if and only if rank$(A) = n$ and rank$(W \quad A) = m$. Then the augmented system gives the solution to the two problems

6

1. *Generalized linear least squares problem* (GLLS)

$$\min_x (Ax - b)^T W^\dagger (Ax - b) + 2c^T x, \qquad b - Ax \perp \mathcal{N}(W). \tag{4.2}$$

If W is positive definite then $y = W^{-1}(b - Ax)$, and W is the covariance matrix of the error vector.

2. *Equality constrained quadratic optimization* (ECQO)

$$\min_y -2b^T y + y^T W y, \qquad A^T y = c.$$

Now suppose that we can factor W into $W = BB^T$. When W is positive definite B can be the Cholesky factor of W. If B is nonsingular, then (4.1) is equivalent to the *weighted* system

$$\begin{pmatrix} \alpha I & B^{-1}A \\ (B^{-1}A)^T & 0 \end{pmatrix} \begin{pmatrix} \alpha^{-1}B^T y \\ x \end{pmatrix} = \begin{pmatrix} B^{-1}b \\ \alpha^{-1}c \end{pmatrix},$$

which is the standard case for the matrix $B^{-1}A$ and vector $B^{-1}b$.

In particular, when the components of the error vector are *uncorrelated*, then W is *diagonal* and if $W > 0$ we can write $W = D^{-2}$,, where

$$B^{-1} = D = W^{-1/2} = \operatorname{diag}(\mu_1, \ldots, \mu_m). \tag{4.3}$$

If B is ill-conditioned, then $\kappa_2(B^{-1}A)$ will be large even when the problem is well-conditioned. Therefore using the weighted problem approach cannot be recommended in general. The system (4.1). also has the advantage that the matrix A can be assumed to be *row equilibrated*, i.e.

$$\max_{1 \le j \le n} |a_{ij}| = 1, \qquad i = 1, \ldots, m,$$

since *any row scaling of A can be included in W*. Note that a row scaling will influence the choice of pivots when the Bunch-Kaufman method is applied. With this more general formulation (4.1) linear constraints can be treated simply by putting $e_i^T W = 0$.

In order to determine an optimal value of α in (4.1), we consider the inverse of the system matrix in (4.1). This can be written in the form

$$M_\alpha^{-1} = \begin{pmatrix} \alpha^{-1}P & (A^g)^T \\ A^g & -\alpha A^g W (A^g)^T \end{pmatrix}, \tag{4.4}$$

where $A^g A = I_n$, i.e., A^g is a generalized inverse of A. The optimal value of α should now be determined by minimizing an error bound based on using (4.4). Unfortunately it is in general not possible to find a *simple* representation of the matrix P, see Wedin [25] who gives an extensive analysis of the structure of the inverse (4.4).

Assume now that $W = BB^T$ is positive definite. Then we can write

$$M_\alpha = \begin{pmatrix} B & 0 \\ 0 & I \end{pmatrix} \begin{pmatrix} \alpha I & B^{-1}A \\ (B^{-1}A)^T & 0 \end{pmatrix} \begin{pmatrix} B^T & 0 \\ 0 & I \end{pmatrix} \tag{4.5}$$

where all three matrices in the product are nonsingular. Using (3.4) it follows that in this case the inverse can be written

$$M_\alpha^{-1} = \begin{pmatrix} \frac{1}{\alpha} B^{-T} P_{\mathcal{N}(A^T B^{-T})} B^{-1} & B^{-T}(A^T B^{-T})^\dagger \\ (B^{-1}A)^\dagger B^{-1} & -\alpha(A^T W^{-1}A)^{-1} \end{pmatrix}. \tag{4.6}$$

Proceeding as in the standard case we obtain the optimal α, and minimizing the error bound for $\bar{x}$, we obtain the optimal α

$$\alpha_{opt} = \Big(\frac{\|(B^{-1}A)^\dagger B^{-1}\|_2 \|y\|_2}{\|(A^T W^{-1}A)^{-1}\|_2 \|x\|_2} \Big)^{1/2}. \tag{4.7}$$

Similarly, minimizing instead the error bound for $\bar{y}$

$$\alpha_{opt} = \Big(\frac{\|B^{-T} P_{\mathcal{N}(A^T B^{-T})} B^{-1}\|_2 \|y\|_2}{\|(B^{-1}A)^\dagger B^{-1}\|_2 \|x\|_2} \Big)^{1/2} \tag{4.8}$$

In general, these two values of α are not equal. It is also not clear how to estimate these from the LDL^T factorization of M_α.

5   Inertia of Augmented Systems and Pivoting

In this section we review some results on the inertia for augmented systems, which are relevant to pivoting strategies. Let B be a nonsingular leading submatrix of the block matrix

$$M = \begin{pmatrix} B & A \\ A^T & C \end{pmatrix}. \tag{5.1}$$

Then the Schur complement of B in M is

$$M/B = C - A^T B^{-1} A.$$

A very nice and complete exposition of properties of the Schur complement is given by Cottle in [11]. In particular the Schur complements ar useful in computing the inertia of a real symmetric matrix. The inertia of M is denoted by $\text{in}(M)$ and defined as the triple

$$\text{in}(M) = (\pi, \nu, \delta)$$

where π, ν, δ are the number of positive, negative and zero eigenvalues of M. Sylvester's famous law of inertia states that inertias of congruent matrices are the same. This can be used to show the *inertia formula*

$$\text{in}(M) = \text{in}(B) + \text{in}(M/B). \tag{5.2}$$

For matrices of the special form (5.1) where $C = 0$, B is arbitrary but symmetric, and A square and nonsingular it holds that $\text{in}(M) = (n, n, 0)$. This result is due to Carlson and Schneider [10]. A new proof is given by Cottle [11] based on the inertia formula and the result that for a 2×2 pivot the inertia equals $(1, 1, 0)$ since if $a_{rs} \neq 0$ we have

$$\lambda_1 \lambda_2 = \begin{vmatrix} b_{rr} & a_{rs} \\ a_{rs} & 0 \end{vmatrix} = -a_{rs}^2 < 0.$$

We now consider the case when $A \in \mathbf{R}^{m \times n}$, $m \geq n$ in (5.1). For $B = I$, we have from (2.10) that $\mathrm{in}(M_\alpha) = (m, r, n - r)$, $r = \mathrm{rank}(A)$. The normal equations (2.9), which results if the first m pivots are chosen from the $(1,1)$-block, equals the Schur complement $M/I = 0 - A^T I A = -A^T A$ which is negative semidefinite. This result also follows from the inertia formula since

$$\mathrm{in}(M/I) = (m, r, n - r) - (m, 0, 0) = (0, r, n - r).$$

Obviously the same result holds when $B = W > 0$, and diagonal.

Now partition $A = \begin{pmatrix} A_1 \\ A_2 \end{pmatrix}$ and assume that $A_1 \in \mathbf{R}^{n \times n}$ is nonsingular. After a symmetric permutation of blocks the augmented matrix (1.3) has the form

$$\begin{pmatrix} W_1 & A_1 & 0 \\ A_1^T & 0 & A_2^T \\ 0 & A_2 & W_2 \end{pmatrix} \tag{5.3}$$

After n steps using 2×2 pivots from the leading blocks the reduced matrix equals

$$M / \begin{pmatrix} W_1 & A_1 \\ A_1^T & 0 \end{pmatrix} = W_2 - \begin{pmatrix} 0 & A_2 \end{pmatrix} \begin{pmatrix} W_1 & A_1 \\ A_1^T & 0 \end{pmatrix}^{-1} \begin{pmatrix} 0 \\ A_2^T \end{pmatrix}.$$

This Schur complement is positive definite, since from the inertia formula we have $(m, n, 0) - (n, n, 0) = (m - n, 0, 0)$, and hence only 1×1 pivots will be used in the remaining steps.

For constrained problems the system matrix can be written in the form

$$M = \begin{pmatrix} 0 & A_1 & 0 \\ A_1^T & 0 & A_2^T \\ 0 & A_2 & W_2 \end{pmatrix}, \tag{5.4}$$

where $A_1 \in \mathbf{R}^{p \times n}$, $p \leq n$, and $W_2 > 0$. If rank $(A_1) = p$ then p elimination steps using 2×2 pivots from A_1 can be taken. The reduced matrix then has the form

$$\begin{pmatrix} 0 & \tilde{A}_2 \\ \tilde{A}_2^T & \tilde{W}_2 \end{pmatrix}.$$

where $\tilde{A}_2 \in \mathbf{R}^{(m-p) \times (n-p)}$. If A has full rank then the inertia for the reduced matrix equals $(m, n, 0) - (p, p, 0) = (m - p, n - p, 0)$ and hence rank $(\tilde{A}_2) = n - p$.

Consider now the case when $A \in \mathbf{R}^{n \times n}$ is square, and assume that block diagonal 2×2 pivots are chosen using elements $a_{11}, a_{22}, \ldots, a_{nn}$. Then it is easily shown that the factorization obtained is of the form

$$\begin{pmatrix} \alpha I & A \\ A^T & 0 \end{pmatrix} = \begin{pmatrix} L & G \\ 0 & U^T \end{pmatrix} \begin{pmatrix} D_1 & D \\ D^T & 0 \end{pmatrix} \begin{pmatrix} L^T & 0 \\ G^T & U \end{pmatrix}, \tag{5.5}$$

where $A = LDU$ with L and U^T unit lower triangular, D and $D_1 > 0$ are diagonal and G strictly lower triangular. Thus there is a close relation between the factorization obtained from the augmented system method when only 2×2 pivots are used, and the LU factorization of A. Multiplying together and identifying the blocks in (5.5) shows that

$$I - LD_1 L^T = LDG^T + GD^T L^T. \tag{5.6}$$

9

When $A \in \mathbf{R}^{m \times n}$, $m > n$ and the first n pivots are 2×2 the situation is similar. For example, when $m = 3$, $n = 2$, we obtain a factorization of the form $M = \mathcal{L}\mathcal{D}\mathcal{L}^T$, where

$$
\mathcal{D} = \begin{pmatrix} \alpha_1 & 0 & 0 & a_{11} & 0 \\ 0 & \alpha_2 & 0 & 0 & u_{22} \\ 0 & 0 & \alpha_3 & 0 & 0 \\ a_{11} & 0 & 0 & 0 & 0 \\ 0 & u_{22} & 0 & 0 & 0 \end{pmatrix} \quad \mathcal{L} = \begin{pmatrix} 1 & 0 & 0 & 0 & 0 \\ l_{21} & 1 & 0 & -\gamma_{21} & 0 \\ l_{31} & l_{32} & 1 & -\gamma_{31} & -\gamma_{32} \\ 0 & 0 & 0 & 1 & 0 \\ 0 & 0 & 0 & u_{12} & 1 \end{pmatrix}.
$$

Again this factorization is closely related to the unsymmetric factorization $A = LDU$. Hence, it is also related to the Peters-Wilkinson method, a sparse version of which is developed in [7].

6   Numerical Results

The following tests were performed using MATLAB on a SUN 4/60 workstation, with machine precision equal to $2.220 \cdot 10^{-16}$. Since the Bunch-Kaufman factorization [9] is not available in MATLAB it was implemented as an m-file . We used approximately the same two sets of test problems as in [5].

In the first set we take $A \in \mathbf{R}^{6 \times 5}$ to be the first five columns of the inverse Hilbert matrix of order six, scaled so that $\|A\|_2 = 1$,

$$
A = H/\|H\|_2, \qquad H = H_6^{-1} \begin{pmatrix} I_5 \\ 0 \end{pmatrix}, \qquad \sigma_n = 2.13 \cdot 10^{-7}.
$$

We take $x = (1, 1/2, 1/3, 1/4, 1/5)^T$, and $b = b_1 + \gamma b_2$, where

$$
b_1 = Ax, \qquad b_2 = (1/6, 1/7, 1/8, 1/9, 1/10, 1/11)^T.
$$

Since b_2 is orthogonal to the columns of A we then have $r = \gamma b_2$.

In Fig. 6.1 we show the errors using formulation (2.6)

$$
\epsilon_x = \|\bar{x} - x\|_2, \qquad \epsilon_r = \|\bar{r} - r\|_2,
$$

when $\gamma = 10^{-6}$ as a function of α. The shape of these error curves is roughly similar to the error bound $f(\alpha)$ in (3.7), which is also plotted. For this problem $\alpha_{opt} = 2.3 \cdot 10^{-7}$. The errors in $\bar{x}$ and $\bar{r}$ are sharply reduced when α is decreased from 1 to 10^{-4}, which is a practical optimum for this problem.

Fig. 6.2 shows ϵ_x for the same problem when $\gamma = 10^{-2} : 10^{-2} : 10^{-8}$, using formulation (2.7). The dependence of the optimal value of α on the residual norm $\|r\|_2 = \gamma$ is very obvious. Formulation (2.6) gave similar, but more irregular results, see Fig. 6.3. This is explained by the fact that (2.7) tends to introduce 2×2 pivots earlier in the LDL^T factorization.

In the second set of test problems the matrix $A \in \mathbf{R}^{21 \times 6}$ is given by

$$
A = VD, \qquad v_{ij} = (i - 1)^{j-1},
$$

where D is a diagonal scaling matrix such that the columns in A has unit norm, $\|a_{\cdot j}\|_2 = 1$, $j = 1, \ldots, 6$. This matrix arises in the fitting of a fifth degree polynomial to data, and is moderately ill-conditioned with $\sigma_n = 1.03 \cdot 10^{-3}$. The exact solution is taken to be $x = (1, 1, 1, 1, 1, 1)^T$, and $b = b_1 + \gamma b_2$, where $b_1 = Ax$, and b_2, $\|b_2\|_2 = 1$, a vector orthogonal to A^T. Weighted test problems were generated by multiplying some rows of A, and b_1 with a weight $\mu \geq 1$, and b_2 with μ^{-1}.

In Fig. 6.4 we show ϵ_x as a function of α for $\gamma = 10^{-6} : 10^{-2} : 10^{-12}$ using (2.7). Again there is a gain in accuracy of about a factor σ_n when α is decreased from 1 to 10^{-2}. Fig. 6.5 shows results when three rows $1 : 10 : 21$ (in MATLAB notation) are weighted with $\mu = 10^4$. Here the results for $\alpha = 1$ deteriorate as μ^2, but taking $\alpha = 10^{-2}$ is sufficient to regain the accuracy. This value of α is sufficiently small to force the use of three 2×2 pivots. Note that the error bound based on $\kappa_2(DA)$ is not relevant for small values of α. The next figure, Fig. 6.6 shows results for the weighted problem, where rows $1 : 5 : 21$ are weighted with $\mu = 10^6$. Here the rows have been equilibrated, the formulation (4.1) used, and the curve for the error bound is based on the condition number for the equilibrated matrix. Now good results are obtained already for $\alpha = 1$, since the small values on the diagonal will force the use of a sufficient number of 2×2 pivots.

7   Conclusions

It is clear from the analysis and the numerical examples given that the Bunch-Kaufman method is stable for augmented systems only if the scaling parameter α is chosen properly. For the unweighted case a new expression for the optimal value α_{opt} is derived. It is interesting to note that although $\alpha = \sigma_n$ always is a stable choice we can have $\alpha_{opt} \gg \sigma_n$. We point out that the optimal value can be estimated a posteriori, provided a condition estimator for the augmented systems is available.

For sparse problems, using a small value of α will often generate too much fill-in during factorization, and the choice of pivots must be a compromise between preserving sparsity and stability. If an unsuitable value for α has been used, it may then be necessary to refactorize M_α using a smaller value of α. A refactorization can often be avoided if iterative refinement of the solution using single precision residuals is used to restore stability. This idea goes back at least to Wilkinson [26], who in an unpublished report remarked that "...when $\bar{x}$ has been determined by a direct method of some *poorer* numerical stability than Gaussian elimination with pivoting ... the use of $\delta^{(1)}$ as an actual correction should yield substantial dividends...and may be of great value in the solution of sparse systems when pivoting requirements have been relaxed."

Arioli, Duff and de Rijk [2] use this device in their implementation of the augmented systems method for sparse least squares problems. They report that taking $\alpha \approx \|A\|_\infty$ often is satisfactory in practice, provided that *iterative refinement* of the computed solution is performed. Using the residual $s = d - M_\alpha \bar{z}_\alpha$ in single precision they solve for a correction Δz_α. Simultaneously the component-wise backward error (see Oettli and Prager [23]) is computed. Only in case this backward error is not sufficiently small, or the iterative improvement diverges, is the matrix M refactorized. The analysis presented here shows that the use of iterative refinement is crucial for the reliable use of the augmented system method.

Acknowledgements

The author gratefully acknowledges helpful comments and advice from Michael Saunders, who carefully read the manuscript and suggested numerous improvements.

References

[1] F. L. Alvarado. Manipulating and visualization of sparse matrices. *ORSA Journal on Computing*, 2:186–207, 1990.

[2] M. Arioli, I. S. Duff, and P.P.M. de Rijk. On the augmented system approach to sparse least-squares problems. *Numer. Math.*, 55:667–684, 1989.

[3] R. H. Bartels, G. H. Golub, and M. A. Saunders. Numerical techniques in mathematical programming. In J. B. Rosen, O. L. Mangasarian, and K. Ritter, editors, *Nonlinear programming*, pages 123–176. Academic Press, New York, 1970.

[4] Å. Björck. Iterative refinement of linear least squares solutions I. *BIT*, 7:257–278, 1967.

[5] Å. Björck. Stability analysis of the method of semi-normal equations for least squares problems. *Linear Algebra Appl. 88/89, 31–48.*, 88/89:31–48, 1987.

[6] Å. Björck. Component-wise perturbation analysis and errors bounds for linear least square solutions. *BIT*, 31:238–244, 1991.

[7] Å. Björck and I. S. Duff. A direct method for the solution of sparse linear least squares problems. *Linear Algebra and Appl.*, 34:43–67, 1980.

[8] Å. Björck and C. C. Paige. Loss and recapture of orthogonality in the modified Gram-Schmidt algorithm. *SIAM J. Matrix Anal. Appl.*, 13:??–??, 1992.

[9] J. R. Bunch and L. Kaufman. Some stable methods for calculating inertia and solving symmetric linear systems. *Mathematics of Computation*, 31:162–179, 1977.

[10] D. Carlson and H. Schneider. Inertia theorems for matrices: the positive semidefinite case. *J. Math. Anal. Appl.*, 6:430–446, 1963.

[11] R. W. Cottle. Manifestations of the Schur complement. *Linear Algebra Appl.*, 8:189–211, 1974.

[12] I. S. Duff, N. I. M. Gould, J. K. Reid, J. A. Scott, and K. Turner. Factorization of sparse symmetric indefinite matrices. Technical Report RAL-90-066, Nov. 1990, Rutherford Appleton Laboratory, 1990.

[13] I. S. Duff, N. Munksgaard, H. B. Nielsen, and J. K. Reid. Direct solution of sets of linear equations whose matrix is sparse symmetric and indefinite. *J. Inst. Maths. Applics.*, 23:235–250, 1979.

[14] I. S. Duff and J. K. Reid. A comparison of some methods for the solution of sparse overdetermined systems of linear equations. *J. Inst. Maths. Applics.*, 17:267–280, 1976.

[15] I. S. Duff and J. K. Reid. The multifrontal solution of indefinite sparse symmetric linear systems. *ACM Trans. Math. Software*, 9:302–325, 1983.

[16] R. Fletcher. Factorizing symmetric indefinite matrices. *Linear Algebra Appl.*, 14:257–272, 1976.

[17] P. E. Gill, W. Murray, M. A. Saunders, and M. H. Wright. A Schur-complement method for sparse quadratic programming. In M.G. Cox and S. Hammarling, editors, *Reliable Numerical Computation*, pages 113–138, Oxford, 1990. Clarendon Press.

[18] G. H. Golub. *personal communication*, 1966.

[19] G. H. Golub and C. F. Van Loan. *Matrix Computations. 2nd ed.* Johns Hopkins Press, Baltimore, MD., 1989.

[20] G. D. Hachtel. Extended applications of the sparse tableau approach—finite elements and least squares. In W. Spillers, editor, *Basic questions in design theory*. North-Holland, Amsterdam, 1974.

[21] G. D. Hachtel. The sparse tableau approach to finite element assembly. In J.R. Bunch and D. J. Rose, editors, *Sparse Matrix Computations*. Academic Press, New York, 1976.

[22] N. J. Higham. How accurate is gaussian elimination? In D. F. Griffiths and G. A. Watson, editors, *Numerical Analysis 1989: Proceedings of the 13th Dundee Conference*, Pitman Research Notes in Mathematics 228, pages 137–154. Longman Scientific and Technical, 1990.

[23] W. Oettli and W. Prager. Compatibility of approximate solution of linear equations with given error bounds for coefficients and right-hand sides. *Numer. Math.*, 6:405–409, 1964.

[24] P.-Å. Wedin. Perturbation theory and condition numbers for generalized and constrained linear least squares problems. Technical Report UMINF–125.85, Institute of Information Processing, University of Umeå, 1985.

[25] J. H. Wilkinson. The use of single precision-residuals in the solution of linear systems. unpublished report, NPL, 1977.

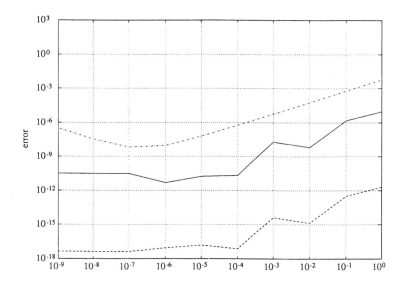

Figure 6.1: Hilbert matrix $\gamma = 10^{-6}$: $\epsilon_x, \epsilon_r, f(\alpha)$, (2.6)

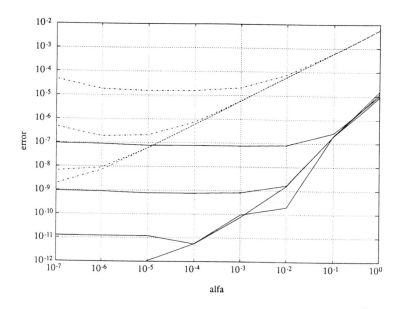

Figure 6.2: Hilbert matrix $\gamma = 10^{-2} - -10^{-8}$: $\epsilon_x, f(\alpha)$, (2.7)

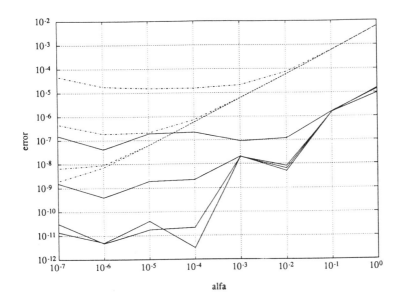

Figure 6.3: Hilbert matrix $\gamma = 10^{-2} - -10^{-8}$: ϵ_x, $f(\alpha)$, (2.6)

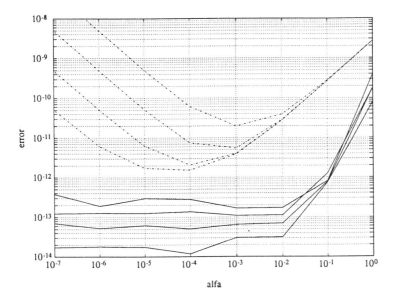

Figure 6.4: Vandermonde matrix $\mu = 1$, $\gamma = 10^{-6} - -10^{-12}$: ϵ_x, $f(\alpha)$, (2.7)

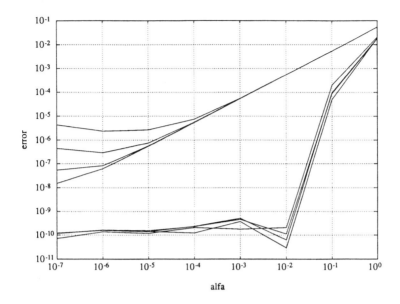

Figure 6.5: Vandermonde matrix $\mu = 10^4$, $\gamma = 10^{-6} - -10^{-12}$: ϵ_x, $f(\alpha)$, (2.7)

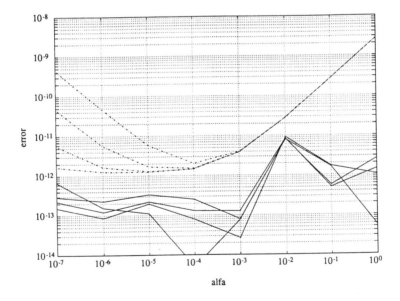

Figure 6.6: Vandermonde matrix $\mu = 10^6$, $\gamma = 10^{-6} - -10^{-12}$: ϵ_x, $f(\alpha)$, (4.1)

M BUHMANN AND A ISERLES

Numerical analysis of functional equations with a variable delay

1 Introduction

The theme of this paper is the *generalized pantograph equation* (GPE)

$$\mathbf{y}'(t) = A\mathbf{y}(t) + B\mathbf{y}(qt) + C\mathbf{y}'(qt), \quad \mathbf{y}(0) = \mathbf{y}_0 \in \mathcal{C}^d, \tag{1}$$

where A, B and C are $d \times d$ complex matrices and q is a given number in $(0, 1)$.

The neutral equation (1) features in a large number of applications – collection of current by electric locomotives [22] (hence the name 'pantograph equation'), number theory [20], probability theory on algebraic structures [23], nonlinear dynamical systems [6], Cherenkov radiation [21], absorption of light by interstellar matter [1], theory of dielectric materials [5] – and this motivates a closer look at both its analytic features and its numerical solution. Moreover, (1) represents a departure from the usual paradigm of delay and neutral equations, namely constant delay [2], [14], and it is a useful starting point for the treatment of more general time-dependent delays [8].

Although our goal is a numerical investigation of (1), we begin with brief description of some analytic features of the GPE, because it is of little use or purpose to investigate how well we can hit a target, unless we know where and what the target is!

2 Analytic features of GPE

The first thing to learn about functional equations is that, being infinite-dimensional dynamical systems, they significantly differ from ordinary differential equations [14]. In particular, we cannot take it for granted that (1) is well-posed. The present section reviews briefly the analytic theory of GPE, addressing itself to well-posedness, stability and behaviour on the stability boundary. A more extensive treatment of the analytic theory will appear in [16].

Let

$$\mathbf{y}(t) = \sum_{k=0}^{\infty} \frac{1}{k!} \mathbf{y}_k t^k$$

be the Taylor expansion of $\mathbf{y}$. Substitution into (1) yields the infinite set of linear equations

$$(I - q^k C)\mathbf{y}_{k+1} = (A + q^k B)\mathbf{y}_k, \quad k \in \mathcal{Z}^+. \tag{2}$$

17

It follows at once that, subject to the existence of a nonnegative integer p such that $q^{-p} \in \sigma(C)$, where $\sigma(C)$ is the spectrum of C, at least one of the equations (2) becomes singular. In that case there exists a linear subspace $\mathcal{K}$ of $\mathcal{C}^d$ such that $\mathbf{y}_0 \in \mathcal{K}$ implies the existence of infinitely many solutions to the GPE, whereas if $\mathbf{y}_0 \notin \mathcal{K}$, no solution exists. On the other hand, if there is no such eigenvalue, then (2) is solvable for all $k \in \mathcal{Z}^+$. Moreover, in that case it is easy to affirm that there exists a positive α such that

$$\|\mathbf{y}_{k+1}\|_2 \le \alpha \|\mathbf{y}_k\|_2, \quad k \in \mathcal{Z}^+.$$

Consequently

$$\|\mathbf{y}(t)\|_2 \le e^{\alpha t} \|\mathbf{y}_0\|_2, \quad t \ge 0,$$

and (1) is well-posed. We henceforth assume that this is indeed the case.

We note in passing that the choice of the starting point at the origin is crucial, otherwise (1) may cease to be well-posed altogether. This fact is a consequence of the infinite dimensionality.

To analyse the asymptotic behaviour of the GPE, we abandon the Taylor expansion in favour of the Dirichlet expansion

$$\mathbf{y}(t) = \sum_{\ell=0}^{\infty} D_\ell e^{q^\ell t A} \mathbf{v}, \tag{3}$$

where D_ℓ, $\ell \in \mathcal{Z}^+$, are matrices, independent of t, and where the rôle of the vector $\mathbf{v}$ is solely to ensure conformity with the initial value. The matrix exponential is defined in the usual way *via* the Taylor series

$$e^F := \sum_{k=0}^{\infty} \frac{1}{k!} F^k.$$

Substitution into (1) produces an infinite set of *commutator equations*

$$\begin{aligned} AD_0 - D_0 A &= 0, \\ AD_\ell - q^\ell D_\ell A &= -BD_{\ell-1} - q^{\ell-1} C D_{\ell-1} A, \quad \ell = 1, 2, \ldots. \end{aligned} \tag{4}$$

Clearly, the first equation is obeyed by the choice $D_0 = I$, the identity matrix. To investigate (4) we note that the commutator equation

$$FX - XG = H,$$

where F, G and H are given $d \times d$ matrices, possesses a unique solution if and only if F and G have no common eigenvalues [11]. In the present case $F = A$, $G = q^\ell A$, $\ell = 1, 2, \ldots$, hence (4) produces a unique solution if and only if no two eigenvalues of A have a ratio of the form q^p with p a positive integer. We assume that this is indeed the case.[1]

In order to investigate convergence of (3), we note that, for $\ell \gg 1$, (4) becomes

$$AD_\ell = -BD_{\ell-1} + \mathcal{O}(q^\ell),$$

---

[1]Existence of 'wrong' ratios of eigenvalues leads to resonance and it will be debated in [16].

hence it is easy to affirm that there exists a positive c such that

$$\|D_\ell\|_2 \le c\left(\rho(A^{-1}B)\right)^\ell, \quad \ell \in \mathcal{Z}^+, \tag{5}$$

where $\rho(F)$ denotes the spectral radius of the matrix F. In other words, if $\rho(A^{-1}B) < 1$ then the series (3) converges.

The exposition here falls short of proving that we can choose $\mathbf{v}$ so that (3) is consistent with any initial value $\mathbf{y}_0$. The reader is referred to [16] for a more complete exposition.

The Dirichlet series representation provides an easy sufficient stability condition which is superior to the more familiar Taylor series representation from [9] and [18]:

Theorem 1 *Letting $\mu[\cdot]$ be the* logarithmic norm [24]

$$\mu[F] = \max\{\operatorname{Re}\lambda : \lambda \in \sigma(F)\},$$

the two conditions $\rho(A^{-1}B) < 1$ and $\mu[A] < 0$ imply that $\lim_{t\to\infty}\mathbf{y}(t) = 0$.

Proof. Given a positive integer n, we split the representation (3)

$$\mathbf{y}(t) = \mathbf{u}_n(t) + \mathbf{w}_n(t),$$

where

$$\mathbf{u}_n(t) = \sum_{\ell=0}^{n-1} D_\ell e^{q^\ell tA}\mathbf{v},$$

$$\mathbf{w}_n(t) = \sum_{\ell=n}^{\infty} D_\ell e^{q^\ell tA}\mathbf{v}.$$

An important property of the logarithmic norm is

$$\|e^F\|_2 \le e^{\mu[F]}.$$

This, together with $\mu[A] < 0$ and (5), proves that

$$\|\mathbf{w}_n(t)\|_2 \le c\|\mathbf{v}\|_2 \sum_{\ell=n}^{\infty}\left(\rho(A^{-1}B)\right)^\ell = \frac{c\|\mathbf{v}\|_2}{1 - \rho(A^{-1}B)}\left(\rho(A^{-1}B)\right)^n.$$

Consequently, given $\varepsilon > 0$, we can choose a sufficiently large n such that

$$\|\mathbf{w}_n(t)\|_2 \le \tfrac{1}{2}\varepsilon \tag{6}$$

uniformly in $t \ge 0$.

In order to bound $\|\mathbf{u}_n(t)\|_2$ we majorize

$$\|e^{q^\ell tA}\|_2 \le e^{q^\ell t\mu[A]} \le e^{q^{n-1}t\mu[A]}, \quad \ell = 0, 1, \ldots, n-1.$$

Therefore,

$$\|\mathbf{u}_n(t)\|_2 \le ce^{q^{n-1}t\mu[A]}\|\mathbf{v}\|_2 \sum_{\ell=0}^{n-1}\left(\rho(A^{-1}B)\right)^\ell = ce^{q^{n-1}t\mu[A]}\|\mathbf{v}\|_2\frac{1 - (\rho(A^{-1}B))^n}{1 - \rho(A^{-1}B)}.$$

19

Consequently, since $\mu[A] < 0$, there exists a t_ε such that for all $t \geq t_\varepsilon$

$$\|\mathbf{u}_n(t)\|_2 \leq \tfrac{1}{2}\varepsilon.$$

The last inequality and (6) imply that

$$\lim_{t \to \infty} \mathbf{y}(t) = \mathbf{0},$$

completing the proof. □

It will be proved in [16] that the above conditions are necessary, as well as sufficient, for $\lim_{t \to \infty} \mathbf{y}(t) = \mathbf{0}$. The most significant part of the proof of that fact is in arguing that $\rho(A^{-1}B) > 1$, $\mu[A] < 0$, are inconsistent with stability. More specifically, one proceeds by differentiating (1), which gives a GPE for $\mathbf{y}'$ and it follows that $\lim_{t \to \infty} \mathbf{y}'(t) = \mathbf{0}$, provided that $\rho(A^{-1}B) < q^{-1}$. However, one can show that in the regime $1 < \rho(A^{-1}B) < q^{-1}$ the convergence to $\mathbf{0}$ occurs logarithmically and

$$\mathbf{y}(t) = \mathbf{y}_0 + \int_0^t \mathbf{y}'(\tau)\mathrm{d}\tau$$

diverges. This argument can be extended by induction to annuli with outer radius q^{-2}, q^{-3} etc.

The behaviour on the stability boundary is also of interest, although it does not impinge on our subsequent analysis. Frederickson [10] proved that the solution of the scalar pantograph equation

$$y'(t) = ay(t) + by(qt), \quad y(0) = y_0,$$

is *almost periodic* when $\mathrm{Re}\, a = 0$, $|b| < |a|$. We recall that a function f is almost periodic if for every $\varepsilon > 0$ there exists a positive T_ε such that

$$|f(t + T_\varepsilon) - f(t)| < \varepsilon, \quad t \geq 0,$$

with the absolute value replaced by norm in the vector case [19]. This can be generalized to (1), provided that $\mu[A] = 0$, $\rho(A^{-1}B) < 1$, and the almost-period T_ε can be explicitly identified for rational values of q [16].

Matters are somewhat less clear for the other portion of the stability boundary, namely $\rho(A^{-1}B) = 1$, $\mu[A] \leq 0$. In that case $\mathbf{y}$ asymptotically approaches a bounded invariant manifold [16]. However, its dimension and the precise nature of the motion of $\mathbf{y}$ therein are still unknown.

3 Stability of recurrences

Conventional ODE solvers can be easily generalized to cater for delay equations. Moreover, as long as the neutral term appears linearly (as is the case with (1)), it can also be accommodated into the standard ODE framework. However, the stability analysis of functional equations is considerably more complicated than that of their ODE counterparts.

The stability of an extended trapezoidal rule, as applied to GPE, has been investigated in detail by the present authors [3]. In this section we proceed along similar lines, albeit with considerably greater generality. The first stage involves a look into general recurrence relations of a specific, nonstationary, form. We henceforth confine our attention to the scalar equation

$$y'(t) = ay(t) + b\left(\tfrac{1}{2}t\right) + cy'\left(\tfrac{1}{2}t\right), \quad y(0) = y_0, \tag{7}$$

where $a, b, c \in \mathcal{C}$. The simplification from vector to scalar systems is merely for ease of exposition. A real loss of generality accrues from the choice $q = \tfrac{1}{2}$. Unfortunately, our method of proof does not generalize to $q \in (0,1)$,[2] although extensive computer experiments hint that the outcome of our analysis remains valid in the general setting.

The stability conditions of Section 2 reduce for (7) to

Condition A$_1$ $\operatorname{Re} a < 0$,

Condition A$_2$ $|b| < |a|$.

The main feature of (7) is the coupling between the state of the dynamical system at times t and $\tfrac{1}{2}t$. The general pattern of recurrence relations in this section retains this characteristic, up to discretization. We now explore

$$y_{2n} = \sum_{\ell=0}^{M} R_\ell y_{2n-1-\ell} + \sum_{\ell=0}^{K} S_\ell y_{n-\ell}, \quad n \geq \max\left\{K, \left[\frac{M}{2}+1\right]\right\}, \tag{8}$$

$$y_{2n+1} = \sum_{\ell=0}^{M} R_\ell y_{2n-\ell} + \sum_{\ell=0}^{K} T_\ell y_{n+1-\ell}, \quad n \geq \max\left\{K-1, \left[\frac{M+1}{2}\right]\right\}. \tag{9}$$

Here R_ℓ, $\ell = 0,1,\ldots,M$, and S_ℓ, T_ℓ, $\ell = 0,1,\ldots,K$, are arbitrary complex numbers, except that we require

$$\sum_{\ell=0}^{K} S_\ell = \sum_{\ell=0}^{K} T_\ell. \tag{10}$$

Starting values to begin the recursion have to be provided.
    We define

$$R(z) := \sum_{\ell=0}^{M} R_\ell z^\ell, \quad S(z) := \sum_{\ell=0}^{K} S_\ell z^\ell, \quad T(z) := \sum_{\ell=0}^{K} T_\ell z^\ell,$$

and we let $\tilde{S}$ and $\tilde{T}$ be the polynomials of degree K and $K-1$ respectively with coefficients

$$\tilde{S}_\ell := \sum_{i=0}^{\ell} T_i - \sum_{i=0}^{\ell-1} S_i, \quad \ell = 0,1,\ldots,K, \qquad \tilde{T}_\ell := \sum_{i=0}^{\ell}(S_i - T_i), \quad \ell = 0,1,\ldots,K-1.$$

Letting finally $W(z) := \tilde{S}(z^2) + z\tilde{T}(z^2)$, we claim that the following assertion is true:

---

[2]It can be extended to $q = 1/L$, $L \geq 2$ an integer [3].

Theorem 2 *Suppose that the following three conditions hold:*

Condition B$_1$ *The polynomial $1 - zR(z)$ does not vanish for $|z| \leq 1$,*

Condition B$_2$ *$|S(1)| < |1 - R(1)|$, and*

Condition B$_3$ *$|T_0 \prod_{|\gamma_j|>1} \gamma_j| < 1$, where γ_j^{-1} are the nonzero roots of W.*

Then $\lim_{n\to\infty} y_n = 0$.

Proof. Let $\{x_n\}_{n=0}^{\infty}$ denote the sequence of first differences of $\{y_n\}_{n=0}^{\infty}$, i.e.

$$x_n = y_{n+1} - y_n, \quad n \in \mathcal{Z}^+.$$

Subtracting (8) from (9) gives

$$x_{2n} = \sum_{\ell=0}^{M} R_\ell x_{2n-1-\ell} + \sum_{\ell=0}^{K} (T_\ell y_{n+1-\ell} - S_\ell y_{n-\ell}).$$

It follows from (10) and from the simple identity

$$y_m = y_0 + \sum_{j=0}^{m-1} x_j, \quad m \in \mathcal{Z}^+,$$

that we have

$$
\begin{aligned}
\sum_{\ell=0}^{K} (T_\ell y_{n+1-\ell} - S_\ell y_{n-\ell}) &= \sum_{j=0}^{n} \left(\sum_{\ell=0}^{\min\{n-j,K\}} T_\ell - \sum_{\ell=0}^{\min\{n-j-1,K\}} S_\ell \right) x_j \\
&= \sum_{j=0}^{n-K-1} (T(1) - S(1))x_j + \sum_{j=n-K}^{n} \left(\sum_{\ell=0}^{n-j} T_\ell - \sum_{\ell=0}^{n-j-1} S_\ell \right) x_j \\
&= \sum_{j=0}^{K} \left(\sum_{\ell=0}^{j} T_\ell - \sum_{\ell=0}^{j-1} S_\ell \right) x_{n-j}.
\end{aligned}
$$

An identical calculation affirms that

$$x_{2n+1} = \sum_{\ell=0}^{M} R_\ell x_{2n-\ell} + \sum_{j=0}^{K-1} \left(\sum_{\ell=0}^{j} S_\ell - \sum_{\ell=0}^{j} T_\ell \right) x_{n-j}.$$

Thus, we derive the recurrences

$$x_{2n} = \sum_{\ell=0}^{M} R_\ell x_{2n-1-\ell} + \sum_{\ell=0}^{K} \tilde{S}_\ell x_{n-\ell}, \quad n \geq \max\left\{ K, \left[\frac{M}{2}\right] + 1 \right\}, \tag{11}$$

$$x_{2n+1} = \sum_{\ell=0}^{M} R_\ell x_{2n-\ell} + \sum_{\ell=0}^{K-1} \tilde{T}_\ell x_{n-\ell}, \quad n \geq \max\left\{ K - 1, \left[\frac{M+1}{2}\right] \right\}. \tag{12}$$

22

We formally introduce the generating function of the sequence $\{x_n\}_{n=0}^{\infty}$,

$$X(z) := \sum_{n=0}^{\infty} x_n z^n, \quad z \in \mathcal{C}.$$

Multiplying (11) by z^{2n} and (12) by z^{2n+1}, summing up and adding the contribution from the starting values yields

$$
\begin{aligned}
X(z) \;=\; & \sum_{n=0}^{\max\{2K-2,M\}} x_n z^n + \sum_{n=\max\{K,[M/2]+1\}}^{\infty} \left(\sum_{\ell=0}^{M} R_\ell x_{2n-1-\ell} \right) z^{2n} \\
& + \sum_{n=\max\{K-1,[(M+1)/2]\}}^{\infty} \left(\sum_{\ell=0}^{M} R_\ell x_{2n-\ell} \right) z^{2n+1} \\
& + \sum_{n=\max\{K,[M/2]+1\}}^{\infty} \left(\sum_{\ell=0}^{K} \tilde{S}_\ell x_{n-\ell} \right) z^{2n} \\
& + \sum_{n=\max\{K-1,[(M+1)/2]\}}^{\infty} \left(\sum_{\ell=0}^{K-1} \tilde{T}_\ell x_{n-\ell} \right) z^{2n+1} \\
\;=\; & \sum_{n=0}^{\max\{2K-2,M\}} x_n z^n + z \sum_{\ell=0}^{M} R_\ell z^\ell \left(X(z) - \sum_{n=0}^{\max\{2K-3,M-1\}-\ell} x_\ell z^\ell \right) \\
& + \sum_{\ell=0}^{K} \tilde{S}_\ell z^{2\ell} \left(X(z^2) - \sum_{n=0}^{\max\{K-1,[M/2]\}-\ell} x_n z^{2n} \right) \\
& + \sum_{\ell=0}^{K-1} \tilde{T}_\ell z^{2\ell+1} \left(X(z^2) - \sum_{n=0}^{\max\{K-2,[(M-1)/2]\}-\ell} x_n z^{2n} \right).
\end{aligned}
$$

Consequently,

$$X(z) = \frac{1}{1 - zR(z)} \left\{ U(z) + \left(\tilde{S}(z^2) + z\tilde{T}(z^2) \right) X(z^2) \right\}, \tag{13}$$

where U is a polynomial of degree $\max\{2K - 2, M\}$ that depends solely on the starting values.

We want to prove first that $x_n = o(1)$ as $n \to \infty$ and we shall show this by looking at the Fourier series associated with $\{x_n\}_{n=0}^{\infty}$. In order to do this we require that the identity (13) is valid in an open disc of radius exceeding unity, and this is true if and only if condition $\mathbf{B_1}$ holds. Assuming that this condition holds, we let $z = e^{i\vartheta}$ in (13), to obtain the Fourier transform

$$\hat{x}(\vartheta) = \frac{1}{1 - e^{i\vartheta} R(e^{i\vartheta})} \left\{ U(e^{i\vartheta}) + W(e^{i\vartheta})\hat{x}(2\vartheta) \right\}, \quad 0 \le \vartheta \le 2\pi.$$

Successively iterating the equation for $\hat{x}$ yields the explicit expression

$$\hat{x}(\vartheta) = \sum_{m=0}^{\infty} \frac{\prod_{\ell=0}^{m-1} W(e^{i2^\ell \vartheta})}{\prod_{\ell=0}^{m} \left(1 - e^{i2^\ell \vartheta} R(e^{i2^\ell \vartheta}) \right)} U(e^{i2^m \vartheta}), \quad 0 \le \vartheta \le 2\pi. \tag{14}$$

23

The latter formula represents $\hat{x}$ as an infinite sum of analytic functions. Our next goal is to prove that $\hat{x}$ is in $L_2[0, 2\pi]$, which implies that $\{x_n\}_{n=0}^{\infty}$ is square-summable and hence $x_n = o(1)$ as $n \to \infty$. This goal will be accomplished by demonstrating that, subject to two extra conditions, the analytic functions in (14) are bounded almost everywhere by a geometric sequence.

It follows from (14) that

$$|\hat{x}(\vartheta)| \le \frac{1}{|1 - e^{i\vartheta} R(e^{i\vartheta})|} \|U\|_{\infty} \sum_{m=0}^{\infty} \prod_{\ell=0}^{m-1} |F(e^{i2^{\ell}\vartheta})|, \tag{15}$$

where

$$F(z) := \frac{W(z)}{1 - z^2 R(z^2)} = \tilde{S}(0) \frac{\prod_{j=1}^{2K}(1 - \gamma_j z)}{\prod_{j=1}^{2M+2}(1 - \delta_j z)},$$

is a rational function. We set

$$g_m(\alpha) := \prod_{\ell=0}^{m-1} \left(1 - \alpha e^{i2^{\ell}\vartheta}\right), \quad \alpha \in \mathcal{C},$$

hence

$$\mathcal{F}_m(\vartheta) := \prod_{\ell=0}^{m-1} F(e^{i2^{\ell}\vartheta}) = \tilde{S}(0) \frac{\prod_{j=1}^{2K} g_m(\gamma_j)}{\prod_{j=1}^{2M+2} g_m(\delta_j)}.$$

The asymptotic behaviour of g_m as $m \to \infty$ can be explicitly determined by using elementary ergodic theory. We have the identity

$$\log |g_m(\alpha)| = \sum_{\ell=0}^{m-1} \log \left|1 - \alpha e^{i2^{\ell}\vartheta}\right| = \sum_{\ell=0}^{m-1} \log \left|1 - \alpha \exp \left(i T^{\ell}(\vartheta)\right)\right|,$$

where T is the *doubling map*

$$T(\vartheta) := 2\vartheta \mod 2\pi.$$

Thus, according to the *mean ergodic theorem* [15]

$$\frac{2\pi}{m} \log |g_m(\alpha)| = \int_0^{2\pi} \log \left|1 - \alpha e^{i\phi}\right| d\phi + e_m(\vartheta)o(1), \qquad |e_m(\vartheta)| = \mathcal{O}(1), \tag{16}$$

almost everywhere for $\vartheta \in [0, 2\pi]$. Consequently, $|\mathcal{F}_m|$ is, asymptotically, determined almost everywhere by the configuration of its zeros and poles. Since

$$\int_0^{2\pi} \log \left|1 - \alpha e^{i\phi}\right| d\phi = \begin{cases} 0 & : |\alpha| \le 1, \\ 2\pi \log |\alpha| & : |\alpha| \ge 1, \end{cases}$$

zeros and poles *outside* the open unit disc do not matter. According to condition $\mathbf{B}_1$, all the poles of F are outside the disc. Therefore, and because $\tilde{S}(0) = T_0$, we have

$$\log |\mathcal{F}_m(\vartheta)| = m \log \left| T_0 \prod_{|\gamma_j| > 1} \gamma_j \right| + o(m)$$

almost everywhere along the unit circle. Hence we have almost everywhere

$$|\mathcal{F}_m(\vartheta)| = \left|T_0 \prod_{|\gamma_j|>1} \gamma_j\right|^m e^{o(m)}, \quad m \to \infty, \tag{17}$$

which implies that (15) is dominated almost everywhere by a geometric series, therefore finite. We need an extra morsel of information to deduce that (17) holds *uniformly* almost everywhere so that we may deduce $\hat{x} \in L_2[0, 2\pi]$, namely that the 'correct' ratio of the constants e_m from (16) is uniformly bounded almost everywhere. This point is of purely technical nature and we follow here the approach from [3], iterating the recurrence *backwards* (i.e. halving, rather than doubling, the angle). An arbitrarily small open neighbourhood of $\vartheta = 0$ can be reached in a finite number of steps and it follows from the analysis in [3] that, to ensure uniform boundedness of the error constants, (14) must be summable for $\vartheta = 0$. But

$$\hat{x}(0) = \frac{U(1)}{1 - R(1)} \sum_{m=0}^{\infty} \left(\frac{W(1)}{1 - R(1)}\right)^m$$

which is a geometric series. Moreover, $W(1) = S(1)$, consequently summability at the origin reduces to the inequality in condition $\mathbf{B}_2$. Stipulating that this condition holds, we have the estimate (17) and the series representation of $\hat{x}$ is dominated term-by-term by a geometric series uniformly almost everywhere in $[0, 2\pi]$.[3] Further, subject to condition $\mathbf{B}_3$ the geometric series converges and we can use Lebesgue's dominated convergence theorem to argue that $\hat{x}$ is in $L_2[0, 2\pi]$. Since the Fourier transform is an isometry, $\hat{x} \in L_2[0, 2\pi]$ is equivalent to ℓ_2-boundedness of $\{x_n\}_{n=0}^{\infty}$, and this, in turn, implies that $\lim_{n\to\infty} x_n = 0$. Unfortunately, this result falls short of the required, since we wish to deduce stability of the sequence $\{y_n\}_{n=0}^{\infty}$, rather than of its first differences $\{x_n\}_{n=0}^{\infty}$. In order to infer from $\{x_n\}_{n=0}^{\infty}$ to $\{y_n\}_{n=0}^{\infty}$ we use a technique from [3]. Since

$$\hat{y}(\vartheta) = \frac{y_0 + e^{i\vartheta}\hat{x}(\vartheta)}{1 - e^{i\vartheta}}, \quad 0 \le \vartheta \le 2\pi,$$

$\hat{y}$ being the Fourier series associated with $\{y_n\}_{n=0}^{\infty}$, integrability of $\hat{x}$ implies integrability of $\hat{y}$ if and only if the polar singularity at the origin is removable. This requires $\hat{x}(0) = -y_0$. However, recall that the series (14) is summable for $\vartheta = 0$, and it is trivial to deduce by telescoping series that

$$\hat{x}(0) = X(1) = \sum_{n-0}^{\infty}(y_{n+1} - y_n) = -y_0,$$

as required. Hence we infer that, subject to conditions $\mathbf{B}_1$, $\mathbf{B}_2$ and $\mathbf{B}_3$, it is true that $\lim_{n\to\infty} y_n = 0$ and the numerical solution sequence is stable. □

Recall conditions $\mathbf{A}_1$ and $\mathbf{A}_2$ which are equivalent to stability of the equation (7). Clearly, it is important to explore the connection between them and the three numerical

---

[3]The phrase 'almost everywhere' is far from an idle mathematical nitpicking, since (14) fails for all *rational* ϑ.

stability conditions. This is done in the next section, where the recursive scheme is identified with a continuous-output multistep ODE solver.

Note that we have proved the conditions $\mathbf{B_1}$, $\mathbf{B_2}$ and $\mathbf{B_3}$ to be sufficient for stability, but nothing is known about their necessity. This is just one of the many outstanding open questions regarding recurrence relations of this section. Clearly, exploration of general values of $q \in (0,1)$ is important, as is the behaviour on the stability boundary.[4] Another intriguing aspect of our recurrence relations is their similarity to systems that are encountered in computer-aided geometric design [12] and in the investigation of wavelets [7].

An alternative approach to stability analysis is to explore monotonically decaying numerical solutions. Results shedding light on this type of behaviour will be reported (for all $q \in (0,1)$) in a forthcoming paper [4].

4   Multistep methods

Let $h > 0$ be the step-length. Integration of (7) yields

$$y(Nh) = y_0 + a \int_0^{Nh} y(\tau)\mathrm{d}\tau + 2b \int_0^{\frac{1}{2}Nh} y(\tau)\mathrm{d}\tau + 2c \left(y \left(\tfrac{1}{2}Nh \right) - y_0 \right), \quad N \in \mathcal{Z}^+. \quad (18)$$

Let (ρ, σ) be a given $(M+1)$-step zero-stable ODE method of order $p \geq 1$. As applied to $\mathbf{y}' = \mathbf{f}(\mathbf{y})$, the method reads

$$\sum_{\ell=0}^{M+1} \rho_{M+1-\ell} \mathbf{y}_{N+1-\ell} = h \sum_{\ell=0}^{M+1} \sigma_{M+1-\ell} \mathbf{f}(\mathbf{y}_{N+1-\ell}), \quad N \geq M.$$

Consistency and zero-stability require $\rho(1) = 0$ and $\rho'(1) = \sigma(1) \neq 0$. We denote by $\mathcal{S}_{\rho,\sigma}$ its *linear stability domain*, i.e. the set of all $z \in C$ such that the sequence that results from the application of the underlying multistep method to the linear test equation $y' = zy$, $y(0) = 1$, $h = 1$, is stable [17].

We deduce from (18) and $\rho(1) = 0$ that

$$\sum_{\ell=0}^{M+1} \rho_{M+1-\ell} y((N+1-\ell)h) = a \sum_{\ell=0}^{M+1} \rho_{M+1-\ell} \int_{(N-M)h}^{(N+1-\ell)h} y(\tau)\mathrm{d}\tau \qquad (19)$$

$$+ 2b \sum_{\ell=0}^{M+1} \rho_{M+1-\ell} \int_{\frac{1}{2}(N-M)h}^{\frac{1}{2}(N+1-\ell)h} y(\tau)\mathrm{d}\tau + 2c \sum_{\ell=0}^{M+1} \rho_{M+1-\ell} y \left(\tfrac{1}{2}(N+1-\ell)h \right).$$

It is obvious how to approximate the first integral, that corresponds to the 'ODE component' of the equation, viz.

$$\sum_{\ell=0}^{M+1} \rho_{M+1-\ell} \int_{(N-M)h}^{(N+1-\ell)h} y(\tau)\mathrm{d}\tau = h \sum_{\ell=0}^{M+1} \sigma_{M+1-\ell} y((N+1-\ell)h) + \mathcal{O}(h^{p+1}).$$

---

[4]The latter leads to some quite striking fractal sets [3], whose fine structure is unknown.

In other words, the R_ℓs of the previous section are

$$R_\ell = -\frac{\rho_{M+1-\ell} - ha\sigma_{M+1-\ell}}{1 - ha\sigma_{M+1}}, \quad \ell = 0, 1, \ldots, M.$$

It follows that

$$1 - zR(z) = z^{M+1}\left(\rho(z^{-1}) - ha\sigma(z^{-1})\right). \tag{20}$$

Recall condition $\mathbf{B_1}$, which requires that $1 - zR(z) \neq 0$ for $|z| \leq 1$. Given that we have

$$1 - zR(z) = \prod_{\ell=1}^{M+1}(1 - \omega_\ell(ha)z),$$

this is equivalent to $|\omega_\ell(ha)| < 1$, $\ell = 1, 2, \ldots, M + 1$. The identity (20) can now be used to argue that this is *precisely* the condition that ha belongs to $\mathcal{S}_{\rho,\sigma}$. In other words, we can rephrase condition $\mathbf{B_1}$ into

Condition $\mathbf{C_1}$ $ha \in \mathcal{S}_{\rho,\sigma}$.

Also condition $\mathbf{B_2}$ can be rephrased in a more helpful manner in the present terminology. Firstly, note that $\rho(1) = 0$ implies

$$1 - R(1) = -\frac{ha\sigma(1)}{1 - ha\sigma_{M+1}}. \tag{21}$$

The coefficients S_ℓ and T_ℓ, $\ell = 0, 1, \ldots, K$, are derived by approximating the remaining terms on the right-hand side of (19) for odd and even values of N respectively. In particular, letting $N = 2n - 1$ yields

$$\sum_{\ell=0}^{K} S_\ell y((n - \ell)h) \approx \frac{2}{1 - ah\sigma_{M+1}}\left\{b\sum_{\ell=0}^{M+1}\rho_{M+1-\ell}\int_{(n-\frac{1}{2}(M+1))h}^{(n-\frac{1}{2}\ell)h} y(\tau)d\tau \right.$$
$$\left. + c\sum_{\ell=0}^{M+1}\rho_{M+1-\ell}y\left(\left(n - \tfrac{1}{2}\ell\right)h\right)\right\}.$$

Consistency requires equality when $y(t) \equiv 1$. This, together with $\rho(1) = 0$, implies that

$$S(1) = \frac{hb}{1 - ha\sigma_{M+1}}\sum_{\ell=0}^{M+1}\ell\rho_\ell.$$

But

$$\sum_{\ell=0}^{M+1}\ell\rho_\ell = \rho'(1) = \sigma(1),$$

and therefore

$$S(1) = \frac{hb\sigma(1)}{1 - ah\sigma_{M+1}}.$$

Since $\sigma(1) \neq 0$, condition $\mathbf{B_2}$ now reduces to

Condition $\mathbf{C_2}$ $|b| < |a|$,

which is precisely condition $\mathbf{A}_2$, which was necessary for stability of the exact equation (7).[5]

Recall that we have derived in Section 3 three conditions for stability. The first two have been rephrased by exploiting properties of the multistep method, whereas the third will be explored in the sequel on a case-by-case basis. It is instructive to compare the $\mathbf{A}$ conditions (on stability of the exact equation), $\mathbf{B}$ conditions (on the formal recurrence) and $\mathbf{C}$ conditions (on the multistep method).

	A	B	C
1	$\mathrm{Re}\,a < 0$	$1 - zR(z) \neq 0,\ |z| \leq 1$	$ha \in \mathcal{S}_{\rho,\sigma}$
2	$|b| < |a|$	$|S(1)| < |1 - R(1)|$	$|b| < |a|$
3	none	$|T_0|\prod_{|\gamma_j|>1}|\gamma_j| < 1$	?

In other words, the first condition translates the left half-plane to the linear stability domain (vindicating basic ODE intuition), whereas the second condition survives intact. The third condition has no exact counterpart and it leads to a limitation on the step-length h. Fortunately, this upper bound is very mild, as will be evident in the remainder of this section.

To explore further the choice of S and T, we let E stand for the *shift operator*, $Ey(t) := y(t + h)$. It is easy to verify that

$$\int_{(n-\alpha)h}^{(n-\beta)h} y(\tau)\mathrm{d}\tau = h\frac{E^{-\beta} - E^{-\alpha}}{\log E}y(nh),$$

therefore

$$\sum_{\ell=0}^{K} S_\ell E^{-\ell}y(nh) \approx \frac{2h}{1 - ah\sigma_{M+1}}\left\{\frac{b}{\log E}\sum_{\ell=0}^{M+1} \rho_{M+1-\ell}\left(E^{-\frac{\ell}{2}} - E^{-\frac{M+1}{2}}\right)\right.$$
$$\left. + c\sum_{\ell=0}^{M+1} \rho_{M+1-\ell}E^{-\frac{\ell}{2}}\right\}y(nh).$$

Thus, to maintain order p, we need

$$\sum_{\ell=0}^{K} S_\ell z^{-\ell} = \frac{2h}{1 - ah\sigma_{M+1}}\left(\frac{b}{\log z} + c\right)\sum_{\ell=0}^{M+1} \rho_{M+1-\ell}z^{-\frac{\ell}{2}} + \mathcal{O}(|1 - z|^{p+1}),$$

where $z = \mathcal{O}(h)$. Note that we have exploited $\rho(1) = 0$. Reworking the last result, we obtain

$$S(z) = \frac{2hz^{\frac{M+1}{2}}}{1 - ah\sigma_{M+1}}\frac{\rho(z^{-\frac{1}{2}})}{\log z}(-b + c\log z) + \mathcal{O}(|1 - z|^{p+1}). \tag{22}$$

An identical calculation (this time expanding about $(n + 1)h$) affirms that

$$T(z) = \frac{2hz^{\frac{M}{2}+1}}{1 - ah\sigma_{M+1}}\frac{\rho(z^{-\frac{1}{2}})}{\log z}(-b + c\log z) + \mathcal{O}(|1 - z|^{p+1}). \tag{23}$$

---

[5]Incidentally, similar calculation affirms that consistency implies $T(1) = S(1)$.

Order p of the original ODE method means that

$$\rho(z) - \sigma(z)\log z = \mathcal{O}(|1 - z|^{p+1}),$$

hence

$$\frac{\rho(z^{-\frac{1}{2}})}{\log z} = -\tfrac{1}{2}\sigma(z^{-\frac{1}{2}}) + \mathcal{O}(|1 - z|^{p}).$$

Substitution into (22) and (23) gives

$$S(z) = \frac{2hz^{\frac{M+1}{2}}}{1 - ah\sigma_{M+1}}\left(\tfrac{1}{2}b\sigma(z^{-\frac{1}{2}}) + c\rho(z^{-\frac{1}{2}})\right) + \mathcal{O}(|1 - z|^{p+1}), \tag{24}$$

$$T(z) = z^{\frac{1}{2}}S(z) + \mathcal{O}(|1 - z|^{p+1}). \tag{25}$$

Expressions (24) and (25) provide a means to derive K-degree polynomials S and T consistently with the order of the numerical method. In particular, it follows that the least (hence the optimal) choice of K is $K = p - 1 \le 2[\tfrac{1}{2}M] + 1$ (the latter inequality is a consequence of the first Dahlquist barrier [17], [24]). We henceforth adopt the practice of selecting $K = p - 1$ and choosing S and T as the least $(p-1)$-degree polynomials that obey (24) and (25) respectively.

Our first example revisits the trapezoidal rule that has been already analysed in the less general formalism of [3]. Now

$$\rho(z) = z - 1, \qquad \sigma(z) = \tfrac{1}{2}(z + 1), \qquad p = 2.$$

Hence $M = 0$. We choose $K = 1$ and truncate

$$\begin{aligned}
S(z) &= \frac{2h\sqrt{z}}{1 - \tfrac{1}{2}ha}\left(c\left(\frac{1}{\sqrt{z}} - 1\right) + \tfrac{1}{4}b\left(\frac{1}{\sqrt{z}} + 1\right)\right) + \mathcal{O}(|1 - z|^{3}) \\
&= \frac{1}{1 - \tfrac{1}{2}ha}\left(\tfrac{1}{4}hb(3 + z) + hc(1 - z)\right)
\end{aligned}$$

and

$$\begin{aligned}
T(z) &= \sqrt{z}S(z) \\
&= \frac{1}{1 - \tfrac{1}{2}ha}\left(\tfrac{1}{4}hb(1 + 3z) + hc(1 - z)\right).
\end{aligned}$$

Therefore

$$W(z) = \frac{h}{1 - \tfrac{1}{2}ha}(1 + z)\left(\left(\tfrac{1}{4}b + c\right) + \left(\tfrac{1}{4}b - c\right)z\right),$$

and straightforward manipulation yields condition $\mathbf{C}_3$

$$h\max\left\{\left|\tfrac{1}{4}b + c\right|, \left|\tfrac{1}{4}b - c\right|\right\} < \left|1 - \tfrac{1}{2}ha\right|,$$

which has been already derived in [3]. Since condition $\mathbf{C}_1$ means that $\operatorname{Re} a < 0$, this restriction on the step-size is usually quite mild.

29

Our next example is the two-step BDF method [13]

$$\rho(z) = z^2 - \tfrac{4}{3}z + \tfrac{1}{3}, \qquad \sigma(z) = \tfrac{2}{3}z^2, \qquad p = 2.$$

We derive

$$S(z) = \frac{\tfrac{2}{3}h}{1 - \tfrac{2}{3}ha}(b + c - cz),$$

$$T(z) = \frac{\tfrac{1}{3}h}{1 - \tfrac{2}{3}ha}(b + 2c + (b - 2c)z),$$

hence

$$W(z) = \frac{\tfrac{1}{3}h}{1 - \tfrac{2}{3}ha}(1 + z)(b + 2c - 2cz).$$

It is now easy to verify that condition $\mathbf{C_3}$ boils down to

$$\tfrac{1}{3}h \max\{|b + 2c|, 2|c|\} \le \left|1 - \tfrac{2}{3}ha\right|.$$

Finally, we consider general BDF methods [13], restricting our discussion to the pure delay case $c = 0$. We consider only zero-stable methods, therefore $M \le 5$. Since

$$z^{M+1}\rho\left(\frac{1}{z}\right) = -\sigma_{M+1}\sum_{\ell=1}^{M+1}\frac{(1 - z)^\ell}{\ell},$$

$$z^{M+1}\sigma\left(\frac{1}{z}\right) = \sigma_{M+1} = \frac{1}{\sum_{\ell=1}^{M+1}\frac{1}{\ell}},$$

we obtain

$$S(z) \equiv \frac{\sigma_{M+1}hb}{1 - \sigma_{M+1}ha},$$

$$T(z) = \frac{\sigma_{M+1}hb}{1 - \sigma_{M+1}ha}\sum_{k=0}^{M+1}\frac{\left(-\frac{1}{2}\right)_k}{k!}(1 - z)^k.$$

Here $(x)_m := x(x + 1)(x + 2)\cdots(x + m - 1)$ is the *Pochhammer symbol*. It is possible to demonstrate with only minor effort that

$$W(z) = \frac{\sigma_{M+1}hb}{1 - \sigma_{M+1}ha}(1 + z)\left(\frac{1}{2} + (1 - z)^2\sum_{k=0}^{M-1}\frac{\left(-\frac{1}{2}\right)_{k+2}}{(k + 2)!}(1 - z^2)^k\right)$$

and -1 is a zero of W. It is, in fact, a zero of multiplicity $M+1$. To see this, let $s^* = S(z)$. Then, according to (25),

$$T(z) = s^*\sqrt{z} + \mathcal{O}(|1 - z|^{m+2}),$$

and therefore

$$W(z) = s^*\frac{|z| + z + \mathcal{O}(|1 - z^2|^{M+2})}{1 + z}.$$

Letting z be real affirms that

$$W(z) = \mathcal{O}(|1 + z|^{M+1}), \qquad z \to -1,$$

and the result can be extended to complex values by analytic continuation.

Since $K = M + 1$, we expect $\deg W = 2M + 2$, but, S being a constant, the degree decreases by one. Consequently

$$W(z) = (1 + z)^{M+1} W_M^\star(z), \qquad \deg W_M^\star = M.$$

The polynomials $W_M^\star$ can be easily calculated from the recurrence

$$(1 + z)W_M^\star(z) = W_{M-1}^\star(z) - \frac{1}{2}\frac{\left(\frac{1}{2}\right)_M}{(M+1)!}(1 - z)^{M+1}.$$

This gives

$$
\begin{aligned}
W_0^\star(z) &\equiv \tfrac{1}{2}, \\
W_1^\star(z) &= \tfrac{1}{8}(3 - z), \\
W_2^\star(z) &= \tfrac{1}{6}(5 - 4z + z^2), \\
W_3^\star(z) &= \tfrac{1}{128}(35 - 47z + 25z^2 - 5z^3).
\end{aligned}
$$

It is easy to verify that all the zeros of $W_M^\star$ reside outside the unit disc for $M \leq 5$, the zero-stable range. Thus, and since moreover

$$|T_0| < \frac{\sigma_{M+1} h |b|}{|1 - \sigma_{M+1} ha|},$$

it is enough for condition $\mathbf{B}_3$ to be true that

$$\frac{\sigma_{M+1} h |b|}{|1 - \sigma_{M+1} ha|} < 1.$$

Reworking the last inequality produces a sufficient condition for the condition $\mathbf{C}_3$ to be true when $c = 0$, namely

$$\sigma_{M+1}^2 h^2 |b|^2 < 1 - 2\sigma_{M+1} h \operatorname{Re} a + \sigma_{M+1}^2 h^2 |a|^2. \tag{26}$$

Note that, as long as $\operatorname{Re} a < 0$, the inequality (26) is subsumed into $|b| < |a|$, i.e. condition $\mathbf{C}_2$, and it produces no restriction on the step-size.

Acknowledgement

We are grateful to Charles Micchelli for studying a draft of this paper and contributing helpful remarks.

References

[1] V.A. Ambartsumian (1944), "On the fluctuation of the brightness of the Milky Way", *Doklady Akad. Nauk USSR* **44**, 223–226 (in Russian).

[2] R. Bellman and K. Cooke (1963), *Differential-Difference Equations*, Academic Press, New York.

[3] M.D. Buhmann and A. Iserles (1990), "On the dynamics of a discretized neutral equation", Technical Report DAMTP 1990/NA8, University of Cambridge.

[4] M.D. Buhmann and A. Iserles, paper in preparation.

[5] C.J. Budd (1990), personal communication.

[6] G.A. Derfel (1990), "Kato problem for functional-differential equations and difference Schrödinger operators", in *Operator Theory: Advances and Applications* **46**, 319–321.

[7] N. Dyn (1990), personal communication.

[8] A. Feldstein, A. Iserles and D. Levin, paper in preparation.

[9] L. Fox, D.F. Mayers, J.R. Ockendon and A.B. Tayler (1971), "On a functional differential equation", *J. Inst. Maths Applics* **8**, 271–307.

[10] P.O. Frederickson (1971), "Dirichlet series solution for certain functional differential equations", *Japan–United States Seminar on Ordinary Differential and Functional equations,* M. Urabe, ed., Springer Lecture Notes in Mathematics 243, Springer-Verlag, Berlin, 247–254.

[11] F.R. Gantmacher (1977), *Matrix Theory*, Chelsea, New York.

[12] J. Gregory (1992), "A review of uniform subdivision", these proceedings.

[13] E. Hairer, S.P. Nørsett and G. Wanner (1987), *Solving Ordinary Differential Equations I: Nonstiff Problems*, Springer-Verlag, Berlin.

[14] J. Hale (1977), *Theory of Functional Differential Equations*, Springer-Verlag, New York.

[15] H. Helson (1983), *Harmonic Analysis*, Addison-Wesley, London.

[16] A. Iserles, paper in preparation.

[17] A. Iserles and S.P. Nørsett (1991), *Order Stars*, Chapman and Hall, London.

[18] T. Kato and J.B. McLeod (1971), "The functional-differential equation $y'(x) = ay(\lambda x) + by(x)$", *Bull. Amer. Math. Soc.* **77**, 891–937.

[19] Y. Katznelson (1968), *An Introduction to Harmonic Analysis*, John Wiley and Sons, Chichester.

[20] K. Mahler (1940), "On a special functional equation", *J. London Math. Soc.* **15**, 115–123.

[21] J.R. Ockendon (1991), personal communication.

[22] J.R. Ockendon and A.B. Tayler (1971), "The dynamics of a current collection system for an electric locomotive", *Proc. Royal Soc. A* **322**, 447–468.

[23] E.Yu. Romanenko and A.N. Sharkovskĭ (1978), "Asymptotic solutions of differential-functional equations", in *Asymptotic Behaviour to Solutions of Differential Difference Equations*, Inst. Math. Akad. Nauk UkrSSR, 5–39 (in Russian).

[24] H.J. Stetter (1973), *Analysis of Discretization Methods for Ordinary Differential Equations*, Springer-Verlag, Berlin.

M.D. Buhmann and A. Iserles
Department of Applied Mathematics and Theoretical Physics
University of Cambridge
Silver Street, Cambridge CB3 9EW.

M P CALVO AND J M SANZ-SERNA
Variable steps for symplectic integrators

1. Introduction. In Mechanics, Optics, Chemistry etc..., situations where dissipation does not play a significant role may be modelled by means of Hamiltonian systems of, ordinary or partial, differential equations [2, 9]. Hamiltonian systems of ODEs are of the form

$$\dot{p}^I = -\partial H/\partial q^I, \qquad \dot{q}^I = \partial H/\partial p^I, \qquad 1 \leq I \leq N, \tag{1.1}$$

where the integer N is the number of degrees of freedom, the Hamiltonian $H = H(\mathbf{p}, \mathbf{q}) = H(p^1, \dots, p^N, q^1, \dots, q^N)$ is a, sufficiently smooth, real function of $2N$ real variables and a dot represents differentiation with respect to t (time). There has been much recent interest in the numerical integration of (1.1) by means of so-called symplectic or canonical integrators (see e.g. [1, 3, 4, 7, 9–16, 18] and references therein). In order to explain in simple terms the meaning and relevance of symplecticness, it is advisable to consider first the question of how to tell, from the knowledge of the *solutions* of a system of ODEs, whether the *system* is of Hamiltonian form or otherwise. More precisely, let S be an autonomous system of ODEs for the dependent variables $(\mathbf{p}, \mathbf{q})$ and let us introduce the $\mathcal{R}^{2N}$-valued function $\Phi(\mathbf{p}_0, \mathbf{q}_0; t)$ such that, for fixed $\mathbf{p}_0, \mathbf{q}_0$ and varying t, $(\mathbf{p}(t), \mathbf{q}(t)) = \Phi(\mathbf{p}_0, \mathbf{q}_0; t)$ is the solution of S with initial condition $\mathbf{p}(0) = \mathbf{p}_0$, $\mathbf{q}(0) = \mathbf{q}_0$. If we now see t as a parameter and $\mathbf{p}_0, \mathbf{q}_0$ as variables, $\Phi(\mathbf{p}_0, \mathbf{q}_0; t)$ defines a transformation in the space $\mathcal{R}^{2N}$ (the phase space). This transformation is the *flow* of the differential system S. If we were given Φ and at the same time S were concealed from us, could we tell whether S is a Hamiltonian system or otherwise? The answer to this question is affirmative. The system S is Hamiltonian *if and only if, for each t, Φ is a symplectic transformation.* Now a transformation T in phase space is said to be symplectic [2, 9] if for any bounded two-dimensional surface D in phase space the sum of the two-dimensional areas of the N-projections of D onto the planes (p^I, q^I) is the same as the sum of the two-dimensional areas of the N-projections of $T(D)$ onto the planes (p^I, q^I). Thus the symplectic character of the flow is the hallmark of Hamiltonian systems. Hamiltonian problems have many specific features not shared by other systems of differential equations. All such specific features (absence of attractors, recurrence, etc...) directly derive from the symplecticness of the corresponding flow [2, 9].

A one-step numerical method used with step-length h defines a transformation in phase space $\Psi(\mathbf{p}_0, \mathbf{q}_0; h)$ that advances the solution h units of time, starting from $(\mathbf{p}_0, \mathbf{q}_0)$. Of course $\Psi(\mathbf{p}_0, \mathbf{q}_0; h)$ is an approximation to $\Phi(\mathbf{p}_0, \mathbf{q}_0; h)$ and the numerical method approximates $\Phi(\mathbf{p}_0, \mathbf{q}_0; nh) = \Phi(\mathbf{p}_0, \mathbf{q}_0; h)^n$ by iterating n times $\Psi(\mathbf{p}_0, \mathbf{q}_0; h)$. For Hamiltonian problems integrated by classical methods, such as explicit Runge-Kutta methods, the transformation Ψ turns out to be *nonsymplectic*. Then the numerical method misses the important specific features associated with symplectic transformations. However there are *symplectic* methods for which Ψ is guaranteed to be symplectic for Hamiltonian problems. It can be shown that for a symplectic method $\Psi(\mathbf{p}_0, \mathbf{q}_0; h)$ is, for each fixed h and except for a negligible remainder [9], identical to the h-flow $\tilde{\Phi}(\mathbf{p}_0, \mathbf{q}_0; h)$ of an autonomous Hamiltonian, whose Hamiltonian function $\tilde{H}_h$ is an approximation to the Hamiltonian H of the system being integrated. Hence, in this case, a numerically computed solution $\Psi(\mathbf{p}_0, \mathbf{q}_0; h)^n$, $n = 0, 1, 2, \ldots$ (except for a negligible remainder) coincides with an orbit $\tilde{\Phi}(\mathbf{p}_0, \mathbf{q}_0; nh)$ of a neighbouring *Hamiltonian* problem. In a backward error analysis approach we are solving exactly an approximate *Hamiltonian problem*. This interpretation is not possible for nonsymplectic integrators: the numerical solution is an exact solution of a problem in which the Hamiltonian structure has been lost.

Numerical experiments have shown that, for Hamiltonian problems, symplectic integrators may well be an improvement on their nonsymplectic counterparts. However the development of symplectic methods has so far been confined to *constant step-size* formulae and, accordingly, numerical tests have used, as reference algorithms, constant step-size implementations of classical methods. Such implementations are, by modern numerical ODE standards very naive, and the question arises of whether, for Hamiltonian problems, a symplectic method with constant step-sizes may actually be more efficient than a modern variable-step code. Before we carried out the experiments reported in this paper, we felt that the answer to that question would be no. On the other hand we suspected that, for Hamiltonian problems, variable step-size symplectic algorithms would improve on standard variable step-size algorithms. Accordingly we decided to develop variable step-size symplectic algorithms.

In this paper we report on our experience with the construction and assessement of *variable step, symplectic, explicit Runge-Kutta-Nyström methods*. We used Runge-Kutta-Nyström methods rather than Runge-Kutta methods because all symplectic Runge-Kutta formulae are implicit [11]. It appears that both our guesses above were wrong: *constant step-size symplectic methods may beat standard variable step-size codes, but variable step-size symplectic codes are not more advantageous than standard variable step-size codes.*

2. Runge-Kutta-Nyström methods.

We restrict our attention to systems of the special form

$$\dot{\mathbf{p}} = \mathbf{f}(\mathbf{q}), \qquad \dot{\mathbf{q}} = \mathbf{p} \tag{2.1}$$

(i.e. to second order systems $\ddot{\mathbf{q}} = \mathbf{f}(\mathbf{q})$). If $\mathbf{f}$ is the gradient of a scalar function $-V(\mathbf{q})$, then (2.1) is a Hamiltonian system with

$$H = H(\mathbf{p}, \mathbf{q}) = T(\mathbf{p}) + V(\mathbf{q}), \qquad T(\mathbf{p}) = \tfrac{1}{2}\mathbf{p}^T\mathbf{p}.$$

In Mechanics, the $\mathbf{q}$ variables represent Lagrangian coordinates, the $\mathbf{p}$ variables the corresponding momenta, $\mathbf{f}$ the forces, T is the kinetic energy, V the potential energy, and H the total energy [2].

An explicit Runge-Kutta-Nyström method for (2.1) takes the form [5, 8]

$$\mathbf{Q}_i = \mathbf{q}_n + h\gamma_i\mathbf{p}_n + h^2 \sum_{j<i} \alpha_{ij}\,\mathbf{f}(\mathbf{Q}_j),$$

$$\mathbf{p}_{n+1} = \mathbf{p}_n + h\sum_{i=1}^{s} b_i\,\mathbf{f}(\mathbf{Q}_i), \tag{2.2}$$

$$\mathbf{q}_{n+1} = \mathbf{q}_n + h\mathbf{p}_n + h^2 \sum_{i=1}^{s} \beta_i\,\mathbf{f}(\mathbf{Q}_i),$$

where, we assume unless otherwise mentioned that the following well-known condition [5, 8] holds

$$\beta_i = b_i(1 - \gamma_i), \qquad 1 \le i \le s. \tag{2.3}$$

As in [5], we consider FSAL methods, i.e. methods with

$$\gamma_1 = 0, \qquad \gamma_s = 1, \tag{2.4a}$$

$$\alpha_{sj} = \beta_j, \qquad 1 \le j \le s - 1. \tag{2.4b}$$

Note that (2.4a) implies, via (2.3), that $\beta_s = 0$ and then the last stage $\mathbf{Q}_s$ of the current step coincides with $\mathbf{q}_{n+1}$, which, in turn, is the first stage of the next step. Therefore, a step of an FSAL s-stage method requires only $s - 1$ evaluations of $\mathbf{f}$.

The method (2.2) is symplectic if [18, 10, 3]

$$\alpha_{ij} = b_j(\gamma_i - \gamma_j), \quad i > j. \tag{2.5}$$

For symplectic methods with s stages, we have s coefficients b_i and s coefficients γ_i as free parameters; the coefficients β_i and α_{ij} are determined by (2.3) and (2.5) respectively. On the other hand (2.3), (2.4a) and (2.5) imply (2.4b), so that a symplectic FSAL method has s coefficients b_i and $s - 2$ coefficients γ_i, $2 \leq i \leq s - 1$, as free parameters.

3. Derivation of a fourth-order symplectic, FSAL method.

For a method (2.2)–(2.3) to have order four, the coefficients should satisfy *seven* order conditions [8]. However, for symplectic methods, not all order conditions are independent [1, 3, 16] and in fact it turns out [3] that it is sufficient to impose only *six* of them. For FSAL symplectic methods, four stages furnish six free coefficients, and, after imposing order four, no room is left for 'tuning' the formula. We then settle for five-stage FSAL, symplectic methods, for which a two-parameter family of order four methods exists. Following a standard practice (see [5, 6]) we choose, amongst the members of this family, the method with 'smallest' truncation error.

The **p**-truncation and **q**-truncation errors of an RKN method are respectively of the form [5, 8]

$$\sum_{i=0}^{\infty} h^{i+1} \sum_j c_j'^{(i+1)} \mathbf{F}_j^{(i+1)} \tag{3.1a}$$

and

$$\sum_{i=1}^{\infty} h^{i+1} \sum_k c_k^{(i+1)} \mathbf{F}_k^{(i)}, \tag{3.1b}$$

where the $\mathbf{F}_j^{(i)}$ are the elementary differential that only depend on the problem (2.1) being integrated and the $c_j'^{(i+1)}$ and $c_k^{(i+1)}$ are polynomials in the method coefficients α_{ij}, γ_i, β_i, b_i. In (3.1a), the sum in j is extended to all special Nyström trees with $i+1$ nodes, whilst in (3.1b) the sum in k is extended to all special Nyström trees with i nodes. For fourth-order methods, $c_j'^{(i)}$ and $c_k^{(i)}$ vanish for $i \leq 4$ and we try to minimize $c_j'^{(5)}$ and $c_k^{(5)}$. We proceed as follows. Let us respectively denote by $\mathbf{c}'^{(i)}$ and $\mathbf{c}^{(i)}$ the vectors with components $c_j'^{(i)}$ and $c_k^{(i)}$ and set

$$A'^{(5)} = \|\mathbf{c}'^{(5)}\|, \qquad A^{(5)} = \|\mathbf{c}^{(5)}\|. \tag{3.2}$$

(The norm is the standard Euclidean norm.) We consider $\phi = A'^{(5)^2} + A^{(5)^2}$ as a function of the eight free coefficients γ_i, $2 \leq i \leq 4$ and b_j, $1 \leq j \leq 5$ and use the NAG subroutine

E04UCF to minimize ϕ, subject to the six equality constraints that impose order four and to bounds $-1.5 \leq \gamma_i, b_j \leq 1.5$. Of course, the minimization subroutine requires an initial guess for the minimum and converges only to a local minimum that depends on the initial guess. A thousand random initial guesses (subject to $-1.5 \leq \gamma_i, b_j \leq 1.5$) were taken and we kept the local minimum with smallest value of ϕ. The method thus obtained does not satisfy to machine precision the conditions for order four, because the NAG routine fails in exactly enforcing the equality constraints. We then kept the values b_1 and b_2 provided by the minimization routine and determined γ_i, $2 \leq i \leq 4$, and b_j, $3 \leq j \leq 5$, by solving the six equations for order four by means of Newton's method in quadruple precision. This of course resulted in a solution, that while being close to that provided by the minimization procedure, satisfies the order conditions to a very high precision. The coefficients are given by

$$
\begin{array}{ll}
\gamma_1 = 0 & b_1 = 0.061758858135626325 \\
\gamma_2 = 0.205177661542286386 & b_2 = 0.338978026553643355 \\
\gamma_3 = 0.608198943146500973 & b_3 = 0.614791307175577566 \qquad (3.3) \\
\gamma_4 = 0.487278066807586965 & b_4 = -0.140548014659373380 \\
\gamma_5 = 1 & b_5 = 0.125019822794526133
\end{array}
$$

along with (2.3) and (2.5).

For this method the quantities in (3.2) are $A'^{(5)} = 6.7E - 4$ and $A^{(5)} = 7.1E - 4$. As a reference method for the numerical tests, we employ the fourth-order, FSAL nonsymplectic formula of Dormand et al. [5, Table 3]. This has four stages (three evaluations) and $A'^{(5)} = 1.8E - 3$, $A^{(5)} = 4.6E - 4$. Thus, per step, the reference method achieves an accuracy comparable to that of the symplectic method (3.3), but is cheaper by a factor 3/4. In general, symplectic integrators require, for the same accuracy, more work than their nonsymplectic counterparts, since to impose symplecticness free parameters are sacrificed that could otherwise be directed at achieving accuracy.

4. Error estimation. The standard way [5, 8] of estimating the errors in a p-th order RKN method (2.2) is to supplement (2.2) with formulae

$$
\hat{p}_{n+1} = p_n + h \sum_{i=1}^{s} \hat{b}_i \, f(Q_i),
$$

$$
\hat{q}_{n+1} = q_n + h p_n + h^2 \sum_{i=1}^{s} \hat{\beta}_i \, f(Q_i),
$$

$$(4.1)$$

in such a way that $(\mathbf{p}_n, \mathbf{q}_n) \mapsto (\hat{\mathbf{p}}_{n+1}, \hat{\mathbf{q}}_{n+1})$ is a RKN method of order $q < p$ (usually $q = p - 1$ or $q = p - 2$). Of course the computation of $(\hat{\mathbf{p}}_{n+1}, \hat{\mathbf{q}}_{n+1})$ employs the same function evaluations $\mathbf{f}(\mathbf{Q}_i)$ that are used to compute $(\mathbf{p}_{n+1}, \mathbf{q}_{n+1})$. The difference between the low order $(\hat{\mathbf{p}}_{n+1}, \hat{\mathbf{q}}_{n+1})$ and high order $(\mathbf{p}_{n+1}, \mathbf{q}_{n+1})$ results is then taken to be an approximation to the local error at the step $n \mapsto n + 1$. For (3.3), we take the order q of the imbedded method to be 3.

The weights $\hat{b}_i$, $1 \le i \le 5$, must satisfy four equations for the local error in $\hat{\mathbf{p}}_{n+1}$ to be $O(h^4)$. These equations are linear in the $\hat{b}_i$'s and it is a simple matter to express $\hat{b}_i$, $1 \le i \le 4$, in terms of $\hat{b}_5$, that remains a free parameter. The value of $\hat{b}_5$ is chosen according to a procedure suggested by Dormand and Prince. The quantities

$$C'^{(5)} = \frac{\|\mathbf{c}'^{(5)} - \hat{\mathbf{c}}'^{(5)}\|}{\|\hat{\mathbf{c}}'^{(4)}\|}, \qquad B'^{(5)} = \frac{\|\hat{\mathbf{c}}'^{(5)}\|}{\|\hat{\mathbf{c}}'^{(4)}\|} \tag{4.2}$$

should be made as small as possible (letters with a hat refer of course to the lower order method). A small $C'^{(5)}$ ensures that, in the (p-component of) the error estimator, the leading $O(h^4)$ term dominates over the next $(O(h^5))$ term of the Taylor expansion. This is benefitial, since the mechanism for stepsize selection assumes an $O(h^4)$ behaviour in the estimator. A small $B'^{(5)}$ ensures that the third-order formula used for estimation is sufficiently different from the fourth-order formula used for time-stepping. We minimize the function $\phi(\hat{b}_5) = B'^{(5)^2} + C'^{(5)^2}$ by the simple procedure of evaluating ϕ at uniformly spaced values of $\hat{b}_5$ (the spacing used was 0.01). This yields $\hat{b}_5 = 0.2$.

The weights $\hat{\beta}_i$, $1 \le i \le 5$, are not assumed to be related via (2.3) to the $\hat{b}_i$. They must satisfy two (linear) equations for the local error in $\hat{\mathbf{q}}_{n+1}$ to be $O(h^4)$. This leaves three free parameters. We arbitrarily set $\hat{\beta}_5 = 0$ and expressed $\hat{\beta}_1$ and $\hat{\beta}_2$ in terms of $\hat{\beta}_3$ and $\hat{\beta}_4$. The free $\hat{\beta}_3$, $\hat{\beta}_4$ are now chosen to minimize $B^{(5)^2} + C^{(5)^2}$, where

$$C^{(5)} = \frac{\|\mathbf{c}^{(5)} - \hat{\mathbf{c}}^{(5)}\|}{\|\hat{\mathbf{c}}^{(4)}\|}, \qquad B^{(5)} = \frac{\|\hat{\mathbf{c}}^{(5)}\|}{\|\hat{\mathbf{c}}^{(4)}\|}. \tag{4.3}$$

The minimization was again performed by sampling the objective function on a grid with 0.01×0.01 spacing. The weights of the third-order formula (4.1) imbedded in (3.3) are as follows

$$\hat{b}_1 = -0.127115143890665440 \qquad \hat{\beta}_1 = 0.110014238746029571$$

$$\hat{b}_2 = 0.698831995430764851 \qquad \hat{\beta}_2 = 0.189985761253970428$$

$$\hat{b}_3 = 0.375269477646788521 \qquad \hat{\beta}_3 = 0.25 \tag{4.4}$$

$$\hat{b}_4 = -0.146986329186887931 \qquad \hat{\beta}_4 = -0.05$$

$$\hat{b}_5 = 0.2 \qquad \hat{\beta}_5 = 0$$

With this choice the quantities in (4.2), (4.3) are

$$C'^{(5)} = 1.06, \qquad B'^{(5)} = 1.06, \qquad C^{(5)} = 0.47, \qquad B^{(5)} = 0.25 .$$

For the fourth-order nonsymplectic scheme used as a reference method, Dormand et al [5] provide an embedded formula with

$$C'^{(5)} = 1.19, \qquad B'^{(5)} = 1.20, \qquad C^{(5)} = 1.02, \qquad B^{(5)} = 1.03 .$$

This shows that the minimizations we carried out above are as successful as those in [5].

5. Numerical results.

The imbedded pair (3.3), (4.4) and the reference imbedded pair were implemented in a standard way following closely the code DOPRIN in [8]. Several test problems were used, but we only report on the results corresponding to the Kepler potential [2] $V(q^1, q^2) = -1/\|q\|$, with initial condition

$$p^1 = 0, \quad p^2 = \sqrt{\frac{1+e}{1-e}}, \quad q^1 = 1 - e, \quad q^2 = 0.$$

Here e is a parameter $0 \le e < 1$. The solution is 2π-periodic and its projection onto the (configuration) q-space is an ellipse with eccentricity e and major semiaxis 1. Initially, the moving mass is at the pericentre of the ellipse (i.e. the closest it can be to the coordinate origin). After half a period (apocentre), its distance r to the origin is $1 + e$. Thus $r_{max}/r_{min} = (1 + e)/(1 - e)$, which is large for large eccentricities. Moreover, the i-th derivatives of the force f behave like $r^{-(i+2)}$, so that, for large eccentricities, the elementary differentials of high order may vary by several orders of magnitude along the orbit. In fact, our well-known test problem with small e (say $e = 0.9$) is often taken as a "severe test for the step-size control procedure" of ODE algorithms [6].

The test equation was integrated combining each of the eccentricities 0.1, 0.3, 0.5, 0.7, 0.9 with each of the final times $10 \times 2\pi$, $30 \times 2\pi$, $90 \times 2\pi$, $270 \times 2\pi$, $810 \times 2\pi$, $2430 \times 2\pi$, $7290 \times 2\pi$, $21870 \times 2\pi$. We were particularly interested in long time intervals, as it is in this sort of simulation that the advantages of symplecticness should be felt (see [12]). For short time intervals, the local error of the formula is of paramount importance and it is as the time interval gets larger that advantages derived from a better qualitative behaviour become more prominent. In Celestial Mechanics very long time integrations are often required with potentials that are small perturbations of the Kepler potential considered here.

In the tests we used the symplectic variable-step code (SV), the nonsymplectic variable-step code (NSV) and also fixed-step implementations of the symplectic (FS) and nonsymplectic (NSF) formulae. The variable-step codes were tried with absolute error tolerances of $1.E-4$, $1.E-5$, ... $1.E-11$ and the fixed-step algorithms were run with step-sizes $2\pi/16$, $2\pi/32$, ... , $2\pi/2048$. Errors were measured in the Euclidean norm of $\mathcal{R}^4$.

Figure 1 gives, for $e = 0.5$ and a final time of 21870 periods, the final error against the computational effort measured by the number of **f**-evaluations. The figure contains information for the runs that yielded errors in the $1.E-1$–$1.E-4$ range, namely

  (i) SV with tolerances $1.E-10$, $1.E-11$ (plus signs joined by a dashed line).
 (ii) NSV with tolerances $1.E-9$, $1.E-10$, $1.E-11$ (circles joined by a solid line).
(iii) SF with time-step $2\pi/256$, $2\pi/512$, $2\pi/1024$ (stars joined by a dashdot line).
(iiii) NSF with time step $2\pi/2048$ (a $\times$ sign).

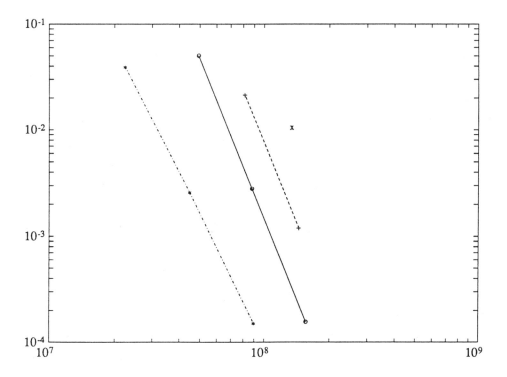

Figure 1. Error against number of function evaluations, after 21870 periods, $e = 0.5$.

Let us first compare the results of SF and NSF. Recall that these RKN formulae roughly have error constants of the same size, but SF has four evaluations per step against three evaluations per step in NSF. Thus *on local error considerations alone,* one would expect that for the same error the numbers of evaluations of the NSF and SF formulae would be in a ratio 3/4. On the contrary, the experimental results show that, for the same error, the symplectic formula is *four times* less expensive than the nonsymplectic process. This shows that there is something in the error propagation mechanism of the symplectic algorithm that gives it a clear advantage over its nonsymplectic counterpart. In fact, it can be proved rigorously [4] that the *global error* of the symplectic algorithm behaves like $O(th^4)$, while for the nonsymplectic method the behaviour is $O(th^4 + t^2h^5)$. For moderate or large values of t the t^2h^5 term dominates over the th^4 term, and NSF behaves like a fifth order method with a very large error constant. Thus, for a given error, the symplectic method needs $h = O(t^{-0.25})$ while the nonsymplectic formula requires $h = O(t^{-0.4})$. This shows that, for large t symplecticness pays. In fact, when $e = 0.5$, SF improves on NSF if t_{final} is larger than, say, 30 periods.

Turning now to a comparison between NSF and NSV, we observe that for the formula of Dormand *et al.* the use of variable step-sizes results in a gain in efficiency by a factor of 2. In the apocentre, the variable step-size code takes step-sizes about 7 times as large as those it takes near the pericentre, with the result that, as expected, NSV saves on function evaluations for a given error. Note that the line that joins the NSV points reveals an h^5 behaviour of the error, in spite of the method having order four. This is again due to the fact that the global error has terms $O(th^4)$ and $O(t^2h^5)$; in the range of t-values and h-values of practical significance the $O(t^2h^5)$ term dominates the $O(th^4)$ term.

On the other hand, for the symplectic formula, going from fixed to variable step-sizes results in a *decrease* in efficiency. We shall return to this point later. For the time being, let us note that, with variable step-sizes, the line joining the points of the symplectic algorithm are in agreement with an h^5 behaviour of the error. In fact SV and NSV show a very similar behaviour. The only difference between them lies in the fact that, for a given error, the costs of NSV and SV are in a ratio 3/4, i.e. in the ratio one would have anticipated from a study of the local errors, without taking symplecticness into account.

As Figure 1, Figure 2 corresponds to a final time $21870 \times 2\pi$, but now $e = 0.3$. Again we have displayed the results corresponding to runs for which the errors lie in the $1.E - 1$–$1.E - 4$ range. These are:

(i) SV with tolerances $1.E - 10$, $1.E - 11$ (plus signs joined by a dashed line).

(ii) NSV with tolerances $1.E - 9$, $1.E - 10$ (circles joined by a solid line).

(iii) SF with time-step $2\pi/128$, $2\pi/256$, $2\pi/512$ (stars joined by a dashdot line).

(iiii) NSF with time-step $2\pi/1024$, $2\pi/2048$ ($\times$ sign, dotted line).

We see that the overall pattern is not changed by changing the eccentricity. The NSV and SV algorithms have efficiencies that are still in the predicted 3/4 ratio. On the other hand, with $e = 0.3$ the advantages of NSV over NSF are less marked as one would have expected. In fact, for $e = 0.3$, both variable-step codes only vary by a factor of 3 the step-size along the orbit. The NSF points, that for $e = 0.5$ were to the right of the SV dashed line, are now exactly on this SV line.

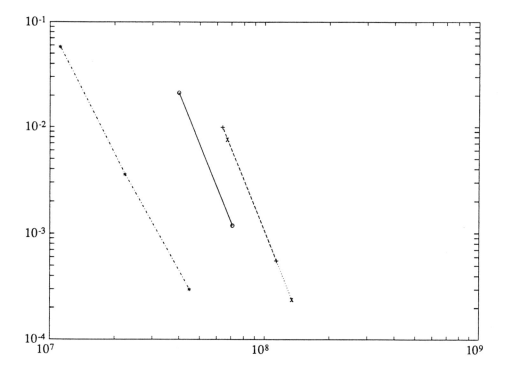

Figure 2. Error against number of function evaluations, after 21870 periods, $e = 0.3$.

Figure 3 corresponds to the same final time with $e = 0.7$. The following runs are represented (results for NSF are not reported as, for step-sizes used, error below $1.E-1$ could not be obtained)

(i) SV with tolerances $1.E - 10$, $1.E - 11$ (plus signs joined by a dashed line).

(ii) NSV with tolerances $1.E - 10$, $1.E - 11$ (circles joined by a solid line).

(iii) SF with time-step $2\pi/512$, $2\pi/1024$, $2\pi/2048$ (stars joined by a dashdot line).

Now SV and NSV become more efficient and change h by a factor of 22. Nevertheless, SF is still the most efficient method: the advantages of symplecticness are not offset by the disadvantages of constant h.

For the smaller values of t_{final} we tried, the picture is very much the same: except if t_{final} is not large and e is large, SF is the most efficient method, NSV is second and SV is 4/3 times worse than NSV. For fixed t_{final}, as e approaches 1, the benefits of variable steps become more prominent and NSV improves on SF. For fixed e, as t_{final} increases, the benefits of symplecticness dominate and SF improves on NSV.

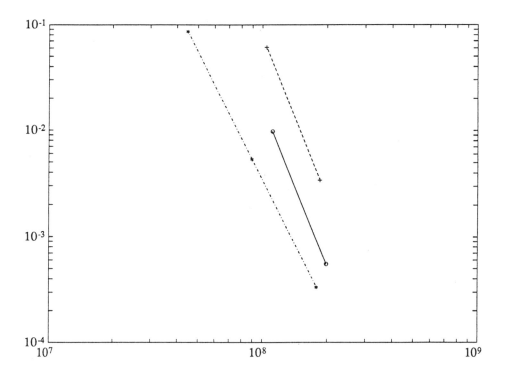

Figure 3. Error against number of function evaluations, after 21870 periods, $e = 0.7$.

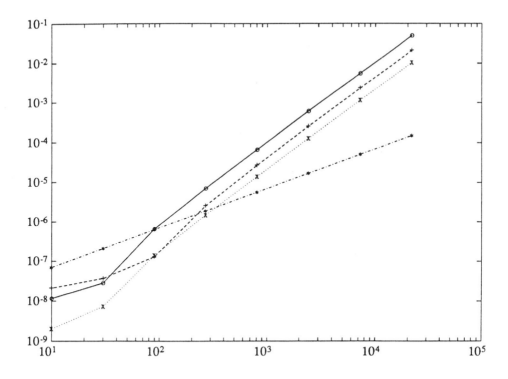

Figure 4. Error against time in periods, $e = 0.5$

Figure 4 gives, for $e = 0.5$, error against time for SV (tolerance $1.E - 10$), NSV (tolerance $1.E - 9$), NF ($h = 2\pi/256$) and NSF ($h = 2\pi/2048$). For SF the error shows a linear behaviour with respect to t, in agreement with earlier considerations. For the other methods the error grows like t^2.

6. Discussion. The experimental results just described and experiments with other potentials (not reported here) have revealed the following points:

(i) Symplectic fixed-step formulae are well suited for the numerical integration of Hamiltonian problems over long time intervals. This agrees with theoretical expectations. Symplectic fixed-step formulae had already been experimentally tested against nonsymplectic fixed-step formulae. Our investigations have now shown that *symplectic fixed-step formulae can beat, not only their nonsymplectic fixed-step counterparts, but also standard variable-step codes.* Obviously the advantages of a symplectic fixed-step code over a variable-step code are smaller than the advantages of a symplectic fixed-step formula over a standard fixed-step formula. While our experiments have been restricted to order four methods, symplectic methods of higher order exist and are currently being studied.

(ii) The advantages of symplecticness and variable-steps cannot be combined. In other words, when a symplectic formula is implemented with variable steps with a view to increasing its efficiency, the advantages of symplecticness disappear. In the experiments above, the performance of SV can always be explained in terms of local errors and cost per step. The only difference between SV and NSV lies in the fact that the first requires four evaluations per step and the second only three.

What is the reason for (ii)? Several explanations could be given. Several authors (see notably [17]) have argued that integrating with variable steps a system

$$dy/dt = \mathbf{F}(\mathbf{y}) \tag{6.1}$$

is to some extent equivalent to integrating with the same method and constant step-sizes a system

$$dy/d\tau = \alpha(\mathbf{y})\mathbf{F}(\mathbf{y}), \tag{6.2}$$

where α is a real-valued function, $\alpha > 0$, determined by the step-changing mechanism and $dt/d\tau = \alpha(\mathbf{y})$. The function α is small (resp. large) in "difficult" (resp. "easy") regions and there, to a given constant step-size in the variable τ, correspond small (resp. large) values of the step-size in t. Now the point is that, when (6.1) is a Hamiltonian system, (6.2) is *not*, in general, Hamiltonian. Then there is no advantage to be expected from the integration of (6.2) with a constant step-size symplectic method and, by implication, there is nothing to recommend integrating (6.1) with a variable-step symplectic integrator.

A different argument would be as follows. We pointed out in the introduction that a step of a symplectic method is (except for a negligible remainder) identical to the h-flow of a Hamiltonian system with perturbed Hamiltonian $\tilde{H}_h$. The key point is that $\tilde{H}_h$ does depend on h (the closer h to 0, the better $\tilde{H}_h$ approximates the original H). Hence for variable steps is not possible to see a computed trajectory for (1.1) as an exact trajectory of a neighbouring Hamiltonian problem.

The experiments performed suggest that fixed-step symplectic algorithms should be recommended for long-time integrations of Hamiltonian problems where the time-scale does not vary dramatically along the solution. When the time-scale undergoes very significant changes a standard code may become competitive. It is also possible that Hamiltonian problems where the time-scale varies dramatically along the solution should first be analytically transformed into Hamiltonian problems amenable to fixed-step integrations and then be integrated by a symplectic numerical method with constant step-sizes.

Acknowledgement. This research has been supported by "Junta de Castilla y León" under project 1031-89 and by "Dirección General de Investigación Científica y Técnica" under project PB89-0351.

References

1. L. Abia and J. M. Sanz-Serna, "Partitioned Runge-Kutta methods for separable Hamiltonian problems," *Applied Mathematics and Computation Reports*, Report 1990/8, December 1990, Universidad de Valladolid.

2. V. I. Arnold, *Mathematical Methods of Classical Mechanics*, 2nd ed., Springer, New York, 1989.

3. M. P. Calvo and J. M. Sanz-Serna, "Order conditions for canonical Runge-Kutta-Nyström methods," *Applied Mathematics and Computation Reports*, Report 1991 /1, January 1991, Universidad de Valladolid.

4. M. P. Calvo and J. M. Sanz-Serna, "Reasons for a failure. The integration of the two-body problem with a symplectic Runge-Kutta-Nyström code with step-changing facilities," (in preparation).

5. J. R. Dormand, M. E. A. El-Mikkawy and P. J. Prince, "Families of Runge-Kutta-Nyström formulae" *IMA J. Numer. Anal.*, vol. 7, pp. 235–250, 1987.

6. J. R. Dormand and P. J. Prince, "A family of embedded Runge-Kutta formulae" *J. Comput. Appl. Math.*, vol. 6, pp. 19–26, 1980.

7. T. Eirola and J. M. Sanz-Serna, "Conservation of integrals and symplectic structure in the integration of differential equations by multistep methods," *Applied Mathematics and Computation Reports*, Report 1990/9, December 1990, Universidad de Valladolid.

8. E. Hairer, S. P. Nørsett and G. Wanner, *Solving Ordinary Differential Equations I, Nonstiff Problems*, Springer, Berlin, 1987.

9. R. S. Mackay, "Some aspects of the dynamics and numerics of Hamiltonian systems," preprint.

10. D. Okunbor and R. D. Skeel, "An explicit Runge-Kutta-Nyström method is canonical if and only if its adjoint is explicit," preprint.

11. J. M. Sanz-Serna, "Runge-Kutta schemes for Hamiltonian systems," *BIT*, vol. 28, pp. 877–883, 1988.

12. J. M. Sanz-Serna, "The numerical integration of Hamiltonian systems," in *Proceedings of the Conference on Computational Differential Equations, Imperial College London, 3rd–7th July 1989*, (to appear).

13. J. M. Sanz-Serna, "Two topics in nonlinear stability," in *Advances in Numerical Analysis, Vol. 1*, W. Light ed., Clarendon, Oxford, 1991.

14. J. M. Sanz-Serna, "Symplectic Runge-Kutta and related methods: recent results," *Applied Mathematics and Computation Reports*, Report 1991 /5, June 1991, Universidad de Valladolid.

15. J. M. Sanz-Serna, "Symplectic integrators for Hamiltonian problems: an overview," (in preparation).

16. J. M. Sanz-Serna and L. Abia, "Order conditions for canonical Runge-Kutta schemes," *SIAM J. Numer. Anal.* (to appear).

17. D. Stoffer and K. Nipp, "Invariant curves for variable step size integrators" *BIT*, vol. 31, pp. 169–180, 1991.

18. Y. B. Suris, "Canonical transformations generated by methods of Runge-Kutta type for the numerical integration of the system $x'' = -\partial U/\partial x$," *Zh. Vychisl. Mat. i Mat. Fiz.*, vol. 29, pp. 202–211, 1989. (in Russian).

M. P. Calvo and J. M. Sanz-Serna

Departamento de Matemática Aplicada y Computación

Facultad de Ciencias

Universidad de Valladolid

Valladolid

Spain

A R CONN, N I M GOULD AND Ph L TOINT

On the number of inner iterations per outer iteration of a globally convergent algorithm for optimization with general nonlinear constraints and simple bounds

1 Introduction

In this paper, we consider the nonlinear programming problem

$$\underset{x \in \Re^n}{\text{minimize}} \quad f(x) \tag{1.1}$$

subject to the general constraints

$$c_i(x) = 0, \quad i = 1, \ldots, m, \tag{1.2}$$

and the simple bounds

$$l \le x \le u. \tag{1.3}$$

We assume that the region $\mathcal{B} = \{x \in \Re^n \mid l \le x \le u\}$ is non-empty and may be infinite. Furthermore, we assume that

AS1. $f(x)$ and the $c_i(x)$ are twice continuously differentiable for all x in $\mathcal{B}$.

The exposition is conveniently simplified by taking the lower bounds as identically equal to zero and the upper bound as infinity. Thus, in most of what follows, $\mathcal{B} = \{x \in \Re^n \mid x \ge 0\}$. The modification required to handle more general bounds is indicated at the end of the paper.

The approach we intend to take is that of Conn et al. (1991a) and is based upon incorporating the equality constraints via an augmented Lagrangian function whilst handling upper and lower bounds directly. The sequential, approximate minimization of the augmented Lagrangian function is performed in a trust region framework such as that proposed by Conn et al. (1988a).

Our aim in this paper is to consider how these two different algorithms mesh together. In particular, we aim to show that ultimately very little work is performed in the iterative sequential minimization algorithm for every iteration of the outer augmented Lagrangian algorithm. This is contrary to most analyses of sequential penalty function and augmented Lagrangian methods in which the effort required to solve the inner iteration subproblems is effectively disregarded, the analysis concentrating on the convergence of the outer iteration (see for instance the book by Bertsekas, 1982. An exception to this is the sequential penalty function method analyzed by Gould, 1989).

This work was primarily motivated by observations that the authors made when testing their large-scale nonlinear programming package LANCELOT (see, Conn et al., 1991b), which is an implementation of the algorithms discussed in this paper. It was often apparent that only a single iteration of the inner iteration subroutine SBMIN was ultimately required for every outer iteration of the sequential augmented Lagrangian program AUGLG. While the conditions required in this paper to turn this observation to a proven result are relatively strong (and we feel probably about as weak as is possible), the package frequently exhibits the same behaviour on problems which violate our assumptions.

We define the concepts and notation that we shall need in section 2. Our algorithm is fully described in section 3 and analyzed in sections 4 and 5.

2   Notation

Let $g(x)$ denotes the gradient $\nabla_x f(x)$ of $f(x)$. Similarly, let $A(x)$ denote the Jacobian of $c(x)$, where

$$c(x) = [c_1(x), \cdots, c_m(x)]^T. \tag{2.1}$$

Thus

$$A(x)^T = [\nabla c_1(x), \cdots \nabla c_m(x)]. \tag{2.2}$$

We define the Lagrangian and augmented Lagrangian functions as

$$\ell(x, \lambda) = f(x) + \sum_{i=1}^{m} \lambda_i c_i(x), \tag{2.3}$$

and

$$\Phi(x, \lambda, \mu) = f(x) + \sum_{i=1}^{m} \lambda_i c_i(x) + \frac{1}{2\mu} \sum_{i=1}^{m} c_i(x)^2 \tag{2.4}$$

respectively. We note that $\ell(x, \lambda)$ is the Lagrangian with respect to the c_i constraints only. Let $g_\ell(x, \lambda)$ and $H_\ell(x, \lambda)$ respectively denote the gradient, $\nabla_x \ell(x, \lambda)$, and Hessian, $\nabla_{xx} \ell(x, \lambda)$, of the Lagrangian.

We denote the non-negativity restrictions by

$$x \in \mathcal{B} = \{x \in \Re^n \mid x \geq 0\}. \tag{2.5}$$

We will make much use of the projection operator defined componentwise by,

$$(P[x, l, u])_i = \begin{cases} l_i & \text{if } x_i \leq l_i \\ u_i & \text{if } x_i \geq u_i \\ x_i & \text{otherwise.} \end{cases} \tag{2.6}$$

This operator projects the point x onto the region defined by the simple bounds (1.3). Let

$$P(x, v, l, u) = x - P[x - v, l, u]. \tag{2.7}$$

Furthermore, define $P[x] = P[x, 0, \infty]$ and $P(x, v) = P(x, v, 0, \infty)$.

Let $x^{(k)} \in \mathcal{B}$ and $\lambda^{(k)}$ be given values of x and λ. If $h(x, \lambda, \ldots)$ is any function of $x, \lambda, \ldots$, we shall write $h^{(k)}$ as a shorthand for $h(x^{(k)}, \lambda^{(k)}, \ldots)$.

Now, let $\mathcal{N} = \{1, \ldots, n\}$. For any $x^{(k)}$ we have two possibilities for each component $x_i^{(k)}, i = 1, \ldots, n$, namely (i) $0 \leq x_i^{(k)} \leq (\nabla_x \Phi^{(k)})_i$ or (ii) $(\nabla_x \Phi^{(k)})_i < x_i^{(k)}$. We shall call all $x_i^{(k)}$ that satisfy (i) *dominated* variables while the remaining $x_i^{(k)}$ are *floating* variables. It is important to notice that, as $x^{(k)} \in \mathcal{B}$

$$(P(x^{(k)}, \nabla_x \Phi^{(k)}))_i = x_i^{(k)} \quad \text{whenever } x_i^{(k)} \text{ is dominated} \tag{2.8}$$

while

$$(P(x^{(k)}, \nabla_x \Phi^{(k)}))_i = (\nabla_x \Phi^{(k)})_i \quad \text{otherwise.} \tag{2.9}$$

50

If x^* is the limit point of the (sub-)sequence $\{x^{(k)}\}, k \in K$, we partition $\mathcal{N}$ into four index sets related to the two possibilities (i) and (ii) above and the corresponding x^*. We define

$$
\begin{aligned}
\mathcal{D}_1 &\overset{\text{def}}{=} \{i \mid x_i^{(k)} \text{ are dominated for all } k \in K \text{ sufficiently large }\}, \\
\mathcal{F}_1 &\overset{\text{def}}{=} \{i \mid x_i^{(k)} \text{ are floating for all } k \in K \text{ sufficiently large and } x_i^* > 0\}, \\
\mathcal{F}_2 &\overset{\text{def}}{=} \{i \mid x_i^{(k)} \text{ are floating for all } k \in K \text{ sufficiently large but } x_i^* = 0\} \text{ and} \\
\mathcal{F}_3 &\overset{\text{def}}{=} \mathcal{N} \setminus (\mathcal{D}_1 \cup \mathcal{F}_1 \cup \mathcal{F}_2).
\end{aligned}
\tag{2.10}
$$

We will use the notation that if $\mathcal{J}_1$ and $\mathcal{J}_2$ are any subsets of $\mathcal{N}$ and H is an n by n matrix, $H_{[\mathcal{J}_1, \mathcal{J}_2]}$ is the matrix formed by taking the rows and columns of H indexed by $\mathcal{J}_1$ and $\mathcal{J}_2$ respectively. Likewise, if A is an m by n matrix, $A_{[\mathcal{J}_1]}$ is the matrix formed by taking the columns of A indexed by $\mathcal{J}_1$.

We denote the (appropriately dimensioned) identity matrix by I; its i-th column is e_i. A vector of ones is denoted by e.

We will use a variety of vector and subordinate matrix norms. We shall only consider norms $\| \cdot \|_z$ which are *consistent* with the two-norm, that is, norms which satisfy the inequalities

$$
\|v\|_z \leq a_0^{\frac{1}{2}} \|v\|_2 \quad \text{and} \quad \|v\|_2 \leq a_0^{\frac{1}{2}} \|v\|_z
\tag{2.11}
$$

for all vectors v and some constant $a_0 \geq 1$, independent of z. It then follows that, for any pair of two-norm-consistent norms $\| \cdot \|_y$ and $\| \cdot \|_z$,

$$
\|v\|_z \leq a_0 \|v\|_y \quad \text{and} \quad \|v\|_y \leq a_0 \|v\|_z.
\tag{2.12}
$$

Following Conn *et al.* (1991a), we now describe an algorithm for solving (1.1), (1.2) and (2.5).

3 Statement of the algorithm

In order to solve the problem (1.1), (1.2) and (2.5), we consider the following algorithmic model.

Algorithm 3.1 [Outer Iteration Algorithm]

step 0 : [Initialization] *The positive constants η_0, ω_0, μ_0, $\tau < 1$, $\gamma_0 < 1$, α_ω, β_ω, α_η, β_η, small positive convergence tolerances ω_* and η_*, and the vector $\lambda^{(0)} \in \Re^m$ are chosen. The two-norm-consistent norms $\| \cdot \|_p$ and $\| \cdot \|_c$ are specified. We require that*

$$
\alpha_\eta < \min(1, \alpha_\omega) \quad \text{and} \quad \beta_\eta < \min(1, \beta_\omega).
\tag{3.1}
$$

Set $\mu^{(0)} = \mu_0$, $\alpha^{(0)} = \min(\mu^{(0)}, \gamma_0)$, $\omega^{(0)} = \omega_0 (\alpha^{(0)})^{\alpha_\omega}$, $\eta^{(0)} = \eta_0 (\alpha^{(0)})^{\alpha_\eta}$ and $k = 0$.

step 1 : [Inner Iteration] *Find $x^{(k)} \in \mathcal{B}$ such that*

$$
\|P(x^{(k)}, \nabla_x \Phi^{(k)})\|_p \leq \omega^{(k)}
\tag{3.2}
$$

If

$$
\|c(x^{(k)})\|_c \leq \eta^{(k)}
\tag{3.3}
$$

execute step 2. Otherwise, execute step 3.

step 2 : [Test for convergence and update Lagrange multipliers] *If*

$$\|P(x^{(k)}, \nabla_x \Phi^{(k)})\|_p \le \omega_* \quad and \quad \|c(x^{(k)})\|_c \le \eta_*, \tag{3.4}$$

stop. Otherwise, set

$$
\begin{aligned}
\alpha^{(k+1)} &= \min(\mu^{(k+1)}, \gamma_0), \\
\eta^{(k+1)} &= \eta^{(k)}(\alpha^{(k+1)})^{\beta_\eta}, \\
\omega^{(k+1)} &= \omega^{(k)}(\alpha^{(k+1)})^{\beta_\omega}, \\
\lambda^{(k+1)} &= \lambda^{(k)} + c(x^{(k)})/\mu^{(k)}, \\
\mu^{(k+1)} &= \mu^{(k)},
\end{aligned}
\tag{3.5}
$$

increment k by one and go to step 1.

step 3 : [Decrease penalty parameter if constraints too large]

$$
\begin{aligned}
\mu^{(k+1)} &= \tau\mu^{(k)}, \\
\alpha^{(k+1)} &= \min(\mu^{(k+1)}, \gamma_0), \\
\eta^{(k+1)} &= \eta_0(\alpha^{(k+1)})^{\alpha_\eta}, \\
\omega^{(k+1)} &= \omega_0(\alpha^{(k+1)})^{\alpha_\omega},
\end{aligned}
\tag{3.6}
$$

increment k by one and go to step 1.

end of Algorithm 3.1

We shall call the vector $P(x^{(k)}, \nabla_x \Phi^{(k)})$ the *projected gradient of the augmented Lagrangian* or the *projected gradient* for short. The norms $\|\cdot\|_p$ and $\|\cdot\|_c$ are normally chosen to be either two or infinity norms.

Our decreasing sequence of $\mu^{(k)}$s is given by $\mu^{(k)} = \mu_0(\tau)^{k_j}$, but any monotonic decreasing sequence of $\mu^{(k)}$'s converging to zero if step 3 is executed an infinite number of times, will suffice. It is also irrelevant, in theory, as to how we find a suitable point $x^{(k)}$ satisfying (3.2). However, from a practical perspective, a suitable point is found by an iterative procedure. In our algorithm, it is normal to start this inner iteration from, or close to, the solution to the last one. Indeed, from the point of view of the results we are about to establish, this is crucial. Such a starting point is desirable as function and derivative information from the conclusion of one inner iteration may be passed as input to the next.

The main purpose of this article is to show that asymptotically we take one inner iteration per outer iteration. More specifically, under certain assumptions, we first show that (3.3) is eventually satisfied at each outer iteration. We then show that, under additional assumptions, it is possible to satisfy the convergence test (3.2) after a single iteration of the algorithm given in Conn *et al.* (1988a).

The specific inner iteration algorithm we shall consider is as follows:

Algorithm 3.2 [Inner Iteration Algorithm]

step 0 : [Initialization] *The positive constants $\mu < \eta < 1$ and $\gamma_0 \le \gamma_1 < 1 \le \gamma_2$ are given. The starting point, $x^{(k,0)}$, a nonnegative convergence tolerance, $\omega^{(k)}$, an initial trust region radius, $\Delta^{(k,0)}$, a symmetric approximation, $B^{(k,0)}$, to the Hessian of the Lagrangian and a two-norm-consistent norm $\|\cdot\|_p$ are specified. Compute $\Phi(x^{(k,0)}, \lambda^{(k)}, \mu^{(k)})$ and its gradient. Set the inner iteration counter $j = 0$.*

step 1 : [Test for convergence] *If*

$$\|P(x^{(k,j)}, \nabla_x \Phi^{(k,j)})\|_p \le \omega^{(k)} \tag{3.7}$$

set $x^{(k)} = x^{(k,j)}$ and stop.

step 2 : [Significantly reduce a model of the augmented Lagrangian function] *Construct a quadratic model,*

$$m^{(k,j)}(x+s) \stackrel{\text{def}}{=} \begin{array}{l} \Phi(x, \lambda^{(k)}, \mu^{(k)}) + s^T \nabla_x \Phi(x, \lambda^{(k)}, \mu^{(k)}) \\ + \tfrac{1}{2} s^T (B^{(k,j)} + \tfrac{1}{\mu} A(x)^T A(x)) s, \end{array} \tag{3.8}$$

of $\Phi(x+s, \lambda^{(k)}, \mu^{(k)})$. Compute a step $s^{(k,j)}$ which significantly reduces the value of the model, $m^{(k,j)}(x^{(k,j)} + s)$.

step 3 : [Compute a measure of the effectiveness of the step] *Compute $\Phi(x^{(k,j)} + s^{(k,j)}, \lambda^{(k)}, \mu^{(k)})$ and the ratio*

$$\rho^{(k,j)} = \frac{\Phi(x^{(k,j)}, \lambda^{(k)}, \mu^{(k)}) - \Phi(x^{(k,j)} + s^{(k,j)}, \lambda^{(k)}, \mu^{(k)})}{m^{(k,j)}(x^{(k,j)}) - m^{(k,j)}(x^{(k,j)} + s^{(k,j)})}. \tag{3.9}$$

step 4 : [Accept or reject the step] *Set*

$$x^{(k,j+1)} = \begin{cases} x^{(k,j)} + s^{(k,j)} & \text{if } \rho^{(k,j)} > \mu \\ x^{(k,j)} & \text{otherwise,} \end{cases} \tag{3.10}$$

and

$$\Delta^{(k,j+1)} = \begin{cases} \gamma_0^{(k,j)} \Delta^{(k,j)} & \text{if } \rho^{(k,j)} \le \mu \\ \Delta^{(k)} & \text{if } \mu < \rho^{(k,j)} < \eta \\ \gamma_2^{(k,j)} \Delta^{(k,j)} & \text{otherwise,} \end{cases} \tag{3.11}$$

where $\gamma_0^{(k,j)} \in [\gamma_0, 1)$ and $\gamma_2^{(k,j)} \in [1, \gamma_2]$.

step 5 : [Updating] *If necessary, compute the gradient of $\Phi(x^{(k,j+1)}, \lambda^{(k)}, \mu^{(k)})$ and a further approximation to the Hessian of the Lagrangian $B^{(k,j+1)}$. Increment the inner iteration counter j by one and go to step 1.*

end of Algorithm 3.2

There are a number of possible ways of choosing $\gamma_0^{(k,j)}$ and $\gamma_2^{(k,j)}$ in step 4. The simplest is merely to pick $\gamma_0^{(k,j)} = \gamma_0$ and $\gamma_2^{(k,j)} = \gamma_2$; other alternatives are discussed by Conn *et al.* (1991b).

It remains to give a description of the starting point, initial trust region radius and approximation to the Hessian of the Lagrangian, and of the calculation, that is performed in step 2 of Algorithm 3.2.

Let $0 < \theta < 1$. We choose

$$x_i^{(k,0)} = \begin{cases} 0 & \text{if } 0 \leq x_i^{(k-1)} \leq \theta(\nabla_x \Phi^{(k-1)})_i; \\ x_i^{(k-1)} & \text{otherwise.} \end{cases} \tag{3.12}$$

Thus variables which are significantly dominated at the end of the $(k-1)$-st iteration are set to their bounds while the remainder are left unaltered. This choice is made since, under a suitable non-degeneracy assumption (AS7 in section 4), the set of dominated variables is asymptotically the same as the set of variables which lie on their bounds (see, Conn *et al.*, 1991a, Theorem 5.4). Our choice of $x^{(k,0)}$ then encourages subsequent iterates to encounter their asymptotic state as soon as possible. We also pick $\Delta^{(k,0)}$ so that

$$\Delta^{(k,0)} \geq \kappa \|P(x^{(k,0)}, \nabla_x \Phi^{(k,0)})\|_p^\zeta \tag{3.13}$$

for some positive constants κ and $\zeta < 1$ (typical values might be $\kappa = 1$ and $\zeta = 0.9$). This value is chosen so that the trust region does not interfere with the asymptotic convergence of the algorithm, while providing a reasonable starting value in the earlier stages of the method. Finally $B^{(k,0)}$ is taken to be any sufficiently good symmetric approximation to the Hessian of the Lagrangian function at $x^{(k)}$. We qualify what we mean by "sufficiently good" in the next section but suffice it to say that exact second derivatives satisfy this property and are often to be recommended.

The calculation in step 2 is performed in two stages.

1. Firstly, the so-called *generalized Cauchy point*, $x^{C(k,j)} \equiv x^{(k,j)} + s^{C(k,j)}$, is determined. This is merely an approximation to the first local minimizer of the quadratic model, $m^{(k,j)}(x+s)$, along the Cauchy arc. The *Cauchy arc* is the path $x + s$, where

$$s = s^{(k,j)}(t) \stackrel{\text{def}}{=} P[x^{(k,j)} - t\nabla_x \Phi(x^{(k,j+1)}, \lambda^{(k)}, \mu^{(k)}), l, u] - x^{(k,j)}, \tag{3.14}$$

as the parameter t increases from 0, which finishes when the path first intersects the boundary of the trust region,

$$\|s\|_t \leq \Delta^{(k,j)}, \tag{3.15}$$

for some two-norm-consistent norm $\|\cdot\|_t$. Thus the Cauchy arc is simply the path which starts in the steepest descent direction for the model but which is subsequently "bent" to follow the boundary of the "box" region defined by the feasible region (2.5) (or, in general, (1.3)) and which stops on the boundary of the trust region (3.15). The two or infinity norm is normally picked, the latter having some advantages as the trust region is then aligned with the feasible region (2.5).

The method proposed by Conn *et al.* (1988a) calculates the exact generalized Cauchy point by marching along the Cauchy arc until either the trust region boundary is encountered or the model starts to increase. An alternative method by Moré (1988) finds an approximation $s^{C(k,j)} = s^{(k,j)}(t^{C(k,j)})$ which is required to lie within the trust-region and to satisfy the Goldstein-Armijo type conditions

$$m^{(k,j)}(x^{(k,j)} + s^{(k,j)}(t^{C(k,j)})) \leq m^{(k,j)}(x^{(k,j)}) + \mu_1 s^{(k,j)}(t^{C(k,j)})^T \nabla_x \Phi(x^{(k,j)}, \lambda^{(k)}, \mu^{(k)}) \tag{3.16}$$

and
$$t^{C(k,j)} \geq \nu_1 \quad \text{or} \quad t^{C(k,j)} \geq \nu_2 t^{L(k,j)}, \tag{3.17}$$

where $t^{L(k,j)} > 0$ is any value for which

$$m^{(k,j)}(x^{(k,j)} + s^{(k,j)}(t^{L(k,j)})) \geq m^{(k,j)}(x^{(k,j)}) + \mu_2 s^{(k,j)}(t^{L(k,j)})^T \nabla_x \Phi(x^{(k,j)}, \lambda^{(k)}, \mu^{(k)}) \tag{3.18}$$

or

$$\|s^{(k,j)}(t^{L(k,j)})\| \geq \nu_3 \Delta^{(k,j)}, \tag{3.19}$$

and the positive constants μ_1, μ_2, ν_1, ν_2 and ν_3 satisfy the restrictions $\mu_1 < \mu_2 < 1$, $\nu_2 < 1$ and $\nu_3 < 1$. Condition (3.16) ensures that a sufficient reduction in the model takes place at each iteration while condition (3.17) is needed to guarantee that every step taken is non-negligible. Moré shows that it is always possible to pick such a value of $t^{C(k,j)}$ using a backtracking linesearch, starting on or near to the trust region boundary. Similar methods have been proposed by Calamai and Moré (1987), Burke and Moré (1988), Toint (1988) and Burke *et al.* (1990).

2. Secondly, we pick $s^{(k,j)}$ so that $x^{(k,j)} + s^{(k,j)}$ lies within (2.5), $\|s^{(k,j)}\|_t \leq \beta_1 \Delta^{(k,j)}$ and

$$\begin{aligned} m^{(k,j)}(x^{(k,j)}) &- m^{(k,j)}(x^{(k,j)} + s^{(k,j)}) \\ &\geq \beta_2(m^{(k,j)}(x^{(k,j)}) - m^{(k,j)}(x^{(k,j)} + s^{C(k,j)})) \geq 0 \end{aligned} \tag{3.20}$$

for some positive $\beta_1 \geq 1$ and $\beta_2 \leq 1$. In fact, we typically choose $\beta_1 = \beta_2 = 1$, in which case we are merely requiring that the computed step gives a value of the model which is no larger than the value at the generalized Cauchy point.

In order to accelerate the convergence of the method, it is normal to try to bias the computed step towards the Newton direction.

The convergence analysis given by Conn *et al.* (1988a) for Algorithm 3.1 indicates that it is desirable to construct improvements beyond the Cauchy point only in the subspace of variables which are free from their bounds at the Cauchy point. In particular, with such a restriction and with a suitable non-degeneracy assumption, it is then shown that the set of variables which are free from their bounds at the solution is determined after a finite number of iterations. This has the advantage of allowing one to analyze the asymptotic convergence rate of the method purely as if it were an unconstrained calculation, merely by focusing on the set of free variables.

Let $\mathcal{F}$ be a subset of $\mathcal{N}$ and let $\mathcal{D} = \mathcal{N} \setminus \mathcal{F}$. Furthermore, let

$$H^{(k,j)} \stackrel{\text{def}}{=} B^{(k,j)} + \frac{1}{\mu^{(k)}} A(x^{(k,j)})^T A(x^{(k,j)}) \tag{3.21}$$

denote the composite approximation to the Hessian of the augmented Lagrangian. The specific algorithm we shall consider may be summarized as follows:

Algorithm 3.3 [Algorithm to significantly reduce the model]

step 0 : [Initialization] *Select positive constants $\nu < 1$, $\xi < 1$, $\beta_1 \geq 1$ and $\beta_2 \leq 1$.*

step 1 : [Calculate the generalized Cauchy point] *Calculate an approximation to the the generalized Cauchy point $x^{C(k,j)} = x^{(k,j)} + s^{C(k,j)}$ using one of the previously mentioned techniques. Compute the set of variables, $\mathcal{F}^{C(k,j)}$, which are free from their bounds at $x^{C(k,j)}$. Set $x = x^{C(k,j)}$, $s = s^{C(k,j)}$ and $\mathcal{F} = \mathcal{F}^{C(k,j)}$.*

step 2 : [Further improve the model] *Let* $\mathcal{C}(\beta_1) = \mathcal{S} \bigcap \mathcal{T}(\beta_1)$, *where*

$$\mathcal{S} = \{s_{[\mathcal{F}]} \mid x^{(k,j)} + s \in \mathcal{B} \quad and \quad s_{[\mathcal{D}]} = s_{[\mathcal{D}]}^{C(k,j)}\} \tag{3.22}$$

and

$$\mathcal{T}(\beta_1) = \{s_{[\mathcal{F}]} \mid \|s\|_t \le \beta_1 \Delta^{(k,j)} \quad and \quad s_{[\mathcal{D}]} = s_{[\mathcal{D}]}^{C(k,j)}\}. \tag{3.23}$$

If $s_{[\mathcal{F}]}$ lies on the boundary of $\mathcal{T}(\beta_1)$, set $s^{(k,j)} = s$ and stop[1]. Otherwise, recompute $s_{[\mathcal{F}]}$ so that (3.20) is satisfied and either $s_{[\mathcal{F}]}$ lies strictly interior to $\mathcal{C}(\beta_1)$ with

$$\begin{aligned}
\|H_{[\mathcal{F},\mathcal{F}]}^{(k,j)} s_{[\mathcal{F}]} &+ (\nabla_x \Phi_{[\mathcal{F}]}^{(k,j)} + H_{[\mathcal{F},\mathcal{D}]}^{(k,j)} s_{[\mathcal{D}]})\|_p \\
&\le \min(\nu, \|P(x^{(k,j)}, \nabla_x \Phi^{(k,j)})\|_p^\xi) \cdot \|P(x^{(k,j)}, \nabla_x \Phi^{(k,j)})\|_p
\end{aligned} \tag{3.24}$$

or $s_{[\mathcal{F}]}$ lies on the boundary of $\mathcal{C}(\beta_1)$. Reset $x_{[\mathcal{F}]}$ to $x_{[\mathcal{F}]} + s_{[\mathcal{F}]}$.

step 3 : [Test for convergence] *If $s_{[\mathcal{F}]}$ lies strictly interior to $\mathcal{C}(\beta_1)$ and (3.24) is satisfied or if it is decided that sufficient passes have been made, set $s^{(k,j)} = s$ and stop. Otherwise remove all of the indices in $\mathcal{F}$ for which $|s_{[\mathcal{F}]i}|$ lies on the boundary of $\mathcal{S}$ and perform another pass by returning to step 2.*

end of Algorithm 3.3

In step 2 of this method, the value of $s_{[\mathcal{F}]}$ would normally be computed as the aggregate step after a number of Conjugate Gradient (CG) iterations, where CG is applied to minimize the model in the subspace defined by the free variables. The CG process will end when either a new bound is encountered or the convergence test (3.24) is satisfied. Algorithm 3.3 is itself finite as the number of free variables at each pass of step 2 is strictly monotonically decreasing. See the paper by Conn *et al.* (1988b) for further details.

4 Convergence analysis

We wish to analyze the asymptotic behaviour of Algorithm 3.1, that is in the case where $\omega_* = \eta_* = 0$. We require the following additional assumptions.

AS2. The iterates $x^{(k)}$ generated by Algorithm 3.1 all lie within a closed bounded domain Ω.

AS3. The matrix $A(x^*)_{[\mathcal{F}_1]}$ is of full rank at any limit point x^* of the sequence $x^{(k)}$ generated by Algorithm 3.1.

Under these assumptions we have the following result.

Theorem 4.1 [Conn *et al.*, 1991a, Theorem 4.4] *Assume that AS1–AS3 hold, that x^* is a limit point of the sequence $\{x^{(k)}\}$ generated by Algorithm 3.1 and that*

$$\bar{\lambda}^{(k)} \stackrel{\text{def}}{=} \lambda^{(k)} + c(x^{(k)})/\mu^{(k)}. \tag{4.1}$$

Then x^* is a Kuhn-Tucker (first order stationary) point for (1.1), (1.2) and (2.5) and the corresponding subsequences of $\{\bar{\lambda}^{(k)}\}$ and $\{\nabla_x \Phi^{(k)}\}$ converge to a set of Lagrange multipliers, λ^*, and the gradient of the Lagrangian, $g_\ell(x^*, \lambda^*)$, for the problem, respectively.

---

[1] If $\|\cdot\|_t$ is the infinity norm, it is possible to transfer components of $\mathcal{F}$ which lie on the trust-region boundary to $\mathcal{D}$ and to continue.

Now consider the following further assumptions.

AS4. The second derivatives of the functions $f(x)$ and $c_i(x)$ are Lipschitz continuous at all points within Ω.

AS5. Suppose that (x^*, λ^*) is a Kuhn-Tucker point for the problem (1.1), (1.2) and (2.5), and

$$
\begin{aligned}
\mathcal{J}_1 &= \{i \mid (g_\ell(x^*, \lambda^*))_i = 0 \text{ and } x_i^* > 0\}, \\
\mathcal{J}_2 &= \{i \mid (g_\ell(x^*, \lambda^*))_i = 0 \text{ and } x_i^* = 0\}.
\end{aligned}
\tag{4.2}
$$

Then we assume that the Kuhn-Tucker matrix

$$
\begin{bmatrix}
H_\ell(x^*, \lambda^*)_{[\mathcal{J}, \mathcal{J}]} & A(x^*)_{[\mathcal{J}]}^T \\
A(x^*)_{[\mathcal{J}]} & 0
\end{bmatrix}
\tag{4.3}
$$

is non-singular for all sets $\mathcal{J}$ made up from the union of $\mathcal{J}_1$ and any subset of $\mathcal{J}_2$.

AS6. Algorithm 3.1 has a single limit point, x^*.

Under these additional assumptions, we are able to infer the following result.

Theorem 4.2 [Conn *et al.*, 1991a, Theorems 5.3 and 5.5] *Assume that AS1, AS3—AS6 hold. Then there is a constant $\mu^* > 0$ such that the penalty parameter $\mu^{(k)}$ generated by Algorithm 3.1 satisfies $\mu^{(k)} = \mu^*$ for all k sufficiently large. Furthermore, $x^{(k)}$ and $\bar{\lambda}^{(k)}$ satisfy the bounds*

$$
\|x^{(k)} - x^*\|_p \le a_x(\alpha^*)^{\alpha_\eta + k\beta_\eta} \quad \text{and} \quad \|\bar{\lambda}^{(k)} - \lambda^*\|_p \le a_\lambda(\alpha^*)^{\alpha_\eta + k\beta_\eta},
\tag{4.4}
$$

where

$$
\alpha^* \stackrel{\text{def}}{=} \min(\mu^*, \gamma_0) \le \mu^*,
\tag{4.5}
$$

for the two-norm-consistent norm $\|.\|_p$ and some positive constants a_x and a_λ.

We shall now investigate the behaviour of Algorithm 3.1 once the penalty parameter has converged to its asymptotic value, μ^*. There is no loss of generality in assuming that we restart the algorithm from the point which is reached when the penalty parameter is reduced for the last time. We shall call this iteration $k = 0$ and will start with $\mu^{(0)} = \mu^*$. By construction, (3.3) is satisfied for all k and the updates (3.5) are always performed. Moreover,

$$
\omega^{(k)} = \omega_0(\alpha^*)^{\alpha_\omega + k\beta_\omega} \quad \text{and} \quad \eta^{(k)} = \eta_0(\alpha^*)^{\alpha_\eta + k\beta_\eta}.
\tag{4.6}
$$

We require the following extra assumptions.

AS7. The set

$$
\mathcal{J}_2 = \{i \mid g_\ell(x^*, \lambda^*)_i = 0 \quad \text{and} \quad x_i^* = 0\} = \emptyset.
\tag{4.7}
$$

AS8. If $\mathcal{J}_1$ is defined by (4.2), the approximations $B^{(k,0)}$ satisfy

$$
\|(B^{(k,0)} - \nabla_{xx}\ell(x^*, \lambda^*))_{[\mathcal{J}_1, \mathcal{J}_1]} s_{[\mathcal{J}_1]}^{(k,0)}\|_p \le v\|s_{[\mathcal{J}_1]}^{(k,0)}\|_p^{1+\varsigma},
\tag{4.8}
$$

for some positive constants v and ς and all k sufficiently large.

AS9. Suppose that (x^*, λ^*) is a Kuhn-Tucker point for the problem (1.1), (1.2) and (2.5), and that $\mathcal{J}_1$ is defined by (4.2). Then we assume that the second derivative approximations $B^{(k,0)}$ have a single limit, B^* and that the perturbed Kuhn-Tucker matrix

$$\begin{bmatrix} B^*_{[\mathcal{J}_1, \mathcal{J}_1]} & A(x^*)^T_{[\mathcal{J}_1]} \\ A(x^*)_{[\mathcal{J}_1]} & -\mu^* I \end{bmatrix} \tag{4.9}$$

is non-singular and has precisely m negative eigenvalues.

Assumption AS7 is often known as the strict complementary slackness condition. We observe that AS8 is closely related to the necessary and sufficient conditions for superlinear convergence of the inner iterates given by Dennis and Moré (1974). We also observe that AS9 is entirely equivalent to requiring that the matrix

$$B^*_{[\mathcal{J}_1, \mathcal{J}_1]} + \frac{1}{\mu^*} A(x^*)^T_{[\mathcal{J}_1]} A(x^*)_{[\mathcal{J}_1]} \tag{4.10}$$

is positive definite (see, for instance, Gould, 1986). The uniqueness of the limit point in AS9 can also be relaxed by requiring that (4.10) has its smallest eigenvalue uniformly bounded from below by some positive quantity for all limit points B^* of the sequence $B^{(k,0)}$. Moreover, Bertsekas (1982, Proposition 2.4) has shown that AS5 and AS7 guarantee AS9 provided that μ^* is sufficiently small. Although we shall merely assume that AS9 holds in this paper, it is of course possible to try to encourage this eventuality. We might, for instance, insist that step 3 of Algorithm (3.1) is executed rather than step 2 so long as the matrix $H^{(k,0)}$ is not positive definite. This is particularly relevant if exact second derivatives are used.

We now show that if we perform the step calculation for Algorithm 3.2 using Algorithm 3.3, a single iteration of Algorithm 3.2 suffices to complete an iteration of Algorithm 3.1 when k is sufficiently large. Moreover, the solution of one inner-iteration subproblem, $x^{(k-1)}$ and the shifted starting point for the next inner iteration (3.12) are asymptotically identical. We do this by showing that, after a finite number of iterations,

(i) moving to the new starting point does not significantly alter the norms of the projected gradient or constraints. Furthermore, the status of each variable (floating or dominated) is unchanged by the move;

(ii) the generalized Cauchy point $x^{C(k,0)}$ occurs before the first "breakpoint" along the Cauchy arc — the breakpoints are the values of $t > 0$ at which the Cauchy arc changes direction as problem or trust region bounds are encountered. Thus the set of variables which are free at the start of the Cauchy arc $x^{(k,0)}$ and those which are free at the generalized Cauchy point are identical;

(iii) any step which satisfies (3.24) also satisfies $s_{[\mathcal{F}_1]}$ lies strictly interior to $\mathcal{C}(\beta_1)$. Thus a single pass of step 2 of Algorithm 3.3 is required;

(iv) the step $s^{(k,0)}$ is accepted in step 4 of Algorithm 3.1;

(v) the new point $x^{(k,1)}$ satisfies the convergence test (3.7); and

(vi) $x^{(k+1,0)} = x^{(k)}$.

We have the following theorem.

Theorem 4.3 *Assume that assumptions AS1,AS3–AS9 hold and that the convergence tolerances β_ω and β_η satisfy the extra condition*

$$\beta_\omega < (1 + \min(\xi, \varsigma))\beta_\eta. \tag{4.11}$$

Then for all k sufficiently large, a single inner iteration of Algorithm 3.2, with the step computed from Algorithm 3.3, suffices to complete an iteration of Algorithm 3.1. Moreover, the solution to one inner iteration subproblem provides the starting point for the next without further adjustment, for all k sufficiently large.

Proof. Recall, we have used Theorem 4.2 to relabel the sequence of iterates so that

$$\|P(x^{(k)}, \nabla_x \Phi(x^{(k)}, \lambda^{(k)}, \mu^*))\|_p \le \omega_0(\alpha^*)^{\alpha_\omega + k\beta_\omega} \tag{4.12}$$

and

$$\|c(x^{(k)})\|_c \le \eta_0(\alpha^*)^{\alpha_\eta + k\beta_\eta} \tag{4.13}$$

for all $k \ge 0$.

We shall follow the outline given above.

(i) Status of the starting point. The non-degeneracy assumption AS7 ensures that for all k sufficiently large, each variable belongs exclusively to one of the sets $\mathcal{F}_1$ and $\mathcal{D}_1$ (see Conn *et al.*, 1991a, Theorem 5.4); moreover,

$$g_\ell(x^*, \lambda^*)_i = 0 \quad \text{and} \quad x_i^* > 0 \quad \text{for all} \quad i \in \mathcal{F}_1 \tag{4.14}$$

and

$$x_i^* = 0 \quad \text{and} \quad g_\ell(x^*, \lambda^*)_i > 0 \quad \text{for all} \quad i \in \mathcal{D}_1. \tag{4.15}$$

As one of $x_i^{(k)}$ and $\nabla_x \Phi_i^{(k)}$ converges to zero while its partner converges to a strictly positive limit for each i (assumption AS7), we may define nontrivial regions which separate the two sequences for all k sufficiently large. Let

$$\epsilon = \frac{\theta}{1+\theta} \min_{j \in \mathcal{N}} \max(x_j^*, g_\ell(x^*, \lambda^*)_j) > 0, \tag{4.16}$$

where θ is as in (3.12). Then there is an iteration k_0 such that for variables in $\mathcal{F}_1$,

$$|x_i^{(k)} - x_i^*| \le \epsilon \quad \text{and} \quad |\nabla_x \Phi_i^{(k)}| \le \epsilon, \tag{4.17}$$

while for those in $\mathcal{D}_1$,

$$|x_i^{(k)}| \le \epsilon \quad \text{and} \quad |\nabla_x \Phi_i^{(k)} - g_\ell(x^*, \lambda^*)_i|, \le \epsilon \tag{4.18}$$

for all $k \ge k_0$. Hence, for those variables in $\mathcal{D}_1$, (4.16) and (4.18) give that

$$\begin{aligned} x_i^{(k)} &\le \epsilon \le \theta[\min_{j \in \mathcal{N}} \max[x_j^*, g_\ell(x^*, \lambda^*)_j] - \epsilon] \\ &\le \theta[g_\ell(x^*, \lambda^*)_i - \epsilon] \le \theta(\nabla_x \Phi^{(k)})_i. \end{aligned} \tag{4.19}$$

Thus, by definition (3.12), $x_i^{(k+1,0)} = 0$ for each $i \in \mathcal{D}_1$ when $k \ge k_0$. Similarly, when $i \in \mathcal{F}_1$ and $k \ge k_0$, $x_i^{(k)} > \theta(\nabla_x \Phi^{(k)})_i$ and hence, using (3.12), $x_i^{(k+1,0)} = x_i^{(k)}$.

We now consider the starting point $x^{(k+1,0)}$ for the next inner iteration in detail. Firstly, combining (2.8), (2.11) and (3.12), we have that

$$\|x^{(k+1,0)} - x^{(k)}\|_z \le a_0 \|P(x^{(k)}, \nabla_x \Phi(x^{(k)}, \lambda^{(k)}, \mu^*))\|_p \tag{4.20}$$

59

for any two-norm-consistent norm $\|.\|_z$. We may bound $\|c(x^{(k+1,0)})\|_p$ using the integral mean value theorem (see, eg, Dennis and Schnabel, 1983, page 74), the boundedness of $A(x)$ (assumptions AS1 and AS2) and inequalities (2.12), (3.1), (4.12), (4.13) and (4.20) to obtain

$$
\begin{aligned}
\|c(x^{(k+1,0)})\|_p &\leq a_0\|c(x^{(k)})\|_c + \|\int_0^1 A(x(s))ds\|_p\|x^{(k+1,0)} - x^{(k)}\|_p \\
&\leq a_0\eta_0(\alpha^*)^{\alpha_\eta+k\beta_\eta} + a_1a_0\omega_0(\alpha^*)^{\alpha_\omega+k\beta_\omega} \\
&\leq a_0(1 + a_1\omega_0/\eta_0)\eta_0(\alpha^*)^{\alpha_\eta+k\beta_\eta}
\end{aligned}
\tag{4.21}
$$

where $x(s) \stackrel{\text{def}}{=} x^{(k)} + s(x^{(k+1,0)} - x^{(k)})$ and a_1 is an upper bound on $\|A(x)\|$ within Ω.

Now consider the variables whose indices i lie in $\mathcal{F}_1$ for $k \geq k_0$. Firstly, (3.12), (4.16) and (4.17) show that

$$
x_i^{(k+1,0)} = x_i^{(k)} \geq \frac{x_i^*}{1+\theta} > 0.
\tag{4.22}
$$

We next bound $|\nabla_x\Phi(x^{(k+1,0)}, \lambda^{(k)}, \mu^*)_i|$. Again using the integral mean value theorem, the convergence of $\bar{\lambda}^{(k)} \equiv \lambda^{(k+1)}$ to λ^* (Theorem 4.1), the boundedness of the Hessian of the Lagrangian (with bounded multiplier estimates) and the constraint Jacobian within Ω (assumptions AS1 and AS2) and the inequalities (2.12), (4.12) and (4.20), we obtain

$$
\begin{aligned}
|\nabla_x\Phi(x^{(k+1,0)}, \lambda^{(k)}, \mu^*)_i| &\leq |\nabla_x\Phi(x^{(k)}, \lambda^{(k)}, \mu^*)_i| + \|x^{(k+1,0)} - x^{(k)}\|_2 \cdot \\
&\quad |e_i^T \int_0^1 (H_\ell(x(s), \lambda^{(k)}) + \frac{1}{\mu^*}A(x(s))^T A(x(s)))ds| \\
&\leq a_0(1 + a_2 + a_1^2/\mu^*)\omega_0(\alpha^*)^{\alpha_\omega+k\beta_\omega},
\end{aligned}
\tag{4.23}
$$

where a_2 is an upper bound on the norm of the Hessian of the Lagrangian (with bounded multiplier estimates) within Ω. We now combine the identity

$$
\nabla_x\Phi(x^{(k+1,0)}, \lambda^{(k+1)}, \mu^*) = \nabla_x\Phi(x^{(k+1,0)}, \lambda^{(k)}, \mu^*) + A(x^{(k+1,0)})^T c(x^{(k+1,0)})/\mu^*
\tag{4.24}
$$

with (2.12), (3.1), (4.5), (4.21) and (4.23) to derive the inequality

$$
\begin{aligned}
|\nabla_x\Phi(x^{(k+1,0)}, \lambda^{(k+1)}, \mu^*)_i| &\leq |\nabla_x\Phi(x^{(k+1,0)}, \lambda^{(k)}, \mu^*)_i| \\
&\quad +a_0\|\nabla_x c_i(x^{(k+1,0)})\|_p\|c(x^{(k+1,0)})\|_p/\mu^* \\
&\leq a_0(1 + a_2 + a_1^2/\mu^*)\omega_0(\alpha^*)^{\alpha_\omega+k\beta_\omega} \\
&\quad +a_0^2a_1(1 + a_1\omega_0/\eta_0)\eta_0(\alpha^*)^{\alpha_\eta-1+k\beta_\eta} \\
&\leq a_3(\alpha^*)^{\alpha_\eta-1+k\beta_\eta},
\end{aligned}
\tag{4.25}
$$

where $a_3 \stackrel{\text{def}}{=} a_0(1 + a_2 + a_1^2(1 + a_0))\omega_0 + a_0^2a_1\eta_0$. As k increases, the right-hand-side of the inequality (4.25) converges to zero. Thus for k sufficiently large, $x_i^{(k+1,0)}$ is floating for each $i \in \mathcal{F}_1$ and (2.9) and (4.25) imply that

$$
|P(x^{(k+1,0)}, \nabla_x\Phi(x^{(k+1,0)}, \lambda^{(k+1)}, \mu^*))_i| = |\nabla_x\Phi(x^{(k+1,0)}, \lambda^{(k+1)}, \mu^*)_i| \leq a_3(\alpha^*)^{\alpha_\eta-1+k\beta_\eta}.
\tag{4.26}
$$

Conversely, consider the variables which lie in $\mathcal{D}_1$ for $k \geq k_0$. We then have that

$$
\begin{aligned}
&|\nabla_x\Phi(x^{(k+1,0)}, \lambda^{(k)}, \mu^*)_i - \nabla_x\Phi(x^{(k)}, \lambda^{(k)}, \mu^*)_i| \\
&\leq \|x^{(k+1,0)} - x^{(k)}\|_2 \cdot |\int_0^1 (H_\ell(x(s), \lambda^{(k)}) + \frac{1}{\mu^*}A(x(s))^T A(x(s)))_i ds| \\
&\leq a_0(a_2 + a_1^2/\mu^*)\omega_0(\alpha^*)^{\alpha_\omega+k\beta_\omega}
\end{aligned}
\tag{4.27}
$$

using the same tools that we used to obtain (4.23). Then, combining (2.12), (3.1), (4.5), (4.21), (4.24) and (4.27) we obtain the inequality

$$
\begin{aligned}
&|\nabla_x\Phi(x^{(k+1,0)}, \lambda^{(k+1)}, \mu^*)_i - \nabla_x\Phi(x^{(k)}, \lambda^{(k)}, \mu^*)_i| \\
&\leq a_0(a_2 + a_1^2/\mu^*)\omega_0(\alpha^*)^{\alpha_\omega+k\beta_\omega} + a_0a_1(1 + a_1\omega_0/\eta_0)\eta_0(\alpha^*)^{\alpha_\eta-1+k\beta_\eta} \leq a_4(\alpha^*)^{\alpha_\eta-1+k\beta_\eta},
\end{aligned}
\tag{4.28}
$$

where $a_4 \stackrel{\text{def}}{=} a_0(a_2 + 2a_1^2)\omega_0 + a_0a_1\eta_0$. Thus, for sufficiently large k the right-hand-side of (4.28) can be made arbitrarily small. Combining this result with (4.18) and the identity $x_i^{(k+1,0)} = 0$, we see that $x_i^{(k+1,0)}$ is dominated for each $i \in \mathcal{D}_1$ and (2.8) and (4.28) imply that

$$P(x^{(k+1,0)}, \nabla_x \Phi(x^{(k+1,0)}, \lambda^{(k+1)}, \mu^*))_i = x_i^{(k+1,0)} = 0. \tag{4.29}$$

Therefore, using (2.9), (2.12), (4.26) and (4.29), we have

$$\|P(x^{(k+1,0)}, \nabla_x \Phi(x^{(k+1,0)}, \lambda^{(k+1)}, \mu^*))\|_p = \|\nabla_x \Phi(x^{(k+1,0)}, \lambda^{(k+1)}, \mu^*)_{[\mathcal{F}_1]}\|_p \leq a_5(\alpha^*)^{\alpha_\eta - 1 + k\beta_\eta}, \tag{4.30}$$

for all k sufficiently large, where $a_5 \stackrel{\text{def}}{=} a_0 a_3 \|e_{[\mathcal{F}_1]}\|_2$.

(ii) The generalized Cauchy point. We consider the Cauchy arc emanating from $x^{(k+1,0)}$. We have shown that the variables in $\mathcal{D}_1$ are on their bounds; the relationships (4.15), (4.18) and (4.28) imply that $\nabla_x \Phi(x^{(k+1,0)}, \lambda^{(k+1)}, \mu^*)_i > 0$ and hence that $s^{(k+1,0)}(t)_i = 0$ for all $t > 0$ and $i \in \mathcal{D}_1$. Thus the variables in $\mathcal{D}_1$ remain fixed on the bounds throughout the first inner iteration and

$$s_{[\mathcal{D}_1]}^{(k+1,0)} = 0 \tag{4.31}$$

for all k sufficiently large.

The remaining variables, those indexed by $\mathcal{F}_1$, are free from their bounds. The set $\mathcal{J}_1$ in assumption AS9 is identical to $\mathcal{F}_1$ and under this assumption the matrix (4.10) is positive definite with extreme eigenvalues $0 < \lambda_{\min} \leq \lambda_{\max}$, say. The definition (3.12) and inequalities (2.8), (2.9) and (3.2) imply that $x^{(k+1,0)}$ converges to x^*. Thus the matrix

$$H_{[\mathcal{F}_1,\mathcal{F}_1]}^{(k+1,0)} = B_{[\mathcal{F}_1,\mathcal{F}_1]}^{(k+1,0)} + \frac{1}{\mu^*} A(x^{(k+1,0)})_{[\mathcal{F}_1]}^T A(x^{(k+1,0)})_{[\mathcal{F}_1]} \tag{4.32}$$

is also positive definite with extreme eigenvalues satisfying

$$0 < \tfrac{1}{2}\lambda_{\min} \leq \lambda_{\min}^{(k+1,0)} \leq \lambda_{\max}^{(k+1,0)} \leq 2\lambda_{\max}, \tag{4.33}$$

say, for all sufficiently large k. Hence the model (3.8) is a strictly convex function in the subspace of free variables during the first inner iteration.

We now show that the set

$$\mathcal{L} \stackrel{\text{def}}{=} \{s_{[\mathcal{F}_1]} \mid m^{(k+1,0)}(x^{(k+1,0)} + s) \leq m^{(k+1,0)}(x^{(k+1,0)}) \text{ and } s_{[\mathcal{D}_1]} = 0\} \tag{4.34}$$

lies strictly interior to the set $\mathcal{C}(1)$ for all k sufficiently large. The diameter d of $\mathcal{L}$, the maximum distance between two members of the set (measured in the two norm), can be no larger than twice the distance from the center of the ellipsoid defined by $\mathcal{L}$ to the point on $\bar{\mathcal{L}}$ (the boundary of $\mathcal{L}$) furthest from the center. The center of $\mathcal{L}$ is the Newton point,

$$s_{[\mathcal{F}_1]}^* = -H_{[\mathcal{F}_1,\mathcal{F}_1]}^{(k+1,0)-1} \nabla_x \Phi(x^{(k+1,0)}, \lambda^{(k+1)}, \mu^*)_{[\mathcal{F}_1]}. \tag{4.35}$$

Let $s_{[\mathcal{F}_1]} \stackrel{\text{def}}{=} s_{[\mathcal{F}_1]}^* + v_{[\mathcal{F}_1]} \in \bar{\mathcal{L}}$. Then, combining (3.8), (4.32), (4.34) and (4.35), we have that

$$\begin{aligned}
\tfrac{1}{2} v_{[\mathcal{F}_1]}^T &H_{[\mathcal{F}_1,\mathcal{F}_1]}^{(k+1,0)} v_{[\mathcal{F}_1]} \\
&= \tfrac{1}{2} s_{[\mathcal{F}_1]}^{*T} H_{[\mathcal{F}_1,\mathcal{F}_1]}^{(k+1,0)} s_{[\mathcal{F}_1]}^* + (m^{(k+1,0)}(x^{(k+1,0)} + s^* + v) - m^{(k+1,0)}(x^{(k+1,0)})) \\
&\quad -(s^* + v)_{[\mathcal{F}_1]}^T (H_{[\mathcal{F}_1,\mathcal{F}_1]}^{(k+1,0)} s_{[\mathcal{F}_1]}^* + \nabla_x \Phi(x^{(k+1,0)}, \lambda^{(k+1)}, \mu^*)_{[\mathcal{F}_1]}) \\
&= \tfrac{1}{2} s_{[\mathcal{F}_1]}^{*T} H_{[\mathcal{F}_1,\mathcal{F}_1]}^{(k+1,0)} s_{[\mathcal{F}_1]}^* \\
&= \tfrac{1}{2} \nabla_x \Phi(x^{(k+1,0)}, \lambda^{(k+1)}, \mu^*)_{[\mathcal{F}_1]}^T H_{[\mathcal{F}_1,\mathcal{F}_1]}^{(k+1,0)-1} \nabla_x \Phi(x^{(k+1,0)}, \lambda^{(k+1)}, \mu^*)_{[\mathcal{F}_1]}.
\end{aligned} \tag{4.36}$$

Hence, using the extremal properties of the Rayleigh quotient and (4.36), we have

$$
\begin{aligned}
d^2 &\overset{\text{def}}{=}\; 4\|v^*_{[\mathcal{F}_1]}\|_2^2 \le 4v^{*T}_{[\mathcal{F}_1]}H^{(k+1,0)}_{[\mathcal{F}_1,\mathcal{F}_1]}v^*_{[\mathcal{F}_1]}/\lambda^{(k+1,0)}_{\min} \le 8v^{*T}_{[\mathcal{F}_1]}H^{(k+1,0)}_{[\mathcal{F}_1,\mathcal{F}_1]}v^*_{[\mathcal{F}_1]}/\lambda_{\min} \\
&=\; 8\nabla_x\Phi(x^{(k+1,0)},\lambda^{(k+1)},\mu^*)^T_{[\mathcal{F}_1]}H^{(k+1,0)-1}_{[\mathcal{F}_1,\mathcal{F}_1]}\nabla_x\Phi(x^{(k+1,0)},\lambda^{(k+1)},\mu^*)_{[\mathcal{F}_1]}/\lambda_{\min} \\
&\le\; 16\|\nabla_x\Phi(x^{(k+1,0)},\lambda^{(k+1)},\mu^*)_{[\mathcal{F}_1]}\|_2^2/\lambda^2_{\min}
\end{aligned}
\tag{4.37}
$$

where $\|v^*_{[\mathcal{F}_1]}\|_2 = \max_{s^*_{[\mathcal{F}_1]}+v_{[\mathcal{F}_1]}\in\mathcal{L}}\|v_{[\mathcal{F}_1]}\|_2$. Thus, using (2.12), (4.30) and (4.37), any step within $\mathcal{L}$ satisfies the bound

$$
\begin{aligned}
\|s_{[\mathcal{F}_1]}\|_2 &\le\; d \le 4\|\nabla_x\Phi(x^{(k+1,0)},\lambda^{(k+1)},\mu^*)_{[\mathcal{F}_1]}\|_2/\lambda_{\min} \\
&\le\; 4a_0 a_5(\alpha^*)^{\alpha_\eta-1+k\beta_\eta}/\lambda_{\min}.
\end{aligned}
\tag{4.38}
$$

The inequality (4.22) shows that $x^{(k+1,0)}_i$, $i \in \mathcal{F}_1$, is separated from its bound for all k sufficiently large while (4.38) and the two-norm consistency of the infinity norm shows that all steps within $\mathcal{L}$ become arbitrarily small. Thus the problem bounds are excluded from $\mathcal{L}$. Moreover (2.11), (3.13), (4.30), (4.31) and (4.38) combine to give

$$
\|s\|_t = \|s_{[\mathcal{F}_1]}\|_t \le a_0^{\frac{1}{2}}\|s_{[\mathcal{F}_1]}\|_2 \le \Delta^{(k+1,0)}\frac{4a_0\|\nabla_x\Phi(x^{(k+1,0)},\lambda^{(k+1)},\mu^*)_{[\mathcal{F}_1]}\|_p^{1-\varsigma}}{\lambda_{\min}\kappa}.
\tag{4.39}
$$

for all steps on or within $\mathcal{L}$. Inequality (4.30) then combines with (4.39) to show that any such step is shorter than the distance to the trust region boundary for all k sufficiently large.

Thus $\mathcal{L}$ lies strictly interior to $\mathcal{C}(1) \subseteq \mathcal{C}(\beta_1)$ for all k sufficiently large. But, as all iterates generated by Algorithm 3.3 satisfy (3.20) and thus lie in $\mathcal{L}$, it follows that both the generalized Cauchy point and any subsequent improvements are not restricted by the boundaries of $\mathcal{C}$ or $\mathcal{C}(\beta_1)$.

It remains to consider the Cauchy step in more detail. The Cauchy arc starts in the steepest descent direction for the variables in $\mathcal{F}_1$. The minimizer of the model in this direction occurs when

$$
t = t^* = \frac{\nabla_x\Phi(x^{(k+1,0)},\lambda^{(k+1)},\mu^*)^T_{[\mathcal{F}_1]}\nabla_x\Phi(x^{(k+1,0)},\lambda^{(k+1)},\mu^*)_{[\mathcal{F}_1]}}{\nabla_x\Phi(x^{(k+1,0)},\lambda^{(k+1)},\mu^*)^T_{[\mathcal{F}_1]}H^{(k+1,0)}_{[\mathcal{F}_1,\mathcal{F}_1]}\nabla_x\Phi(x^{(k+1,0)},\lambda^{(k+1)},\mu^*)_{[\mathcal{F}_1]}}.
\tag{4.40}
$$

and thus, from the above discussion, gives the generalized Cauchy point proposed by Conn *et al.* (1988a). We use the definition of t^*, (2.11) and the extremal property of the Rayleigh quotient to obtain

$$
\begin{aligned}
m^{(k,0)}(x^{(k+1,0)}) - m^{(k+1,0)}(x^{(k+1,0)}+s^{C(k+1,0)}) &=\; \tfrac{1}{2}t^*\|\nabla_x\Phi(x^{(k+1,0)},\lambda^{(k+1)},\mu^*)_{[\mathcal{F}_1]}\|_2^2 \\
&\ge\; \frac{\|\nabla_x\Phi(x^{(k+1,0)},\lambda^{(k+1)},\mu^*)_{[\mathcal{F}_1]}\|_p^2}{4a_0\lambda_{\max}}
\end{aligned}
\tag{4.41}
$$

for this variant of the generalized Cauchy point. Alternatively, if Moré's (1988) variant is used, the requirement (3.16) and the definition of the Cauchy arc imply that

$$
m^{(k+1,0)}(x^{(k+1,0)}) - m^{(k+1,0)}(x^{(k+1,0)}+s^{C(k+1,0)}) \ge \mu_1 t^{C(k,0)}\|\nabla_x\Phi(x^{(k+1,0)},\lambda^{(k+1)},\mu^*)_{[\mathcal{F}_1]}\|_2^2.
\tag{4.42}
$$

If the first alternative of (3.17) holds, (4.42) implies that

$$
m^{(k+1,0)}(x^{(k+1,0)}) - m^{(k+1,0)}(x^{(k+1,0)}+s^{C(k+1,0)}) \ge \mu_1\nu_1\|\nabla_x\Phi(x^{(k+1,0)},\lambda^{(k+1)},\mu^*)_{[\mathcal{F}_1]}\|_2^2.
\tag{4.43}
$$

Otherwise, we may use the same arguments as above to show that it is impossible for $t^{L(k+1,0)}$ to satisfy (3.19) when k is sufficiently large. Therefore, $t^{L(k+1,0)}$ must satisfy (3.18). Combining (3.8), (3.18), (4.32) and the definition of the Cauchy arc, we have that

$$
\begin{aligned}
&\tfrac{1}{2}(t^{L(k+1,0)})^2 \nabla_x \Phi(x^{(k+1,0)}, \lambda^{(k+1)}, \mu^*)^T_{[\mathcal{F}_1]} H^{(k+1,0)}_{[\mathcal{F}_1,\mathcal{F}_1]} \nabla_x \Phi(x^{(k+1,0)}, \lambda^{(k+1)}, \mu^*)_{[\mathcal{F}_1]} \\
&\geq (1-\mu_2) t^{L(k+1,0)} \|\nabla_x \Phi(x^{(k+1,0)}, \lambda^{(k+1)}, \mu^*)_{[\mathcal{F}_1]}\|^2_2.
\end{aligned}
\tag{4.44}
$$

Hence, combining (4.33) and (4.44) with the extremal properties of the Rayleigh quotient, we have that $t^{L(k,j)} \geq (1-\mu_2)/\lambda_{\max}$. Thus, when the second alternative of (3.17) holds, this result and (4.42) give that

$$
\begin{aligned}
&m^{(k+1,0)}(x^{(k+1,0)}) - m^{(k+1,0)}(x^{(k+1,0)} + s^{C(k+1,0)}) \\
&\geq [\mu_1 \nu_2 (1-\mu_2)/\lambda_{\max}] \|\nabla_x \Phi(x^{(k+1,0)}, \lambda^{(k+1)}, \mu^*)_{[\mathcal{F}_1]}\|^2_2.
\end{aligned}
\tag{4.45}
$$

Therefore, (2.12), (4.43) and (4.45) give the inequality

$$
\begin{aligned}
&m^{(k,0)}(x^{(k,0)}) - m^{(k,0)}(x^{(k,0)} + s^{C(k,0)}) \\
&\geq (\mu_1/a_0)\min(\nu_1, \nu_2(1-\mu_2)/\lambda_{\max})\|\nabla_x \Phi(x^{(k+1,0)}, \lambda^{(k+1)}, \mu^*)_{[\mathcal{F}_1]}\|^2_p.
\end{aligned}
\tag{4.46}
$$

We shall make use of thses results in (iv) below.

(iii) **Improvements beyond the generalized Cauchy point.** We have that $x^{(k+1,0)}_{[\mathcal{D}]} = 0$, and, as a consequence of (4.30), $\|P(x^{(k+1,0)}, \nabla_x \Phi^{(k+1,0)})\|^\xi_p \leq \nu$ for all k sufficiently large. Hence, because we have shown that any s in $\mathcal{L}$ lies strictly interior to $\mathcal{C}$, a single pass of step 2 of Algorithm 3.3 is required. We must pick s to satisfy (3.24) and (3.20) by determining $s^{(k+1,0)}_{[\mathcal{F}_1]}$ so that

$$
\|H^{(k+1,0)}_{[\mathcal{F}_1,\mathcal{F}_1]} s^{(k+1,0)}_{[\mathcal{F}_1]} + \nabla_x \Phi^{(k+1,0)}_{[\mathcal{F}]}\|_p \leq \|\nabla_x \Phi^{(k+1,0)}_{[\mathcal{F}]}\|^{1+\xi}_p.
\tag{4.47}
$$

and

$$
\begin{aligned}
&m^{(k,j)}(x^{(k+1,0)}) - m^{(k+1,0)}(x^{(k+1,0)} + s^{(k+1,0)}) \\
&\geq \beta_2(m^{(k+1,0)}(x^{(k+1,0)}) - m^{(k+1,0)}(x^{(k+1,0)} + s^{C(k+1,0)}))
\end{aligned}
\tag{4.48}
$$

for some $\beta_2 \leq 1$. The set of values which satisfy (4.47) and (4.48) is non-empty as the Newton step (4.35) satisfies both inequalities.

It remains to consider such a step in slightly more detail. Suppose that $s^{(k+1,0)}_{[\mathcal{F}_1]}$ satisfies (4.47). Let

$$
r^{(k+1,0)}_{[\mathcal{F}_1]} = H^{(k+1,0)}_{[\mathcal{F}_1,\mathcal{F}_1]} s^{(k+1,0)}_{[\mathcal{F}_1]} + \nabla_x \Phi^{(k+1,0)}_{[\mathcal{F}]}
\tag{4.49}
$$

Then combining (2.11), (4.33), (4.47) and (4.49), we have

$$
\begin{aligned}
\|s^{(k+1,0)}_{[\mathcal{F}_1]}\|_p &\leq a_0 \|H^{(k+1,0)-1}_{[\mathcal{F}_1,\mathcal{F}_1]}\|_2 (\|r^{(k+1,0)}_{[\mathcal{F}_1]}\|_p + \|\nabla_x \Phi^{(k+1,0)}_{[\mathcal{F}]}\|_p) \\
&\leq 2a_0 \|\nabla_x \Phi^{(k+1,0)}_{[\mathcal{F}]}\|_p (1 + \|\nabla_x \Phi^{(k+1,0)}_{[\mathcal{F}]}\|^\xi_p)/\lambda_{\min}.
\end{aligned}
\tag{4.50}
$$

(iv) **Acceptance of the new point.** We have seen that $s^{(k+1,0)}_{[\mathcal{D}_1]} = 0$ and $s^{(k+1,0)}_{[\mathcal{F}_1]}$ satisfies (4.47). We now wish to show that the quantity

$$
|\rho^{(k+1,0)} - 1| = \frac{|\Phi(x^{(k+1,0)} + s^{(k+1,0)}, \lambda^{(k+1)}, \mu^*) - m^{(k+1,0)}(x^{(k+1,0)} + s^{(k+1,0)})|}{|m^{(k+1,0)}(x^{(k+1,0)}) - m^{(k+1,0)}(x^{(k+1,0)} + s^{(k+1,0)})|}
\tag{4.51}
$$

converges to zero. For then the new point will prove acceptable in step 4 of Algorithm 3.1.

Consider first the denominator on the right-hand-side of (4.51). Combining (4.41), (4.46) and (4.48), we have

$$m^{(k+1,0)}(x^{(k+1,0)}) - m^{(k+1,0)}(x^{(k+1,0)} + s^{(k+1,0)}) \geq a_6 \|\nabla_x \Phi(x^{(k+1,0)}, \lambda^{(k+1)}, \mu^*)_{[\mathcal{F}_1]}\|_p^2, \quad (4.52)$$

where $a_6 = \beta_2 \min(1/(4a_0\lambda_{\max}), \mu_1 \min(\nu_1, \nu_2(1 - \mu_2)/\lambda_{\max})/a_0)$. Turning to the numerator on the right-hand-side of (4.51), we use the integral mean value theorem to obtain

$$
\begin{aligned}
&\Phi(x^{(k+1,0)} + s^{(k+1,0)}, \lambda^{(k+1)}, \mu^*) \\
={}& \Phi(x^{(k+1,0)}, \lambda^{(k+1)}, \mu^*) + s_{[\mathcal{F}_1]}^{(k+1,0)T} \nabla_x \Phi(x^{(k+1,0)}, \lambda^{(k+1)}, \mu^*)_{[\mathcal{F}_1]} \\
&+ \tfrac{1}{2} \int_0^1 s_{[\mathcal{F}_1]}^{(k+1,0)T} \nabla_{xx} \Phi(x(t), \lambda^{(k+1)}, \mu^*)_{[\mathcal{F}_1,\mathcal{F}_1]} s_{[\mathcal{F}_1]}^{(k+1,0)} dt \\
={}& \Phi(x^{(k+1,0)}, \lambda^{(k+1)}, \mu^*) + s_{[\mathcal{F}_1]}^{(k+1,0)T} \nabla_x \Phi(x^{(k+1,0)}, \lambda^{(k+1)}, \mu^*)_{[\mathcal{F}_1]} \\
&+ \tfrac{1}{2} \int_0^1 s_{[\mathcal{F}_1]}^{(k+1,0)T} (\nabla_{xx}\Phi(x(t), \lambda^{(k+1)}, \mu^*) - \nabla_{xx}\Phi(x^{(k+1,0)}, \lambda^{(k+1)}, \mu^*))_{[\mathcal{F}_1,\mathcal{F}_1]} s_{[\mathcal{F}_1]}^{(k+1,0)} dt \\
&+ \tfrac{1}{2} s_{[\mathcal{F}_1]}^{(k+1,0)T} (\nabla_{xx}\Phi(x^{(k+1,0)}, \lambda^{(k+1)}, \mu^*) - H^{(k+1,0)})_{[\mathcal{F}_1,\mathcal{F}_1]} s_{[\mathcal{F}_1]}^{(k+1,0)} \\
&+ \tfrac{1}{2} s_{[\mathcal{F}_1]}^{(k+1,0)T} H_{[\mathcal{F}_1,\mathcal{F}_1]}^{(k+1,0)} s_{[\mathcal{F}_1]}^{(k+1,0)} \\
={}& m^{(k+1,0)}(x^{(k+1,0)} + s^{(k+1,0)}) \\
&+ \tfrac{1}{2} \int_0^1 s_{[\mathcal{F}_1]}^{(k+1,0)T} (\nabla_{xx}\Phi(x(t), \lambda^{(k+1)}, \mu^*) - \nabla_{xx}\Phi(x^{(k+1,0)}, \lambda^{(k+1)}, \mu^*))_{[\mathcal{F}_1,\mathcal{F}_1]} s_{[\mathcal{F}_1]}^{(k+1,0)} dt \\
&+ \tfrac{1}{2} s_{[\mathcal{F}_1]}^{(k+1,0)T} (\nabla_{xx}\Phi(x^{(k+1,0)}, \lambda^{(k+1)}, \mu^*) - H^{(k+1,0)})_{[\mathcal{F}_1,\mathcal{F}_1]} s_{[\mathcal{F}_1]}^{(k+1,0)},
\end{aligned}
$$
$$(4.53)$$

where $x(t) = x^{(k+1,0)} + ts^{(k+1,0)}$. Considering the last two terms in (4.53) in turn, we have the bounds

$$
\begin{aligned}
&|\tfrac{1}{2} \int_0^1 s_{[\mathcal{F}_1]}^{(k+1,0)T} (\nabla_{xx}\Phi(x(t), \lambda^{(k+1)}, \mu^*) - \nabla_{xx}\Phi(x^{(k+1,0)}, \lambda^{(k+1)}, \mu^*))_{[\mathcal{F}_1,\mathcal{F}_1]} s_{[\mathcal{F}_1]}^{(k+1,0)} dt| \\
&\leq \tfrac{1}{4} a_0 a_7 \|s_{[\mathcal{F}_1]}^{(k+1,0)}\|_p^3
\end{aligned}
\quad (4.54)
$$

using (2.11), the convergence (and hence boundedness) of the Lagrange multiplier estimates and the Lipschitz continuity of the second derivatives of the problem functions (assumption AS4) with some composite Lipschitz constant a_7, and

$$
\begin{aligned}
&|\tfrac{1}{2} s_{[\mathcal{F}_1]}^{(k+1,0)T} (\nabla_{xx}\Phi(x^{(k+1,0)}, \lambda^{(k+1)}, \mu^*) - H^{(k+1,0)})_{[\mathcal{F}_1,\mathcal{F}_1]} s_{[\mathcal{F}_1]}^{(k+1,0)}| \\
&\leq a_0(\tfrac{1}{2}v\|s_{[\mathcal{F}_1]}^{(k+1,0)}\|_p^\varsigma + \|(\nabla_{xx}\ell(x^{(k+1,0)}, \lambda^{(k+1)}) - \nabla_{xx}\ell(x^*, \lambda^*))_{[\mathcal{F}_1,\mathcal{F}_1]}\|_p) \|s_{[\mathcal{F}_1]}^{(k+1,0)}\|_p^2
\end{aligned}
\quad (4.55)
$$

using (2.11), (3.21), the definition of the Hessian of the augmented Lagrangian and AS8. Thus, combining (4.50), (4.51), (4.52), (4.53), (4.54) and (4.55), we obtain

$$
|\rho^{(k+1,0)} - 1| \leq 4a_0^3 (1 + \|\nabla_x \Phi_{[\mathcal{F}]}^{(k+1,0)}\|^\varsigma)_p^2 \cdot
\frac{\tfrac{1}{4} a_7 \|s_{[\mathcal{F}_1]}^{(k+1,0)}\|_p + \tfrac{1}{2} v\|s_{[\mathcal{F}_1]}^{(k+1,0)}\|_p^\varsigma + \|(\nabla_{xx}\ell(x^{(k+1,0)}, \lambda^{(k+1)}) - \nabla_{xx}\ell(x^*, \lambda^*))_{[\mathcal{F}_1,\mathcal{F}_1]}\|_p}{a_6 \lambda_{\min}^2}.
\quad (4.56)
$$

As the right-hand-side of (4.56) converges to zero as k increases, $x^{(k+1,1)} = x^{(k+1,0)} + s^{(k+1,0)}$ for all k sufficiently large.

(v) **Convergence of the inner iteration at the new point.** We now show that $x^{(k+1,1)}$ satisfies the inner-iteration convergence test (3.7).

Firstly, in the same vein as (4.27), for $i \in \mathcal{D}_1$ we have that

$$
\begin{aligned}
&|\nabla_x \Phi(x^{(k+1,1)}, \lambda^{(k+1)}, \mu^*)_i - \nabla_x \Phi(x^{(k+1,0)}, \lambda^{(k+1)}, \mu^*)_i| \\
&\leq \|s^{(k+1,0)}\|_2 \cdot |\int_0^1 (H_\ell(x(t), \lambda^{(k)}) + \tfrac{1}{\mu^*} A(x(t))^T A(x(t)))_i dt| \\
&\leq (a_2 + a_1^2/\mu^*)\|s^{(k+1,0)}\|_2,
\end{aligned}
\quad (4.57)
$$

where $x(t) = x^{(k+1,0)} + ts^{(k+1,0)}$. Thus, as the right-hand-side of (4.57) can be made arbitrarily small, by taking k sufficiently large, (4.18) and the identity $x_i^{(k+1,1)} = x_i^{(k+1,0)} = 0$ for each $i \in \mathcal{D}_1$, imply that $x_i^{(k+1,1)}$ is dominated for each $i \in \mathcal{D}_1$ and (2.8) and (4.25) imply that

$$P(x^{(k+1,1)}, \nabla_x \Phi(x^{(k+1,1)}, \lambda^{(k+1)}, \mu^*))_i = x_i^{(k+1,0)} = 0. \tag{4.58}$$

We now consider the components of $P(x^{(k+1,1)}, \nabla_x \Phi(x^{(k+1,1)}, \lambda^{(k+1)}, \mu^*))_i$ for $i \in \mathcal{F}_1$. Using the integral mean value theorem, we have

$$
\begin{aligned}
&\nabla_x \Phi(x^{(k+1,1)}, \lambda^{(k+1)}, \mu^*)_{[\mathcal{F}_1]} \\
&= \nabla_x \Phi(x^{(k+1,0)}, \lambda^{(k+1)}, \mu^*)_{[\mathcal{F}_1]} + \int_0^1 \nabla_{xx} \Phi(x(t), \lambda^{(k+1)}, \mu^*)_{[\mathcal{F}_1,\mathcal{F}_1]} s_{[\mathcal{F}_1]}^{(k+1,0)} dt \\
&= (H_{[\mathcal{F}_1,\mathcal{F}_1]}^{(k+1,0)} s_{[\mathcal{F}_1]}^{(k+1,0)} + \nabla_x \Phi(x^{(k+1,0)}, \lambda^{(k+1)}, \mu^*)_{[\mathcal{F}_1]}) \\
&\quad + \int_0^1 (\nabla_{xx} \Phi(x(t), \lambda^{(k+1)}, \mu^*) - \nabla_{xx} \Phi(x^{(k+1,0)}, \lambda^{(k+1)}, \mu^*))_{[\mathcal{F}_1,\mathcal{F}_1]} s_{[\mathcal{F}_1]}^{(k+1,0)} dt \\
&\quad + (\nabla_{xx} \Phi(x^{(k+1,0)}, \lambda^{(k+1)}, \mu^*) - H^{(k+1,0)})_{[\mathcal{F}_1,\mathcal{F}_1]} s_{[\mathcal{F}_1]}^{(k+1,0)}
\end{aligned} \tag{4.59}
$$

where $x(t) = x^{(k+1,0)} + ts^{(k+1,0)}$. We observe that each of the three terms on the right-hand-side of (4.59) reflects a different aspect of the approximations made. The first corresponds to the approximation to the Newton direction used, the second to the approximation of a nonlinear function by a quadratic and the third to the particular approximation to the second derivatives used. We now bound each of these terms in turn.

The first term satisfies the bound (4.47). Hence, combining (4.30) and (4.47), we obtain

$$\|H_{[\mathcal{F}_1,\mathcal{F}_1]}^{(k+1,0)} s_{[\mathcal{F}_1]}^{(k+1,0)} + \nabla_x \Phi_{[\mathcal{F}]}^{(k+1,0)}\|_p \le a_5^{1+\xi}(\alpha^*)^{(\alpha_\eta - 1)(1+\xi) + k\beta_\eta(1+\xi)}. \tag{4.60}$$

The second term satisfies the bound

$$
\begin{aligned}
&\| \int_0^1 (\nabla_{xx} \Phi(x(t), \lambda^{(k+1)}, \mu^*) - \nabla_{xx} \Phi(x^{(k+1,0)}, \lambda^{(k+1)}, \mu^*))_{[\mathcal{F}_1,\mathcal{F}_1]} s_{[\mathcal{F}_1]}^{(k+1,0)} dt\|_p \\
&\le \tfrac{1}{2} a_0 a_7 \|s_{[\mathcal{F}_1]}^{(k+1,0)}\|_p^2.
\end{aligned} \tag{4.61}
$$

by the same arguments we used to establish inequality (4.54). Picking k sufficiently large that $\|\nabla_x \Phi_{[\mathcal{F}]}^{(k+1,0)}\| \le 1$, we may combine (4.30), (4.50) and (4.61) so that

$$
\begin{aligned}
&\| \int_0^1 (\nabla_{xx} \Phi(x(t), \lambda^{(k+1)}, \mu^*) - \nabla_{xx} \Phi(x^{(k+1,0)}, \lambda^{(k+1)}, \mu^*))_{[\mathcal{F}_1,\mathcal{F}_1]} s_{[\mathcal{F}_1]}^{(k+1,0)} dt\| \\
&\le 8 a_0^3 a_5^2 a_7 (\alpha^*)^{2\alpha_\eta - 2 + k2\beta_\eta} / \lambda_{\min}^2.
\end{aligned} \tag{4.62}
$$

Lastly, using the same arguments as those used to establish (4.55), the definitions (3.21) and of the Hessian of the augmented Lagrangian, the Lipschitz continuity of the second derivatives of the problem functions (assumption AS4) and the accuracy of the second derivative approximations (assumption AS8) imply that

$$
\begin{aligned}
&\|(\nabla_{xx} \Phi(x^{(k+1,0)}, \lambda^{(k+1)}, \mu^*) - H^{(k+1,0)})_{[\mathcal{F}_1,\mathcal{F}_1]} s_{[\mathcal{F}_1]}^{(k+1,0)}\|_p \\
&\le (v\|s_{[\mathcal{F}_1]}^{(k+1,0)}\|_p^\varsigma + \|(\nabla_{xx}\ell(x^{(k+1,0)}, \lambda^{(k+1)}) - \nabla_{xx}\ell(x^*, \lambda^*))_{[\mathcal{F}_1,\mathcal{F}_1]}\|_p)\|s_{[\mathcal{F}_1]}^{(k+1,0)}\|_p \\
&\le (v\|s_{[\mathcal{F}_1]}^{(k+1,0)}\|_p^\varsigma + a_8\|x^{(k+1,0)} - x^*\|_p + a_9\|\lambda^{(k+1)} - \lambda^*\|_p)\|s_{[\mathcal{F}_1]}^{(k+1,0)}\|_p,
\end{aligned} \tag{4.63}
$$

for some composite Lipschitz constants a_8 and a_9. Again picking k sufficiently large that $\|\nabla_x \Phi_{[\mathcal{F}]}^{(k+1,0)}\|_p \le 1$ and recalling that $\lambda^{(k+1)} = \bar{\lambda}^{(k)}$, we may combine (2.12), (4.4), (4.12), (4.20), (4.30), (4.50) and (4.63) so that

$$
\begin{aligned}
&\|(\nabla_{xx} \Phi(x^{(k+1,0)}, \lambda^{(k+1)}, \mu^*) - H^{(k+1,0)})_{[\mathcal{F}_1,\mathcal{F}_1]} s_{[\mathcal{F}_1]}^{(k+1,0)}\|_p \\
&\le [v((4a_0 a_5/\lambda_{\min})(\alpha^*)^{\alpha_\eta - 1 + k\beta_\eta})^\varsigma + a_8(a_x(\alpha^*)^{\alpha_\eta + k\beta_\eta} + a_0\omega_0(\alpha^*)^{\alpha_\omega + k\beta_\omega}) \\
&\quad + a_9 a_\lambda(\alpha^*)^{\alpha_\eta + k\beta_\eta}](4a_0 a_5/\lambda_{\min})(\alpha^*)^{\alpha_\eta - 1 + k\beta_\eta}.
\end{aligned} \tag{4.64}
$$

65

We now combine equation (4.59) with the inequalities (4.60), (4.62) and (4.64), the condition $\xi < 1$ and the definitions of $\alpha_\eta < 1$ and $\beta_\eta > 0$ to obtain the bound

$$\|\nabla_x \Phi(x^{(k+1,1)}, \lambda^{(k+1)}, \mu^*)_{[\mathcal{F}_1]}\| \leq a_{10}(\alpha^*)^{\bar{\alpha}+k\bar{\beta}}, \tag{4.65}$$

where

$$
\begin{aligned}
a_{10} &= a_5^{1+\xi} + 8a_0^3 a_5^2 a_6/\lambda_{\min}^2 + (4a_0 a_5/\lambda_{\min})(v((4a_0 a_5/\lambda_{\min})^\varsigma) + a_8(a_x + a_0\omega_0) + a_9 a_\lambda), \\
\bar{\alpha} &= (\alpha_\eta - 1)(1 + \max(1, \varsigma)) \quad \text{and} \\
\bar{\beta} &= \beta_\eta(1 + \min(\xi, \varsigma)).
\end{aligned}
\tag{4.66}
$$

Firstly, observe that the right-hand-side of (4.65) may be made arbitrarily small. Therefore, (2.9), (4.58) and (4.65) imply that

$$\|P(x^{(k+1,1)}, \nabla_x \Phi(x^{(k+1,1)}, \lambda^{(k+1)}, \mu^*))\|_p = \|\nabla_x \Phi(x^{(k+1,1)}, \lambda^{(k+1)}, \mu^*)_{[\mathcal{F}_1]}\|_p \leq a_{10}(\alpha^*)^{\bar{\alpha}+k\bar{\beta}}. \tag{4.67}$$

Secondly, define $\delta = \log_{\alpha^*}(a_{10}/\omega_0)$. Now let k_1 be any integer for which

$$k_1 \geq \frac{\alpha_\omega + \beta_\omega - \bar{\alpha} - \delta}{\bar{\beta} - \beta_\omega}. \tag{4.68}$$

Then (4.11), (4.67) and (4.68) imply that

$$\|P(x^{(k+1,1)}, \nabla_x \Phi(x^{(k+1,1)}, \lambda^{(k+1)}, \mu^*))\|_p \leq a_{10}(\alpha^*)^{\bar{\alpha}+k\bar{\beta}} \leq \omega_0(\alpha^*)^{\alpha_\omega+(k+1)\beta_\omega} = \omega^{(k+1)} \tag{4.69}$$

for all sufficiently large $k \geq k_1$. Thus, the iterate $x^{(k+1,1)}$ satisfies the inner iteration convergence test (3.2) for all k sufficiently large and we have $x^{(k+1)} = x^{(k+1,1)}$.

(vi) Redundancy of the shifted starting point. Finally, we observe that all the variables $x_i^{(k+1)}$, $i \in \mathcal{D}$, lie on their bounds for sufficiently large k. Therefore, $x^{(k+2,0)} = x^{(k+1)}$ and the perturbed starting point is redundant. ∎

5 The general case

We now turn briefly to the more general problem (1.1)—(1.3). The presence of the more general bounds (1.3) does not significantly alter the conclusions that we are able to draw. The algorithms of section 3 are basically unchanged. We now use the region $\mathcal{B} = \{x \in \Re^n \mid l \leq x \leq u\}$ and replace $P(x, v)$ by $P(x, v, l, u)$ where appropriate. The concept of floating and dominated variables stays essentially the same. Now for each iterate in $\mathcal{B}$ we have three mutually exclusive possibilities, namely, (i) $0 \leq x_i^{(k)} - l_i \leq (\nabla_x \Phi^{(k)})_i$, (ii) $(\nabla_x \Phi^{(k)})_i \leq x_i^{(k)} - u_i \leq 0$ or (iii) $x_i^{(k)} - u_i < (\nabla_x \Phi^{(k)})_i < x_i^{(k)} - l_i$, for each component $x_i^{(k)}$. In case (i) we then have that $(\nabla_x \Phi^{(k)})_i = x_i^{(k)} - l_i$ while in case (ii) $(\nabla_x \Phi^{(k)})_i = x_i^{(k)} - u_i$ and in case (iii) $(\nabla_x \Phi^{(k)})_i = (\nabla_x \Phi^{(k)})_i$. The variables that satisfy (i) and (ii) are said to be the dominated variables, the ones satisfying (i) are *dominated above* while those satisfying (ii) are *dominated below*. Consequently, the sets corresponding to (2.10) are straightforward to define. $\mathcal{D}_1$ is now made up as the union of two sets $\mathcal{D}_{1l}$, whose variables are dominated above for all k sufficiently large, and $\mathcal{D}_{1u}$, whose variables are dominated below for all k sufficiently large. $\mathcal{F}_1$ contains variables which float for all k sufficiently large and which converge to values interior to $\mathcal{B}$. Similarly $\mathcal{F}_2$ is the union of two sets, $\mathcal{F}_{2l}$ and $\mathcal{F}_{2u}$, whose variables are floating for all k sufficiently large but which converge to their lower and upper bounds respectively. We also replace (3.12) by

$$
x_i^{(k,0)} = \begin{cases} l_i & \text{if } 0 \le x_i^{(k-1)} - l_i \le \theta(\nabla_x \Phi^{(k-1)})_i \\ u_i & \text{if } \theta(\nabla_x \Phi^{(k-1)})_i \le x_i^{(k-1)} - u_i \le 0 \\ x_i^{(k-1)} & \text{otherwise.} \end{cases} \tag{5.1}
$$

With such definitions, we may reprove the results of section 4, extending AS5, AS7—AS9 in the obvious way. The only important new ingredient is that Conn *et al.* (1991a) indicate that the non-degeneracy assumption AS7 ensures that the iterates are asymptotically isolated in three sets $\mathcal{F}_1$, $\mathcal{D}_{1l}$ and $\mathcal{D}_{1u}$.

6 Conclusions

We have shown that, under suitable assumptions, a single inner iteration is needed for each outer iteration of the augmented Lagrangian algorithm which lies at the heart of the LANCELOT package. This then places the algorithm in the class of diagonal multiplier methods whose asymptotic behaviour has been studied by Tapia (1977).

References

[Bertsekas, 1982] D.P. Bertsekas. *Constrained Optimization and Lagrange multiplier methods*. Academic Press, London and New York, 1982.

[Burke and Moré, 1988] J. V. Burke and J. J. Moré. On the identification of active constraints. *SIAM Journal on Numerical Analysis*, 25:1197–1211, 1988.

[Burke et al., 1990] J. V. Burke, J. J. Moré, and G. Toraldo. Convergence properties of trust region methods for linear and convex constraints. *Mathematical Programming*, 47:305–336, 1990.

[Calamai and Moré, 1987] P. H. Calamai and J. J. Moré. Projected gradient methods for linearly constrained problems. *Mathematical Programming*, 39:93–116, 1987.

[Conn et al., 1988a] A. R. Conn, N. I. M. Gould, and Ph. L. Toint. Global convergence of a class of trust region algorithms for optimization with simple bounds. *SIAM Journal on Numerical Analysis*, 25:433–460, 1988. See also *ibid.* 26:764-767, 1989.

[Conn et al., 1988b] A. R. Conn, N. I. M. Gould, and Ph. L. Toint. Testing a class of methods for solving minimization problems with simple bounds on the variables. *Mathematics of Computation*, 50:399–430, 1988.

[Conn et al., 1991a] A. R. Conn, N. I. M. Gould, and Ph. L. Toint. A globally convergent augmented Lagrangian algorithm for optimization with general constraints and simple bounds. *SIAM Journal on Numerical Analysis*, 28:545–572, 1991.

[Conn et al., 1991b] A. R. Conn, N. I. M. Gould, and Ph. L. Toint. LANCELOT: *a Fortran package for large-scale nonlinear optimization (Release A)*, 1991.

[Dennis and Moré, 1974] J. E. Dennis and J. J. Moré. A characterization of superlinear convergence and its application to quasi-Newton methods. *Mathematics of Computation*, 28:549–560, 1974.

[Dennis and Schnabel, 1983] J. E. Dennis and R. B. Schnabel. *Numerical methods for unconstrained optimization and nonlinear equations*. Prentice-Hall, Englewood Cliffs, New Jersey, 1983.

[Gould, 1986] N. I. M. Gould. On the accurate determination of search directions for simple differentiable penalty functions. *IMA Journal of Numerical Analysis*, 6:357–372, 1986.

[Gould, 1989] N. I. M. Gould. On the convergence of a sequential penalty function method for constrained minimization. *SIAM Journal on Numerical Analysis*, 26:107–128, 1989.

[Moré, 1988] J. J. Moré. Trust regions and projected gradients. In M. Iri and K. Yajima, editors, *Systems Modelling and Optimization*, pages 1–13. Lecture Notes in Control and Information Science, Vol. 113, Springer, Berlin, 1988.

[Tapia, 1977] R. A. Tapia. Diagonalized multiplier methods and quasi-newton methods for constrained optimization. *Journal of Optimization Theory and Applications*, 22:135–194, 1977.

[Toint, 1988] Ph. L. Toint. Global convergence of a class of trust region methods for nonconvex minimization in Hilbert space. *IMA Journal of Numerical Analysis*, 8:231–252, 1988.

A. R. Conn,
T. J. Watson IBM Research Center,
Yorktown Heights, New York 10598, USA.

Nick Gould,
Atlas Centre, Rutherford Appleton Laboratory,
Chilton, Oxfordshire, England.

Ph. L. Toint,
Department of Mathematics, FUNDP,
Namur, B-5000, Belgium.

J W DEMMEL, J J DONGARRA AND W KAHAN

On designing portable high performance numerical libraries

Abstract

High quality portable numerical libraries have existed for many years. These libraries, such as LINPACK and EISPACK, were designed to be accurate, robust, efficient and portable in a Fortran environment of conventional uniprocessors, diverse floating point arithmetics, and limited input data structures. These libraries are no longer adequate on modern high performance computer architectures. We describe their inadequacies and how we are addressing them in the LAPACK project, a library of numerical linear algebra routines designed to supplant LINPACK and EISPACK. We shall show how the new architectures lead to important changes in the goals as well as the methods of library design.

1   Introduction

The original goal of the LAPACK [14] project was to modernize the widely used LINPACK [11] and EISPACK [33, 23] numerical linear algebra libraries to make them run efficiently on shared memory vector and parallel processors. On these machines, LINPACK and EISPACK are inefficient because their memory access patterns disregard the multilayered memory hierarchies of the machines, thereby spending too much time moving data instead of doing useful floating point operations. LAPACK tries to cure this by reorganizing the algorithms to use block matrix operations. These block operations can be optimized for each architecture to account for the memory hierarchy, and so provide a transportable way to achieve high efficiency on diverse modern machines.

*Supported by the NSF via grants DCR-8552474, ASC-8715728 and ASC-9005933.

†Supported by the NSF via grants ASC-8715728 and ASC-9005933 and by the DOE via grant DE-AC05-84OR21400.

‡Supported by the NSF via grant ASC-9005933.

We say "transportable" instead of "portable" because for fastest possible performance LAPACK requires that highly optimized block matrix operations be already implemented on each machine by the manufacturers or someone else. In other words the correctness of the code is portable, but high performance is not if we limit ourselves to a single (Fortran) source code. Thus we have modified the traditional and honorable goal of portability in use among numerical library designers, where both correctness and performance were retained as the source code was moved to new machines, because it is no longer appropriate on modern architectures.

Portability is just one of the many traditional design goals of numerical software libraries we reconsidered and sometimes modified in the course of designing LAPACK. Other goals are numerical stability (or accuracy), robustness against over/underflow, portability of correctness (in contrast to portability of performance), and scope (which input data structures to support). Recent changes in computer architectures and numerical methods have permitted us to to strengthen these goals in many cases, resulting in a library more capable than before. These changes include the availability of massive parallelism, IEEE floating point arithmetic, new high accuracy algorithms, and better condition estimation techniques. We have also identified tradeoffs among the goals, as well as certain architectural and language features whose presence (or absence) makes achieving these goals easier.

Section 2 reviews traditional goals of library design. Section 3 gives an overview of the LAPACK library. The next three sections discuss how traditional design goals and methods have been modified: Section 4 deals with efficiency, section 5 with stability and robustness, section 6 with portability, and section 7 with scope. Section 8 lists particular architectural and programming language features that bear upon the goals. Section 9 describes future work on distributed memory machines.

We will use the notation $\|x\|$ to refer to the largest absolute component of the vector x, and $\|A\|$ to be the corresponding matrix norm (the maximum absolute row sum). ε will denote the machine roundoff, UNFL the underflow threshold (smallest positive normalized floating point number) and OVFL the overflow threshold (the largest finite floating point number).

The breadth of material we will cover does not permit us to describe or justify all our claims in detail. Instead we give an overview, and relegate details to future papers.

2 Traditional Library Design Goals

The traditional goals of good library design are the following:

- stability and robustness,

- efficiency,

- portability, and

• wide scope.

Let us consider these in more detail in the context of libraries for numerical linear algebra, particularly LINPACK and EISPACK. The terms have traditional interpretations:

In linear algebra, *stability* refers specifically to *backward stability with respect to norms* as developed by Wilkinson [35, 24]. In the context of solving a linear system $Ax = b$, for example, this means that the computed solution $\hat{x}$ solves a perturbed system $(A + E)\hat{x} = b + f$ where $\|E\| = O(\varepsilon)\|A\|$ and $\|f\| = O(\varepsilon)\|b\|$. [1] Similarly, in finding eigenvalues of a matrix A the computed eigenvalues are the exact eigenvalues of $A + E$ where again $\|E\| = O(\varepsilon)\|A\|$. [2] *Robustness* the ability of a computer program to detect and gracefully recover from abnormal situations without unnecessary interruption of the computer run such as in overflows and dangerous underflows. In particular, it means that if the inputs are "far" from over/underflow, and the true answer is far from over/underflow, then the program should not overflow (which generally halts execution) or underflow in such a way that the answer is much less accurate than in the presence of roundoff alone. For example, in standard Gaussian elimination with pivoting, intermediate underflows do not change the bounds for $\|E\|$ and $\|f\|$ above so long as A, b and x are far enough from underflow themselves [13].

Among other things, *efficiency* means that the performance (floating point operations per second, or flops) should not degrade for large problems; this property is frequently called *scalability*. When using direct methods as in LINPACK and EISPACK, it also means that the running time should not vary greatly for problems of the same size (though occasional examples where this occurs are sometimes dismissed as "pathological cases"). Maintaining performance on large problems means, for example, avoiding unnecessary page faults. This was a problem with EISPACK, and was fixed in LINPACK by using column oriented code which accesses matrix entries in consecutive memory locations in columns (since Fortran stores matrices by column) instead of by rows. Running time depends almost entirely on a problem's dimension alone, not just for algorithms with fixed operation counts like Gaussian elimination, but also for routines that iterate (to find eigenvalues). Why this should be so for some eigenroutines is still not completely understood; worse, some nonconvergent examples have been discovered only recently [8].

Portability in its most inclusive sense means that the code is written in a standard language (say Fortran), and that the source code can be compiled on an arbitrary machine with an arbitrary Fortran compiler to produce a program that will run correctly and efficiently. We call this the "mail order software" model of portability, since it reflects the model used by software servers like Netlib [18]. This notion of portability is quite demanding. It demands that all relevant properties of the computer's arithmetic and

---

[1] The constants in $O(\varepsilon)$ depend on dimensionality in a way that is important in practice but not here.

[2] This is one version of backward stability. More generally one can say that an algorithm is backward stable if the answer is scarcely worse than what would be be computed exactly from a slightly perturbed input, even if one cannot construct this slightly perturbed input.

architecture be discovered at runtime within the confines of a Fortran code. For example, if the overflow threshold is important to know for scaling purposes, it must be discovered at runtime *without overflowing*, since overflow is generally fatal. Such demands have resulted in quite large and sophisticated programs [27, 21] which must be modified continually to deal with new architectures and software releases. The mail order software notion of portability also means that codes generally must be written for the worst possible machine expected to be used, thereby often degrading performance on all the others.

Finally, *wide scope* refers to the range of input problems and data structures the code will support. For example, LINPACK and EISPACK deal with dense matrices (stored in a rectangular array), packed matrices (where only the upper or lower half of a symmetric matrix is stored), and band matrices (where only the nonzero bands are stored). In addition, there are some special internally used formats such as Householder vectors to represent orthogonal matrices. Then there are sparse matrices which may be stored in innumerable ways; but in this paper we will limit ourselves to dense and band matrices, the mathematical types addressed by LINPACK, EISPACK and LAPACK.

3 LAPACK Overview

Teams at the University of California at Berkeley, the University of Tennessee, the Courant Institute of Mathematical Sciences, the Numerical Algorithms Group, Ltd., Rice University, Argonne National Laboratory, and Oak Ridge National Laboratory are developing a transportable linear algebra library called LAPACK (short for Linear Algebra Package). The library is intended to provide a coordinated set of subroutines to solve the most common linear algebra problems and to run efficiently on a wide range of high-performance computers.

LAPACK will provide routines for solving systems of simultaneous linear equations, least-squares solutions of linear systems of equations, eigenvalue problems and singular value problems. The associated matrix factorizations (LU, Cholesky, QR, SVD, Schur, generalized Schur) will also be provided, as will related computations such as reordering of the Schur factorizations and estimating condition numbers. Dense and banded matrices will be handled, but not general sparse matrices. In all areas, similar functionality will be provided for real and complex matrices, in both single and double precision. LAPACK will be in the public domain and available from Netlib some time in 1991.

The library is written in standard Fortran 77. The high performance is attained by calls to block matrix operations, such as matrix-multiply, in the innermost loops [14, 2]. These operations are standardized as Fortran subroutines called the Level 3 BLAS (Basic Linear Algebra Subprograms [16]). Although standard Fortran implementations of the Level 3 BLAS are available on Netlib, high performance can generally be attained only by using implementations optimized for each particular architecture. In particular, all parallelism (if any) is embedded in the BLAS and invisible to the user.

Besides depending upon locally implemented Level 3 BLAS, good performance also requires knowledge of certain machine-dependent *block sizes*, which are the sizes of the submatrices processed by the Level 3 BLAS. For example, if the block size is 32 for the Gaussian Elimination routine on a particular machine, then the matrix will be processed in groups of 32 columns at a time. Details of the memory hierarchy determine the block size that optimizes performance [1].

4   New Goals and Methods: Efficiency

The most important fact is that the Level 3 BLAS have turned out to be a satisfactory mechanism for producing fast transportable code for most dense linear algebra computations on high performance *shared memory* machines. (Dealing with distributed memory machines is future work we describe below.) Gaussian elimination and its variants, QR decomposition, and reductions to Hessenberg, tridiagonal and bidiagonal forms (as preparation for finding eigenvalues and singular values) all admit efficient block implementations [1, 2]. Such codes are often nearly as fast as full assembly language implementations for sufficiently large matrices, but approach their asymptotic speeds more slowly. Parallelism, embedded in the BLAS, is generally useful only on sufficiently large problems, and can in fact slow down processing on small problems. This means that the number of processors exercised should ideally be a function of the problem size, something not always taken into account by existing BLAS implementations.

However, the BLAS do not deal with all problems, even in the shared memory world. First, the real nonsymmetric eigenvalue problem involves solving systems with quasi-triangular matrices (block triangular matrices with 1 by 1 and 2 by 2 blocks). These are not handled by the BLAS and so must be written in Fortran. As a result, the real nonsymmetric eigenproblem runs relatively slowly compared to the complex nonsymmetric eigenproblem, which has only standard triangular matrices.

Second, finding eigenvalues of a symmetric tridiagonal matrix, and singular values of a bidiagonal matrix, can not exploit blocking. For these problems, we invented other methods which are potentially quite parallel [20, 4]. However, since the parallelism is not embedded in the BLAS, and since standard Fortran 77 cannot express parallelism, these methods are currently implemented only as serial codes. We intend to supply parallel versions in future releases.

Third, the Hessenberg eigenvalue algorithm has proven quite difficult to parallelize. We have a partially blocked implementation of the QR algorithm but the speedup is modest [5]. There has been quite recent progress [19], but it remains an open problem to produce a highly parallel and reliably stable and convergent algorithm for this problem and for the generalized Hessenberg eigenvalue problem.

Fourth is the issue of performance tuning, in particular choosing the block size parameters. In principal, the optimal block size could depend on the machine, problem dimension, and other problem parameters such as leading matrix dimension. We have a mechanism (subroutine ILAENV) for choosing the block size based on all this information. But we still need a better way to choose the block size. We used brute force during beta testing of LAPACK, running exhaustive tests on different machines, with ranges of block sizes and problem dimensions. This has produced a large volume of test results, too large for thorough human inspection and evaluation.

There appear to be at least three ways to choose block parameters. First, we could take the exhaustive tests we have done, find the optimal block sizes, and store them in tables in subroutine ILAENV; each machine would require its own special tables. Second, we could devise an automatic installation procedure which could run just a few benchmarks and automatically produce the necessary tables. Third, we could devise algorithms which tuned themselves at run-time, choosing parameters automatically [9, 10]. The choice of method depends on the degree of portability we desire; we return to this in Section 6 below.

Finally, we have determined that floating point exception handling impacts efficiency. Since overflow is a fatal exception on some machines, completely portable code must avoid it at all costs. This means extra tests, branches, and scaling must be inserted if spurious overflow is possible at all, and these slow down the code. For example, the condition estimators in LAPACK provide error bounds, and more generally warn about inaccurate answers to ill-conditioned problems. It is therefore important that these routines resist overflow. Since their main operation is (generally) solving a triangular system of equations, we cannot use the standard Level 2 BLAS triangular equation solver [17] because it is unprotected against overflow. Instead, we have another triangular solver written in Fortran including scaling in the inner loop. This gives us a double performance penalty, since we cannot use optimized BLAS, and since we must do many more floating point operations and branches. The same issues arise in computing eigenvectors.

If we could assume we had IEEE arithmetic [3], none of this would be necessary. Instead, we would run with the usual BLAS routine. If an overflow occurred, we could either trap, or else substitute an ∞ symbol, set an "overflow flag" and continue computing using the rules of infinity-arithmetic. If we trapped, we could immediately deduce that the problem is very ill-conditioned, and terminate early returning a large condition number. If we continued with infinity-arithmetic, we could check the overflow flag at the end of the computation and again deduce that the problem is very ill-conditioned. If we use trapping, we need to be able to handle the trap and resume execution, not just terminate. To be fast, this cannot involve an expensive operating system call. Similarly, infinity-arithmetic must be done at normal hardware floating point speed, not via software, lest performance suffer devastation.

Many but not all machines support IEEE arithmetic. Many that claim to do so support neither the user readable overflow flag they should nor user handleable traps. And those that do support these things often use intolerably slow software implementations. Thus, we did not supply IEEE-exploiting routines in the first version of LAPACK. However, we

intend to do so in future versions. This raises the question of portability, which we return
to in Section 6.

5 New Goals and Methods: Accuracy and Robustness

During work on LAPACK we have found better, or at least different, ways to understand
the traditional goals described in Section 2. The first improvement in accuracy and
stability involved replacing the norms traditionally used for backward stability analysis.
For example, consider solving $Ax = b$. As we said before, traditionally we have only
guaranteed that the computed $\hat{x}$ satisfied $(A + E)\hat{x} = b + f$ where E and f were small in
norm compared to A and b, respectively. If A were sparse, there was no guarantee that E
would be sparse. Similarly, if A had both very large and very small entries, some entries
of E could be very large compared to the corresponding entries of A. In other words, the
usual methods did not respect the sparsity or scaling of the original problem.

Instead, LAPACK uses a method which (except for certain rare cases) guarantees
componentwise relative backward stability: this means that $|E_{ij}| = O(\varepsilon)|A_{ij}|$ and $|f_k| =
O(\varepsilon)|b_k|$. This respects both sparsity and scaling, and can result in a much more accu-
rate $\hat{x}$. We have done this for various problems in LAPACK, including the bidiagonal
singular value decomposition and symmetric tridiagonal eigenproblem. Future releases of
LAPACK will extend this to other routines as well [14].

Second, we intend to supply condition estimators (i.e. error bounds) for every quantity computed by the library. This includes, for example, eigenvalues, eigenvectors and
invariant subspaces [6]. Some problems remain for future releases (the generalized nonsymmetric eigenproblem).

Third, we determined that Strassen-based matrix multiplication is adequately accurate
to achieve traditional normwise backward stability [15]. Strassen's method is not as
accurate as conventional matrix multiplication when the matrices are badly row or column
scaled, but if either the matrices are already reasonably scaled or if the bad scaling is first
removed, it is adequate. Thus it may be used in Level 3 BLAS implementations [25, 7].

Fourth, there is possibly a tradeoff between stability and speed in certain algorithms.
Some modern parallel architectures are designed to support particular communication
patterns and so may execute one algorithm, call it Algorithm A, much less efficiently
than another, Algorithm B, even though on conventional computers A may have been as
fast or faster than B. If Algorithm A is stable and Algorithm B is not, this means that the
new architecture will not be able to run simultaneously as fast as possible and correctly
in all cases. Thus one is tempted, in the interest of speed, to use an unstable algorithm.
Since "the fast drives out the slow even if the fast is wrong", many users will prefer
the faster algorithm despite occasional inaccuracy. So we are motivated to find a way
to use unstable algorithm B provided we can check quickly whether it got an accurate

answer, and only occasionally resort to the slower alternative. For example, consider finding the eigenvalues of an n by n symmetric tridiagonal matrix T with diagonal entries $a_1, \ldots, a_n$ and offdiagonal entries $b_1, \ldots, b_{n-1}$. A standard bisection-based method uses the fact that the number of eigenvalues of T less than σ is the number of negative d_i where $d_i = (a_i - \sigma) - b_{i-1}^2/d_{i-1}$ (we take $b_0 = 0$ and $d_0 = 1$) [24]. Evaluating this recurrence straightforwardly requires $O(n)$ time and is stable. Using a parallel-prefix algorithm the d_i can be evaluated in $O(\log n)$ time but no stability proof exists. So we need either a stability proof for the $O(\log n)$ algorithm or a fast way to check the accuracy of the computed eigenvalues at the end of the computation. Similar issues arise with other tree-based algorithms.

Fifth, there is at least one important routine which *requires* double the input precision in some intermediate calculations to compute the answer correctly [34]. This is the solution of the so-called secular equation in the divide and conquer algorithm for the symmetric tridiagonal eigenproblem. This is somewhat surprising, since all other algorithms for this problem require only the input precision in all intermediate calculations. In fact, we must be careful to say what it means to require double precision, since in principal all computations could be done simulating arbitrary precision using integers: We mean in fact that there is an intermediate quantity in the algorithm which must be computed to high relative accuracy despite catastrophic cancellation in order to guarantee stability. (There are other examples where we were able to find an adequate single precision algorithm only after great effort, whereas an algorithm using a little double precision arithmetic was obvious. So even though double precision is not necessary in these cases, it would have made software design much easier.)

This requirement for double the input precision impacts library design as follows. Our original design goal was *not* to use mixed precision arithmetic. This traditional goal arose both because standard Fortran compilers were not required to supply a double precision complex data type, and because of the desire to use the same algorithm whether the input precision were single precision or double precision. (The use of mixed precision would have required quadruple precision for double precision input, and quadruple is rarely available.) An alternative is to simulate double precision using single (and quadruple using double). Provided the underlying arithmetic is accurate enough, there are a number of standard techniques for simulating "doubled precision" arithmetic using a few single precision operations [12, 30, 34, 32]. However, this means that we must either assume the arithmetic is sufficiently accurate, not true on all machines, or decide at run time whether the arithmetic is sufficiently accurate and then either do the simulated precision doubling or return an error flag. Making this decision at run-time is quite challenging, because there is no simple characterization of which arithmetics are sufficiently accurate. The desired simulation works, for example, with IEEE arithmetic, IBM 370 arithmetic, or VAX arithmetic, but requires different correctness proofs in each case. It does not work with Cray arithmetic. Thus it almost appears that we must be able to determine the floating point architecture at run-time in sufficient detail to determine the machine manufacturer.

The routine for determining floating point properties at run time, SLAMCH, has several other difficult tasks. It must also determine the overflow and underflow thresholds OVFL and UNFL, in particular without overflowing. OVFL and UNFL are used for scaling to avoid overflow or harmful underflow during subsequent calculations. Unfortunately, there can be different effective over/underflow thresholds depending on the operation and on the software. For example, the Cray divides a/b essentially using reciprocal approximation and multiplication $a*(1/b)$. If a and b are both tiny, then $1/b$ may overflow even though the true quotient a/b is quite moderate in value. The Cray and NEC machines both implement complex division in the simplest possible way, without branches: $\frac{a+ib}{c+id} = \frac{ac+bd}{c^2+d^2} + i\frac{bc-ad}{c^2+d^2}$. Thus even if the true quotient is modest in size, the computation can overflow if either c or d exceeds $\text{OVFL}^{1/2}$ in magnitude or both are sufficiently less than $\text{UNFL}^{1/2}$ in magnitude. This effectively cuts the exponent range in half. Similarly, there is a Level 1 BLAS routine called SNRM2 [29] which computes the Euclidean length of a vector: $(\sum_i x_i^2)^{1/2}$. The Cray uses this straightforward implementation which can again fail unnecessarily if any $|x_i| > \text{OVFL}^{1/2}$ or all $|x_i| < \text{UNFL}^{1/2}$. As a result of all these and other details, and the fact that new hardware and compilers are constantly appearing, SLAMCH is currently 2000 lines long and growing.

All told, a surprisingly large fraction of the programming effort and lines of code were devoted to clever algorithms using only the input precision to compute various quantities while avoiding overflow, harmful underflow and unacceptable roundoff. In all these cases, there were obvious algorithms based on higher precision *and wider exponent range*. Simulating doubled precision using single can only supply higher precision, not the wider exponent range. This means the extra programming effort to avoid over/underflow by scaling would still remain. By far the best solution would be the availability of a format with higher precision and wider exponent range for the relatively few critical operations. One approach to consider in future libraries is identifying a few high precision and/or wide exponent range primitives from which the ones we need can be built. Like the BLAS, one could supply (at least partly) portable versions which might depend on precision doubling techniques and scaling, but expect the manufacturers to supply more efficient ones for each machine.

6 New Goals and Methods: Portability

As stated above, we can ask for portability of correctness (or of accuracy and robustness), or of performance. We have nearly abandoned portability of performance because of the need for machine dependent BLAS and block sizes. However, we do supply strictly portable Fortran BLAS and default block sizes which may provide adequate performance in some cases, but probably not peak performance on many architectures.

We have tried strictly to maintain portability of correctness. The "mail order software" model described above recognizes that code developed on one machine is often embedded (and hidden) in an application on another machine, and then used on a third. Consequently, it would be unreasonable to expect a user acquiring a code to modify all its subparts to ensure they run correctly on her machine. Since no standard language mechanism exists yet for making environmental enquiries about floating point properties, etc., all this must be done at run time. This explains if not justifies the enormous intellectual effort that has been spent on codes like SLAMCH [27, 21].

However, there is a tradeoff between this kind of portability on the one hand and efficiency, accuracy and robustness on the other. Most machines now supply IEEE arithmetic. As mentioned above, there are numerous places where significantly faster, more accurate and more robust code could have been written had we been able to assume that IEEE arithmetic *and standard high level language access to its exception handling features* were available. Unfortunately, no such standard high level language access exists yet. There have been attempts at such a standard [31, 28] but they fall far short of what is needed and could even make writing efficient portable code harder by mandating a standard environment antagonistic to what we need.

A deleterious by-product of the present situation is the near absence of any payoff for the many manufacturers who have supplied careful and complete IEEE arithmetic implementations, because little software exists that takes advantage of its features. Unless such software is written, manufacturers will have little incentive to implement these features, which then may even disappear from future versions of the standard.

We intend to produce IEEE-exploiting versions of those LAPACK codes which could benefit from special features of IEEE arithmetic. This includes condition estimators, eigenvector algorithms, and others. Not only will this code perform much better than the current portable code, but it will provide incentives to manufacturers to implement IEEE arithmetic with full access to its exception-handling features.

7 New Goals and Methods: Scope

In conventional libraries, as well as in the first version of LAPACK, dense rectangular matrices are stored in essentially one standard data structure: A statement like "DIMENSION A(20,10)" used to indicate that A is a rectangular array stored in consecutive memory locations (or contains a matrix stored in groups of evenly spaced consecutive memory locations). This is no longer a reasonable model on distributed memory machines, because there is no longer any such standard memory mapping. There are a number of competing parallel programming models (SPMD vs. MPMD, SIMD vs. MIMD, explicit message passing vs. implicit message passing, send/receive vs. put/get, etc.) and a large number of ways in which data can be distributed among memories [22, 26]. For example, a one-dimensional array could be laid out in at least four different regular ways, with datum i stored in memory $\lfloor i/b \rfloor \bmod p + 1$, where p is the number of memories used, and

b is a blocking parameter. Various examples are shown below for $0 \le i \le 15$; each box represents a data item, and the number inside is the number of the memory in which it is stored:

All in one ($p = 1, b = 1$)

1	1	1	1	1	1	1	1	1	1	1	1	1	1	1	1

Blocked ($p = 4, b = 4$)

1	1	1	1	2	2	2	2	3	3	3	3	4	4	4	4

Cyclic ($p = 4, b = 1$)

1	2	3	4	1	2	3	4	1	2	3	4	1	2	3	4

Block cyclic ($p = 4, b = 2$)

1	1	2	2	3	3	4	4	1	1	2	2	3	3	4	4

Irregular

1	4	1	3	2	4	1	1	3	2	4	4	1	3	2	3

A multidimensional array may have each dimension stored in a different one of the layouts above, as shown in the following examples, labeled as above:

1212	1212	1212	1212
3434	3434	3434	3434
1212	1212	1212	1212
3434	3434	3434	3434
1212	1212	1212	1212
3434	3434	3434	3434
1212	1212	1212	1212
3434	3434	3434	3434

1	1	1	1	1	1	1	1
2	2	2	2	2	2	2	2
3	3	3	3	3	3	3	3
4	4	4	4	4	4	4	4
1	1	1	1	1	1	1	1
2	2	2	2	2	2	2	2
3	3	3	3	3	3	3	3
4	4	4	4	4	4	4	4

The first version of LAPACK was designed to handle *single problem instances*, e.g. a single system of linear equations to solve. On massively parallel machines one can expect users to want to solve many problems simultaneously. One way to do this is to use a multi-dimensional array, where two of the dimensions are the matrix dimensions and the others index independent problems.

Data layout is closely related to efficiency, because it is related to scalability. To be more precise, let $E(N, P, M, I)$ be the efficiency of the code as a function of problem size $N = n^2$ (n = matrix dimension), number of processors P, memory size per processor M and number of independent problem instances I. Scalability means that as these four parameters grow, E should stay acceptably large, say at least 0.5 (i.e. at least half as fast as the best possible code for that machine). With four parameters, there are several ways they could grow, reflecting different uses of the library. For example, suppose P and N grow with $N = O(P)$, and M and I remain constant. This corresponds to adding more identical processors to the system and letting the problem size grow proportionally to the total memory. This can only be done with a data layout where each memory contains a small constant size submatrix of the whole matrix. A second example is to let M and N grow with $N = O(M)$, and P and I constant. This corresponds to adding memory to each processor, and letting the problem size grow proportionally. Here the data layout requires each memory to hold a constant fraction of the entire matrix, perhaps a growing submatrix or growing number of columns. A third example is to let P, M and N grow with $M = O(P)$ and $N = O(P^2)$, and I constant. This corresponds to keeping a constant number of columns (or rows) per memory. Finally, we can keep N and M constant and let I and $P = O(I)$ grow. This corresponds to solving more independent problem instances of the same size, and keeping the same sized submatrix on each processor. Thus, depending on what kinds of scalability we wish to support, we may have to support many data layouts.

If ever there were a case for semi-automatically generated polyalgorithms, this may be it. The danger in choosing to support only a few of the plethora of possibilities is that the decision may turn into a self-fulfilling prophecy rendering the other memory mappings of little use.

It is still unclear which programming model is best, and how many of these diverse data layouts need to be supported.

8 Suggestions for Architectures and Programming Languages

We have listed a number of suggestions for architectures and programming languages in earlier sections; we summarize them here:

1. Ability to express parallelism in a high level language.

2. Ability to perform floating point operations reasonably efficiently in double the largest input precision, even if only simulated in software using that input precision exclusively. Even better is a doubled precision format with wider exponent range.

3. Access to efficiently implemented exception handling facilities, particularly infinity arithmetic. Trap handlers are a poor substitute.

4. Carefully implemented complex arithmetic and BLAS.

5. A standard set of floating point enquiries sufficiently detailed to describe the features of the last items, and unambiguously. Perhaps NextAfter [3] is the key. We are currently working on a standard for these enquiries.

6. BLAS for dealing with quasi-triangular matrices.

Note that a complete implementation of IEEE arithmetic would satisfy suggestions 2 and 3 above.

9 Future Work

We have recently begun work on a new version of LAPACK. We intend to pursue all the goals listed above, in particular

- Producing a version for distributed memory parallel machines,

- Adding more routines satisfying new componentwise relative stability bounds,

- Adding condition estimators and error bounds for all quantities computable by the library,

- Producing routines designed to exploit exception handling features of IEEE arithmetic, and

- Producing Fortran 90 and C versions of the software.

We hope the insight we gained in this project will influence future developers of hardware, compilers and systems software so that they provide tools to facilitate development of high quality portable numerical software.

10 Acknowledgements

The authors acknowledge the work of the many contributors to the LAPACK project: E. Anderson, Z. Bai, C. Bischof, P. Deift, J. DuCroz, A. Greenbaum, S. Hammarling, E. Jessup, L.-C. Li, A. McKenney, D. Sorensen, P. Tang, C. Tomei, and K. Veselić.

References

[1] E. Anderson and J. Dongarra. Results from the initial release of LAPACK. Computer Science Dept. Technical Report CS-89-89, University of Tennessee, Knoxville, 1989. (LAPACK Working Note #16).

[2] E. Anderson and J. Dongarra. Evaluating block algorithm variants in LAPACK. Computer Science Dept. Technical Report CS-90-103, University of Tennessee, Knoxville, 1990. (LAPACK Working Note #19).

[3] ANSI/IEEE, New York. *IEEE Standard for Binary Floating Point Arithmetic*, Std 754-1985 edition, 1985.

[4] M. Assadullah, J. Demmel, S. Figueroa, A. Greenbaum, and A. McKenney. On finding eigenvalues and singular values by bisection. LAPACK Working Note. in preparation.

[5] Z. Bai and J. Demmel. On a block implementation of Hessenberg multishift QR iteration. *International Journal of High Speed Computing*, 1(1):97–112, 1989. (also LAPACK Working Note #8).

[6] Z. Bai, J. Demmel, and A. McKenney. On the conditioning of the nonsymmetric eigenproblem: Theory and software. Computer Science Dept. Technical Report 469, Courant Institute, New York, NY, October 1989. (LAPACK Working Note #13).

[7] D. H. Bailey, K. Lee, and H. D. Simon. Using Strassen's algorithm to accelerate the solution of linear systems. *J. Supercomputing*, 4:97–371, 1991.

[8] S. Batterson. Convergence of the shifted QR algorithm on 3 by 3 normal matrices. *Num. Math.*, 58:341–352, 1990.

[9] C. Bischof. Adaptive blocking in the QR factorization. *J. Supercomputing*, 3(3):193–208, 1989.

[10] C. Bischof and P. Lacroute. An adaptive blocking strategy for matrix factorizations. In H. Burkhart, editor, *Lecture Notes in Computer Science 457*, pages 210–221, New York, NY, 1990. Springer Verlag.

[11] J. Bunch, J. Dongarra, C. Moler, and G. W. Stewart. *LINPACK User's Guide*. SIAM, Philadelphia, PA, 1979.

[12] T. Dekker. A floating point technique for extending the available precision. *Num. Math.*, 18:224–242, 1971.

[13] J. Demmel. Underflow and the reliability of numerical software. *SIAM J. Sci. Stat. Comput.*, 5(4):887–919, Dec 1984.

[14] J. Demmel. LAPACK: A portable linear algebra library for supercomputers. In *Proceedings of the 1989 IEEE Control Systems Society Workshop on Computer-Aided Control System Design*, Tampa, FL, Dec 1989. IEEE.

[15] J. Demmel and N. J. Higham. Stability of block algorithms with fast Level 3 BLAS. to appear in *ACM Trans. Math. Soft.*

[16] J. Dongarra, J. Du Croz, I. Duff, and S. Hammarling. A set of level 3 basic linear algebra subprograms. *ACM Trans. Math. Soft.*, 16(1):1–17, March 1990.

[17] J. Dongarra, J. Du Croz, S. Hammarling, and Richard J. Hanson. An extended set of fortran basic linear algebra subroutines. *ACM Trans. Math. Soft.*, 14(1):1–17, March 1988.

[18] J. Dongarra and E. Grosse. Distribution of mathematical software via electronic mail. *Communications of the ACM*, 30(5):403–407, July 1987.

[19] J. Dongarra and M. Sidani. A parallel algorithm for the non-symmetric eigenvalue problem. Computer Science Dept. Technical Report CS-91-137, University of Tennessee, Knoxville, TN, 1991.

[20] J. Dongarra and D. Sorensen. A fully parallel algorithm for the symmetric eigenproblem. *SIAM J. Sci. Stat. Comput.*, 8(2):139–154, March 1987.

[21] J. Du Croz and M. Pont. The development of a floating-point validation package. In M. J. Irwin and R. Stefanelli, editors, *Proceedings of the 8th Symposium on Computer Arithmetic*, Como, Italy, May 19-21 1987. IEEE Computer Society Press.

[22] G. Fox, S. Hiranandani, K. Kennedy, C. Koelbel, U. Kremer, C.-W. Tseng, and M.-Y. Wu. Fortran D language specification. Computer Science Department Report CRPC-TR90079, Rice University, Houston, TX, December 1990.

[23] B. S. Garbow, J. M. Boyle, J. J. Dongarra, and C. B. Moler. *Matrix Eigensystem Routines – EISPACK Guide Extension*, volume 51 of *Lecture Notes in Computer Science*. Springer-Verlag, Berlin, 1977.

[24] G. Golub and C. Van Loan. *Matrix Computations*. Johns Hopkins University Press, Baltimore, MD, 2nd edition, 1989.

[25] N. J. Higham. Exploiting fast matrix multiplication within the Level 3 BLAS. *ACM Trans. Math. Soft.*, 16:352–368, 1990.

[26] S. Hiranandani, K. Kennedy, C. Koelbel, U. Kremer, and C.-W. Tseng. An overview of the Fortran D programming system. Computer Science Department Report COMP TR91-154, Rice University, Houston, TX, March 1991.

[27] W. Kahan. Paranoia. available from Netlib[18].

[28] W. Kahan. Analysis and refutation of the International Standard ISO/IEC for Language Compatible Arithmetic. submitted to SIGNUM Newsletter, 1991.

[29] C. Lawson, R. Hanson, D. Kincaid, and F. Krogh. Basic linear algebra subprograms for fortran usage. *ACM Trans. Math. Soft.*, 5:308–323, 1979.

[30] S. Linnainmaa. Software for doubled-precision floating point computations. *ACM Trans. Math. Soft.*, 7:272–283, 1981.

[31] M. Payne and B. Wichmann. Information technology - programming languages - language compatible arithmetic. Project JTC1.22.28, ISO/IEC JTC1/SC22/WG11, 1 March 1991. First Committee Draft (Version 3.1).

[32] D. Priest. Algorithms for arbitrary precision floating point arithmetic. In P. Kornerup and D. Matula, editors, *Proceedings of the 10th Symposium on Computer Arithmetic*, pages 132–145, Grenoble, France, June 26-28 1991. IEEE Computer Society Press.

[33] B. T. Smith, J. M. Boyle, J. J. Dongarra, B. S. Garbow, Y. Ikebe, V. C. Klema, and C. B. Moler. *Matrix Eigensystem Routines – EISPACK Guide*, volume 6 of *Lecture Notes in Computer Science*. Springer-Verlag, Berlin, 1976.

[34] D. Sorensen and P. Tang. On the orthogonality of eigenvectors computed by divide-and-conquer techniques. Mathematics and Computer Science Division MCS-P152-0490, Argonne National Lab, Argonne, IL, May 1990. to appear in SIAM J. Num. Anal.

[35] J. H. Wilkinson. *The Algebraic Eigenvalue Problem*. Oxford University Press, Oxford, 1965.

James Demmel, Computer Science Division and Mathematics Department, University of California, Berkeley, CA 94720

Jack Dongarra, Computer Science Department, University of Tennessee, Knoxville, TN 37996

W. Kahan, Computer Science Division and Mathematics Department, University of California, Berkeley, CA 94720

J J DONGARRA AND M SIDANI

A parallel algorithm for the non-symmetric eigenvalue problem

Abstract

This paper describes a parallel algorithm for computing the eigenvalues and eigenvectors of a non-symmetric matrix. The algorithm is based on a divide-and-conquer procedure and uses an iterative refinement technique.

1 Introduction

A fully parallel algorithm for the symmetric eigenvalue problem was recently proposed in [5]. This algorithm is based on a divide-and-conquer procedure outlined in [2]. The fundamental principle behind this algorithm is that the partitioning by rank-one tearing interlaces the eigenvalues of the modified problem with the eigenvalues of the original problem.

In this paper we propose a parallel algorithm for the solution of the non-symmetric eigenvalue problem that uses some of the features of the divide-and-conquer algorithm for the symmetric case. In particular, the original problem is divided into two smaller and independent subproblems by a rank-one modification of the matrix. Once the eigensystems of the smaller subproblems are known, corrections to these can be computed using Newton's method. Another possible approach consists in using continuation methods [10].

In section 2 we describe the algorithm. Section 3 covers the deflation step required to overcome multiple convergence to a particular eigenvalue. In section 4 the convergence behavior is discussed. In section 5 we discuss the case when the matrix or its rank one modification has a defective system of eigenvectors. Section 6 estimates the amount of work the parallel algorithm requires and compares this to the standard techniques. Section 7 describes the parallel algorithm and its different parallel implementations, and gives numerical results.

2 The Algorithm

Given a matrix H, an eigenpair (x_0, λ_0) of H can be thought of as a solution to the following polynomial system

$$(S) \begin{cases} Hx - \lambda x = 0 \\ e_s^T x = 1 \end{cases}$$

---

*This work was supported in part by the National Science Foundation Science and Technology Center Cooperative Agreement No. CCR-8809615, and in part by the Applied Mathematical Sciences subprogram of the Office of Energy Research, U.S Department of Energy, under Contract DE-AC05-84OR21400.

where e_s is the sth unit vector. Let

$$F_s(x, \lambda) = \begin{pmatrix} Hx - \lambda x \\ e_s^T x - 1 \end{pmatrix}.$$

Then, finding an eigenpair of H reduces to finding a zero of F_s. In what follows, unless otherwise mentioned, H is assumed to be a real, unreduced (no zeros on the subdiagonal), upper-Hessenberg matrix of order n. If H has a zero on the subdiagonal then, finding its eigenvalues reduces to finding those of the blocks on the diagonal. Also as a consequence of our assumption that H is unreduced, an eigenvalue of H can only have geometric multiplicity one. We can write H as

$$H = \left(\begin{array}{c|c} H_{11} & H_{12} \\ \hline \alpha e_1^{(k)} e_k^{(k)T} & H_{22} \end{array} \right),$$

where H_{11} and H_{22} are upper-Hessenberg of dimensions $k \times k$ and $n-k \times n-k$, respectively; $\alpha = h_{k+1,k}$, and $e_i^{(k)}$ is the ith unit vector of length k.

Let $H_0 = H - \alpha e_{k+1}^{(n)} e_k^{(n)T}$, then $H_0 = \left(\begin{array}{c|c} H_{11} & H_{12} \\ \hline 0 & H_{22} \end{array} \right)$, and $\sigma(H_0) = \sigma(H_{11}) \cup \sigma(H_{22})$

(where $\sigma(M)$ is the spectrum of M). The algorithm can then be described as follows: We first find the k eigenpairs of H_{11} and the $n-k$ eigenpairs of H_{22} by some method, perhaps the QR algorithm. These eigenpairs are then used to construct initial approximations to the eigenpairs of H in the following way: if λ is an eigenvalue of H_{11} and x is the corresponding eigenvector, then λ is viewed as an approximate eigenvalue of H with the corresponding approximate eigenvector taken to be $\begin{pmatrix} x \\ 0 \end{pmatrix}$, where $n-k$ zeros are appended to x. On the other hand, if λ is an eigenvalue of H_{22} and x is the corresponding eigenvector, then $\begin{pmatrix} 0 \\ x \end{pmatrix}$, where k zeros are prefixed to x is taken as an approximate eigenvector of H corresponding to the approximate eigenvalue λ.

Having constructed these initial guesses, we now use them as starting points for Newton's method to find the zeros of F_s, i.e., the eigenpairs of H. Hence, given an approximate zero (x, λ) of F_s the next approximation is

$$x' = x + y; \lambda' = \lambda + \mu,$$

where (y, μ) is the solution of the system

$$\begin{pmatrix} H - \lambda I & -x \\ e_s^T & 0 \end{pmatrix} \begin{pmatrix} y \\ \mu \end{pmatrix} = \begin{pmatrix} r \\ 0 \end{pmatrix}, \tag{1}$$

where $r = \lambda x - Hx$.

Newton's method comes into this problem in a rather "natural" way. Indeed, suppose that (x, λ) is an approximate eigenpair of H: $Hx \approx \lambda x$. Then it can be shown that a correction (y, μ) to this approximate eigenpair satisfying

$$H(x + y) = (\lambda + \mu)(x + y),$$

86

is the solution of equation (1), provided second order terms are ignored and x, and $x + y$ are normalized so that their s^{th} component is one.

We use now first order perturbation theory in order to shed some light on the issue of which subdiagonal entry to introduce the zero. Let $E = -\alpha e_{k+1} e_k^T$, where as above: $\alpha = h_{k+1,k}$ (note that $H + E = H_0$, introduced above). Then classical results from function theory ([9], v. 2, pp. 119-134) allow us to state that in a small neighborhood of zero, we have

$$(H + \epsilon E)x(\epsilon) = \lambda(\epsilon)x(\epsilon), \tag{2}$$

for all ϵ in that neighborhood and for differentiable $(x(\epsilon), \lambda(\epsilon))$ corresponding to a simple eigenpair (x, λ) of H. Clearly: $x(0) = x$ and $\lambda(0) = \lambda$. Let y^H be the left eigenvector of H corresponding to λ. Then differentiating both sides in equation (2), pre-multipying by y^H and setting $\epsilon = 0$ we get (with primes indicating derivatives),

$$y^H E x = \lambda'(0) y^H x,$$

and therefore

$$|\lambda'(0)| = \frac{|y^H E x|}{|y^H x|} = \frac{|\alpha| |y_{k+1}| |x_k|}{|y^H x|}. \tag{3}$$

The factors of the numerator are not really independent. Indeed if $\alpha = 0$, then at least one of y_{k+1} or x_k is zero. Hence for α small, we can expect one of y_{k+1} and x_k to be correspondingly small, since the components of the eigenvectors vary continuously. Therefore, we have found it sufficient in practice to look for the smallest subdiagonal entry in a pre-specified range, and accept it as the subdiagonal entry (in that range) with respect to which the eigenvalues of the matrix are least sensitive, and set it equal to zero.

An outline of the algorithm follows:

Algorithm 2.1 *Given an upper-Hessenberg matrix H, the following algorithm computes the eigensystems of two submatrices of H and uses them as initial guesses for starting Newton iterations for determining the eigensystem of H.*

Determine subdiagonal element α where $H = \left(\begin{array}{c|c} H_{11} & H_{12} \\ \hline 0 \ \alpha & H_{22} \end{array} \right)$, should be split;

Determine initial guesses from eigensystems of the 2 diagonal blocks H_{11} and H_{22};

For each initial guess (λ_i, x_i) iterate until convergence:

$$\left(\begin{array}{cc} H - \lambda_i I & -x_i \\ e_s^T & 0 \end{array} \right) \left(\begin{array}{c} y \\ \mu \end{array} \right) = \left(\begin{array}{c} r_i \\ 0 \end{array} \right) ; \lambda_i \leftarrow \lambda_i + \mu \ ; \ x_i \leftarrow x_i + y;$$

end;

Check for duplicates and deflate if necessary;

More will be said about the last step, deflation, in section 3. We end this section with some implementational details. Our algorithm will accept (x, λ) as an eigenpair of H when $\|Hx - \lambda x\|/\|x\|\|H\| < tol$, where tol is some specified tolerance of order ϵ, the machine unit roundoff. Under these conditions ([8]), (x, λ) is an exact eigenpair of a matrix obtained from H by a slight perturbation: Indeed,

$$(H + \frac{1}{x^H x} r x^H)x = \lambda x,$$

where $r = \lambda x - Hx$.

Starting from two complex conjugate initial approximations, Newton's method will converge to two complex conjugate zeros of F_s, or the same real zero. This will allow significant savings in the computations.

Lastly when starting from a real initial guess, only *real* corrections are computed. Therefore the imaginary part of the eigenvalue should be perturbed if convergence to a complex eigenpair is to be made possible. As a case illustrating this situation we mention that of a 2×2 real matrix whose eigenvalues are complex.

3   Deflation

A naive implementation of the algorithm in section 2 may result in many eigenvalues being found multiple times and consequently some eigenvalues not being found at all since the number of initial guesses is n (at most). To avoid this unwanted situation, we included a deflation step in our algorithm which is designed to obtain further zeros using Newton's method. In this section we propose various methods for doing this. Basically, these methods fall into two classes: the methods of one class obtain further zeros by explicitly avoiding a particular eigenvalue; the methods of the other class by avoiding the eigenvector.

A common drawback that these methods have is that they tend to serialize the computation. However, it has been our experience that the need to deflate arises infrequently, less than 5% of the time in our tests.

3.1   Deflation using elementary transformations

We now describe one possible deflating similarity transformation. The assumptions on H, λ and x are as above.

Since we are assuming H to be upper-Hessenberg with no zeros on the subdiagonal, then $x_n \neq 0$ and the elementary transformation

$$M = [e_1, \ldots, e_{n-1}, x]$$

i.e.,

$$
M = \begin{pmatrix} 1 & & & x_1 \\ & \ddots & & \vdots \\ & & 1 & x_{n-1} \\ & \mathbf{0} & & x_n \end{pmatrix} \tag{4}
$$

is non-singular. The inverse of this matrix is

$$
M^{-1} = \begin{pmatrix} 1 & & & -\frac{x_1}{x_n} \\ & \ddots & & \vdots \\ & & 1 & -\frac{x_{n-1}}{x_n} \\ & \mathbf{0} & & \frac{1}{x_n} \end{pmatrix}. \tag{5}
$$

It is easy to see that $M^{-1}x = e_n$. Now we let

$$
\widetilde{H} = M^{-1}HM. \tag{6}
$$

Then the last column of $\widetilde{H}$ is λe_n. Furthermore the leading principal submatrix of order $n-1$ of $\widetilde{H}$, which we will call H', is upper Hessenberg, has the property that

$$
\sigma(H') = \sigma(H) - \{\lambda\}
$$

and differs from the leading $n-1 \times n-1$ principal submatrix of H in the last column only. The strategy we have just described is given in [13] and applies regardless of whether the eigenpair (x, λ) is real or not. However when (x, λ) is not real, the last column of H' alone is not real. Hence the leading principal submatrix of H' of order $n-2$ is real. Also in this case it can be verified that an eigenvector of H' corresponding to $\bar{\lambda}$ can be chosen to be real [4]. We can carry out a deflation that produces a matrix H'' of order $n-2$ with the property that

$$
\sigma(H'') = \sigma(H') - \{\bar{\lambda}\},
$$

in the same way we obtained H' from H. Then the $n-2 \times n-2$ matrix H'' is real.

We remark here, as can be readily realized, that H' (or H'') is quite cheap to obtain in practice once an eigenpair of H is available. It requires $O(n)$ operations consisting of a vector normalization, a scalar-vector multiplication and a vector-vector addition. However the conditioning of the matrix M might raise concern. Indeed:

$$
cond_\infty(M) = \|M\|_\infty \|M^{-1}\|_\infty \approx max(|x_i|) max\left(\frac{|x_i|}{|x_n|}\right)
$$

which can be large if $x_n \ll x_i$. Having noted this, it is clear that the ill-conditioning of M can be easily detected, and therefore one of the more stable (and costlier) methods which we introduce next and in the following sections can be used.

It is possible to prevent the ill-conditioning of M from bearing on the algorithm by avoiding a similarity transformation. More precisely, the eigenvalue problem we want to solve can be thought of as a generalized eigenvalue problem,

$$
Hx = \lambda Bx,
$$

with $B = I$. We want to find $\sigma(H) = \sigma(H, I)$. Given any non-singular M and M',

$$\sigma(H, I) = \sigma(M'HM, M'M).$$

Given a particular eigenpair (x, λ) we would like to choose M and M' in a way that reduces the problem to one where λ is no longer in the spectrum. The similarity transformation introduced above, derived its deflating property from the fact that $M^{-1}x = e_n$. We let $M' = DM^{-1}$ where D is the diagonal matrix with the following entries

$$\begin{cases} d_i = 1, & if \ \frac{|x_i|}{|x_n|} \leq 1 \\ d_i = -\frac{x_n}{x_i}, & if \ \frac{|x_i|}{|x_n|} > 1 \end{cases}$$

i.e., D is chosen so that all the entries in M' are less than or equal to one. Then M' reduces x to a multiple of e_n and we have

$$M'HM = \left(\begin{array}{c|c} H' & 0 \\ \hline 0 & \gamma \end{array} \right), \quad M'M = \left(\begin{array}{c|c} D' & 0 \\ \hline 0 & \delta \end{array} \right), \tag{7}$$

with $\lambda = \gamma/\delta$. Also, $\sigma(H', D') = \sigma(H, I) - \{\lambda\}$ and therefore by this transformation λ has been "removed" from the spectrum. Working on the solution of a generalized eigenvalue problem from this point on will not in general cause any dramatic increase in the cost of the algorithm mainly because D' is diagonal. A Newton step with this problem involves the following computation

$$\left(\begin{array}{cc} H' - \lambda_i D' & -D'x_i \\ e_s^T & 0 \end{array} \right) \left(\begin{array}{c} y \\ \mu \end{array} \right) = \left(\begin{array}{c} r_i \\ 0 \end{array} \right), \quad \lambda_i \leftarrow \lambda_i + \mu \ ; \ x_i \leftarrow x_i + y \tag{8}$$

where

$$r_i = \lambda_i D' x_i - H x_i.$$

This equation is arrived at in a way similar to what we did at the beginning of section 2. Clearly the previous computation involves an $O(n)$ increase in the cost of one step: this comes from the multiplications by D'. If λ is complex, then after deflating λ and its conjugate $\bar{\lambda}$ the resulting matrices H' and D' are complex in general. This is the major drawback of this method.

Finally we mention another approach that can be of interest when the similarity transformation (6) involves a very ill-conditioned M (4). This approach consists of interchanging two components of x and the corresponding columns and rows in H so that the last component of x is large enough. More precisely: Let x_s be the largest component of x (in absolute value) and let P_{ns} be the matrix obtained from the identity matrix by permuting the nth and s^{th} columns. Then $(P_{ns}x, \lambda)$ is an eigenpair of the matrix $P_{ns}HP_{ns}$. Now scale the vector $P_{ns}x$ so that the last component is 1 and call that vector $\tilde{x}$. Then, as above, the elementary transformation

$$
M = \begin{pmatrix} 1 & & \tilde{x}_1 \\ & \ddots & \vdots \\ & & \tilde{x}_{n-1} \\ 0 & & 1 \end{pmatrix}
$$

can be used to deflate the matrix $P_{ns}HP_{ns}$. The fact that here $P_{ns}HP_{ns}$ is not upper-Hessenberg is of no consequence. In fact, we can make the following general statement: Given any matrix A of order n, and any eigenpair (x, λ) of A, then a matrix Q satisfying

$$
Q^{-1}x = e_1 \quad \text{or} \quad Q^{-1}x = e_n,
$$

can be used to deflate A, in the sense that

$$
Q^{-1}AQ = [\lambda e_1, B_1] \quad \text{or} \quad Q^{-1}AQ = [B_2, \lambda e_n],
$$

respectively; where B_1 and B_2 are $n \times n - 1$ matrices.

Having thus deflated the matrix $P_{ns}HP_{ns}$, the leading principal submatrix of order $n-1$ of

$$
M^{-1}P_{ns}HP_{ns}M, \tag{9}
$$

call it H', has all the eigenvalues of H except λ (if λ is simple). However, H' is not upper-Hessenberg in general and therefore will be reduced back to Hessenberg form before applying Newton's iterations; this is meant to save on the cost of factorizing the Jacobian when solving the linear systems arising at each step of Newton's iteration. Note that it is only the trailing diagonal submatrix of order $n - s \times n - s$ of H' that needs to be reduced and that if $s = n - 1$ or $s = n$, then H' is upper-Hessenberg. Moreover s needs not be chosen so that x_s is the largest component of x (in absolute value). Indeed, since the size of the matrix to be reduced to upper-Hessenberg form increases when s approaches 1, it is more advantageous to choose the largest s for which the ratios x_i/x_s are moderate. We wish therefore to define a threshold t for the size of these ratios on the basis of which s will be determined. We propose $\|H\|_\infty$ as a value for t ([4]).

This method also suffers from the fact that when the eigenvalue to be deflated is non-real, the resulting matrix H' is complex.

3.2 Deflation with help from the left eigenvector

The method we introduce now is different in spirit from the ones in the previous section, in that no attempt is made at modifying the matrix.

Assume that (x, λ) is an exact eigenpair of H and that λ is simple: $Hx = \lambda x$. No assumption is made about the remaining eigenvalues of H. Let (x, X) be such that

$$
(x, X)^{-1}H(x, X) = \begin{pmatrix} \lambda & 0 \\ 0 & J \end{pmatrix}
$$

where the right hand side is the Jordan canonical form of H. Now set

$$(x, X)^{-1} = \begin{pmatrix} y^H \\ Y \end{pmatrix}.$$

Then it is clear that y^H is a left eigenvector of H corresponding to λ and furthermore that

$$y^H X = 0.$$

This property can be used to modify Newton's method to avoid converging to the eigenpair (x, λ) a second time. When (x, λ) has been obtained once, our algorithm for avoiding it then consists of the following major steps:

1) Compute the left eigenvector y^H corresponding to λ, with $\|y\|_2 = 1$;

2) Given the current eigenpair (z, μ) compute the Newton correction from algorithm 2.1;

3) Let z' be the approximate eigenvector obtained after adding the Newton correction to z, choose the next eigenvector z'' as: $z'' = (I - yy^H)z'$.

In the last step we are just projecting z' onto the space X.

The algorithm will be adversely affected if the eigenvalue λ is ill-conditioned, i.e., if $y^H x$ is very small. Indeed, in this case, if the current eigenvector $z = \alpha x + Xv$, where v is a vector of length $n - 1$, then

$$(I - yy^H)z = z - (y^H z)y = \alpha x + Xv - (\alpha y^H x)y \approx z,$$

showing that z is hardly modified by the projection.

This method adds $O(n^2)$ work to the cost of finding one eigenpair distinct from (x, λ): this is the cost of computing the left eigenvector corresponding to λ; the cost of a single projection is $O(n)$.

3.3 Deflation with orthogonal transformations

Plane rotations may be used in the deflation process. However, some of the algorithms involving these transformations require high accuracy in the computed eigenvalue [13] for them to remain stable, whereas others requiring only approximate eigenpairs with small residuals, result in the destruction of the structure of the matrix [1, 13]. They are also much more expensive than the methods we introduced earlier. These are not then attractive.

3.4 Remarks on identifying duplicate eigenvalues

This can be a very hard problem [7]. On the one hand, duplicate eigenvalues corresponding to ill-conditioned eigenvalues of the given matrix should not be expected to be recognizable as such. On the other hand, when a matrix has a very large norm compared to the distance of some of its eigenvalues, some of its distinct computed eigenvalues might be mistakenly declared to be duplicates.

Example 3.1 *We illustrate this last case with the following matrix*

$$H = \begin{pmatrix} 2 \cdot 10^{-7} & 0 & 0 \\ 2 & 10^{-7} & 0 \\ 0 & 2 & 10^{10} \end{pmatrix}.$$

The tolerated size of the residuals for this matrix as chosen in our algorithm is tol $\approx$ $\|H\|\epsilon > 10^{-6}$. Therefore if we decide to declare as duplicates those eigenvalues whose difference is less than tol then the other two distinct eigenvalues of H will be declared as duplicates.

4 Convergence

In section 2, we mentioned that computing an eigenpair of H reduces to computing a zero of

$$F_s(x, \lambda) = \begin{pmatrix} Hx - \lambda x \\ e_s^T x - 1 \end{pmatrix}.$$

The Jacobian of F_s at (x, λ) is

$$F_s'(x, \lambda) = \begin{pmatrix} D_x(Hx - \lambda x) & D_\lambda(Hx - \lambda x) \\ D_x(e_s^T - 1) & D_\lambda(e_s^T - 1) \end{pmatrix} = \begin{pmatrix} H - \lambda I & -x \\ e_s^T & 0 \end{pmatrix},$$

where $D_x(F)$ denotes the derivative with respect to x of the function F. In this section, we give sufficient conditions for the convergence of our procedure. The result is a version of the Kantorovich theorem as it applies to our case.

Theorem 4.1 (Wilkinson) *Assume (x, λ) is an exact zero of F_s. Then*

$$\begin{pmatrix} H - \lambda I & -x \\ e_s^T & 0 \end{pmatrix}$$

is singular if and only if λ is multiple.

Proof. See [3]. □

   Remark: Since we are assuming that H is upper-Hessenberg and unreduced, an eigenvalue can only be non-derogatory, i.e., the associated eigenspace has dimension one.

   The previous result applies to the Jacobian at a zero of F_s. We wish to know more about the Jacobian at those approximations arising during Newton's iteration and before convergence is declared.

Theorem 4.2 *Assume (x, λ) is not an exact zero of F_s. Then $\begin{pmatrix} H - \lambda I & -x \\ e_s^T & 0 \end{pmatrix}$ is singular if and only if at least one of the following is true:*

1) λ is an eigenvalue of H and has an eigenvector whose s^{th} component is zero.

2) x belongs to the space generated by $(c_1, \ldots, c_{s-1}, c_{s+1}, \ldots, c_n)$, where c_i is the ith column of $H - \lambda I$ (λ may or may not be an eigenvalue).

Proof. See [4].

Remarks: The theorem tells us that more often than not, a singular Jacobian is an indication that we already have an eigenvalue, and this has been our experience indeed. The singularity of the Jacobian is also an indication of an ill-conditioned eigensystem. In fact, if we accept that the current eigenvector was moving in the "right" direction then if it satisfies condition 2 of the theorem, we can say that the eigenvalue is acting like a multiple one since the eigenvector is also in the range of $(H - \lambda I)$. In practice we have not encountered a situation where condition 1 applied: this is understandable again if we accept that the eigenvector is moving in the right direction since then the eigenvalue would be acting like an eigenvalue of geometric multiplicity more than one, which is impossible since the matrix is unreduced (recall that $x_s = 1$).

Suppose now that our procedure is started with initial guess (x_0, λ_0). We call (x_0, λ_0), ..., (x_k, λ_k) the sequence of iterates produced by the algorithm.

Theorem 4.3 *If $\beta_0 = Kc_1 < 1/4$, then the sequence (x_k, λ_k) converges quadratically starting from (x_0, λ_0); where $K = \|F'^{-1}_s(x_0, \lambda_0)\|$ and $c_1 = \|(x_1, \lambda_1) - (x_0, \lambda_0)\|$.*

See [3] for a proof. The process can be regarded as starting from any of the iterates (x_i, λ_i) and in fact will often converge even when the conditions of the theorem are not satisfied at (x_0, λ_0). These conditions will then be met for some (x_k, λ_k) at which stage convergence becomes quadratic.

5 Defective Case

The connection between Newton's method and inverse iteration is well known [11]. We sketch a proof in order to motivate the subsequent analysis in which the emphasis is placed on the case when H or H_0 is defective: we let J be the Jacobian of the map F_s at (x, λ) defined in section 2,

$$J = \begin{pmatrix} H - \lambda I & -x \\ e_s^T & 0 \end{pmatrix},$$

and we assume that $x_s = 1$. The order of the Jacobian is $n + 1$. Then,

$$J = J_0 + e_{n+1}v^T$$

with

$$J_0 = \begin{pmatrix} H - \lambda I & -x \\ 0 & 1 \end{pmatrix} \quad \& \quad v = e_s - e_{n+1} = \begin{pmatrix} 0 \\ \vdots \\ 0 \\ 0 \\ 1 \\ 0 \\ \vdots \\ -1 \end{pmatrix}.$$

Assume J_0 is not singular; this is true if and only if λ is not an eigenvalue. Assume also that J is not singular. Then it can be shown ([4]) using the Sherman-Morrison-Woodbury formula ([8]) that our scheme reduces to the following: Given (x, λ) compute the next iterate (x_1, λ_1) via

$$(H - \lambda I)\tilde{x} = x$$
$$x_1 = \frac{1}{\tilde{x}_s}\tilde{x} \tag{10}$$
$$\lambda_1 = \lambda + \frac{1}{\tilde{x}_s}.$$

5.1   Case when H is defective

When the algorithm is expressed as in (10) we can readily see some of the difficulties that arise when the matrix H is defective or almost defective, which is more likely in general due to roundoff errors. These problems are similar to the kind of problems that face the application of inverse iteration. Let $\lambda_1, \ldots, \lambda_\ell$ be a cluster of eigenvalues of H and suppose that the initial approximate eigenvalue λ corresponds to one of these. The eigenvectors corresponding to these eigenvalues are then almost linearly dependent. In general, the eigenvector corresponding to λ can be expected to converge to the space generated by these eigenvectors. As the eigenvalue λ approaches the cluster however, continued corrections to the eigenvector cannot be expected to refine it. We refer the reader to the particularly lucid account in [11] for a justification of these claims. Solving as in inverse iteration (see INVIT [12]) is a possible way around this problem. Computing the residual in extended precision arithmetic is also an obvious approach, and has been successful in practice.

When approaching a singular solution, Newton's method loses its quadratic convergence rate. We have proved earlier (Thm. 4.1) that the Jacobian is singular at multiple eigenpairs and therefore we can expect slower convergence when these are the target (see Fig. 1).

The eigenvalues of a defective matrix are not necessarily extremely sensitive to our dividing process. Indeed, as an extreme case which will help illustrate our point, the eigenvalues of a defective matrix can remain virtually unchanged after a zero has been introduced in the subdiagonal. An example, is the 2×2 matrix $\begin{pmatrix} 1 & 0 \\ 1 & 1 \end{pmatrix}$.

Finally, we mention that the deflation process can contribute to the improvement of the condition of the eigenvalues by removing members of clusters.

5.2   Case when H_0 is defective

It can happen in this case (e.g., if H is non-defective) that the initial dividing process would leave us with a number of initial approximations that is smaller than n. This will inhibit the parallelism of the algorithm in that less Newton processes can be started simultaneously. Furthermore, whatever eigenpairs we have can be extremely poor approx-

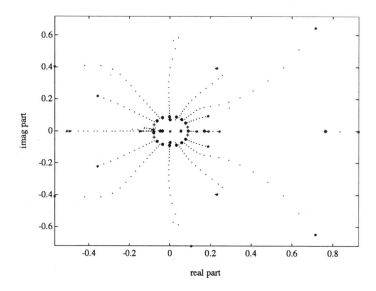

Figure 1: *The behavior of the algorithm for an almost defective 25×25 matrix. The crosses are the eigenvalues of the original matrix; the stars are the initial guesses; the circles are the eigenvalues computed by our algorithm; the dots are the iterates arising in Newton's iterations.*

imations to the desired ones. An extreme situation is illustrated by the matrix

$$H = \begin{pmatrix} 0 & 0 & 0 & 1 \\ 1 & 0 & 0 & 0 \\ 0 & 1 & 0 & 0 \\ 0 & 0 & 1 & 0 \end{pmatrix}.$$

No matter where the zero is introduced on the subdiagonal, the resulting matrix has 0 as its only eigenvalue, and we only have two initial approximations to the four distinct eigenpairs of H, namely:

$$\left(\begin{pmatrix} 0 \\ 1 \\ 0 \\ 0 \end{pmatrix}, 0 \right), \quad \left(\begin{pmatrix} 0 \\ 0 \\ 0 \\ 1 \end{pmatrix}, 0 \right),$$

if the zero is introduced in the $(3, 2)$ position.

5.3 Known failures

Some matrices of the structure mentioned at the end of section 5.2 (companion-like matrices), provided us with the only cases where the algorithm failed in practice to converge to the desired eigenpairs, i.e., failed to produce eigenpairs with small residuals after a

96

fixed number of iterations. Should these matrices be subjected to orthogonal similarity transformations however and then reduced back to upper-Hessenberg form, the dividing process will provide us with much better initial approximations and indeed, will converge for all initial approximations. We are certainly not advocating this as a viable scheme: we wanted to emphasize the fact that it is the structure of the matrix that caused the poor approximations and the failures, and not some inherent difficulty with the spectrum of these matrices.

6   Work Estimates

We assume in this section that we are given a real dense matrix A. The expression for the work done by one processor (assuming among other things that all processors share equally the cost of all stages), is (see [4])

$$W_p = \frac{14n^3}{3p} + 18 \left(\frac{p'}{p}\right) \left(\frac{n}{p'}\right)^3 + \sum_{\ell=0}^{m-1} 6(\frac{n}{p})ks_\ell^2 + 2\frac{n^3}{p}, \tag{11}$$

where n is the order of the matrix; p is the number of processors; k is the average number of Newton iterations needed before an eigenpair is accepted; m is the number of zeros introduced on the subdiagonal in the divide phase; p' is the number of subproblems at the lowest level, $p' = 2^m$; and $s_\ell = \frac{1}{2^\ell}$. As for the terms of expression (11): $\frac{14n^3}{3p}$ is the processor's contribution to the reduction of the original matrix A to upper-Hessenberg form H; $18 \left(\frac{p'}{p}\right) \left(\frac{n}{p'}\right)^3$ is the cost of applying the QR algrithm to $\frac{p'}{p}$ matrices of size $\frac{n}{p'}$; $\sum_{\ell=0}^{m-1} 6(\frac{n}{p})ks_\ell^2$ is our estimate for the cost of solving k linear systems with matrices of size s_ℓ, repeating this for $\frac{n}{p}$ initial guesses and for $\ell = 0, \cdots, m-1$; $2\frac{n^3}{p}$ is the processor's contribution to the computation of the eigenvectors of the original dense matrix A once those of H have been computed.

Expression (11) can be rewritten as

$$W_p = \frac{n^3}{p} \left(\frac{20}{3} + 8k + (18 - 8k)\frac{1}{4^m}\right). \tag{12}$$

The cost of getting the eigenvalues and eigenvectors by the QR algorithm is $25n^3$ [8]. A reasonable value for k is 3, however there are cases when k is 2 or less. There are also cases where k is larger than 3, mostly with matrices of small order or defective matrices. Figures 2, 3, 4 and 5 show plots of the work estimate for various values of the parameters involved.

7   Parallel Algorithms Details and Performance

Our algorithm is inherently parallel. Indeed, the eigensystems of H_{11} and H_{22} can be computed in parallel. After which Newton's method can be started with different initial

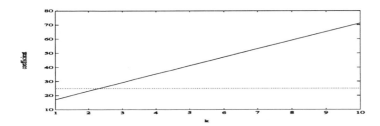

Figure 2: Variation of the Coefficient of n^3 in work estimate model in terms of the number of steps k, with $m = 1$, and $p = 1$ (see expression for work).

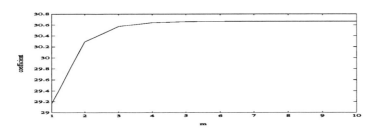

Figure 3: Variation of the Coefficient of n^3 in work estimate model in terms of the number of splits m, with $k = 3$ and $p = 1$ (see expression for work).

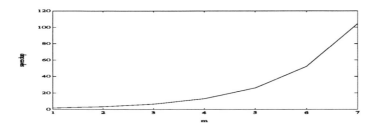

Figure 4: Predicted speedup over QR in terms of number of splits m, with $p = 2^m$ and $k = 3$ (see expression for work).

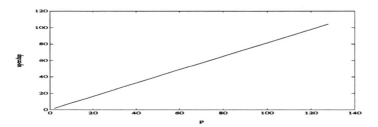

Figure 5: Predicted speedup over QR in terms of the number of processors p, with $m = \log_2 p$ and $k = 3$ (see expression for work).

guesses on different processors. Furthermore, the dividing process can be applied recursively to obtain yet smaller subproblems. This will permit the use of more processors. However, care must be exercised if the algorithm is to be kept efficient since it has been our experience that for small matrices the algorithm requires more steps to converge in general.

In a parallel implementation, the given, generally dense, matrix is first reduced to upper-Hessenberg form using a blocked algorithm. Next comes the partitioning phase or "divide". This phase amounts to constructing a binary tree with each node representing a partition into two subproblems. Each of these problems may be spawned independently without fear of data conflicts; the computation at this level (the lowest) consists of calls to the EISPACK routine HQR2. The tree is then traversed in reverse order with Newton's method applied at each node using the results from the sons as initial approximations. Note here that unlike the symmetric case, the computation at a node does not have to wait for both sons to complete in order to start. However the need to check for duplicates has imposed some synchronization.

7.1 Shared memory implementation

So far we have used SCHEDULE [6] to implement the algorithm on shared memory computers. SCHEDULE is a package of FORTRAN and C subroutines designed to aid in programming explicitly parallel algorithms for numerical calculations. An important part of this package is the provision of a mechanism for dynamically spawning processes even when such a capability is not present within the parallel language extensions provided for a given machine.

7.2 Distributed memory implementation

The current implementation on distributed memory machines requires that the matrix be stored on each processor. This obviously puts constraints on the size of problems that can be solved. With this implementation however, communication is needed only during the deflation phase. This implementation is best described through the contribution of a particular processor. Suppose that we have 4 processors at our disposal, $p_0, \ldots, p_3$ and that accordingly the matrix H has been divided into 4 subproblems: $H_0, \ldots, H_3$, that their common size is $n/4$ and that they occur in this order on the diagonal of the matrix. We describe now the contribution of p_2 by steps:

1) Call HQR2 to solve for the eigensystem of the matrix H_2.

2) Refine the output from step 1 to get $1/2$ the number of (i.e., $n/4$) eigenpairs of the matrix $H_{\frac{1}{2}}$:

$$H_{\frac{1}{2}} = \left(\begin{array}{c|c} H_2 & B \\ \hline 0 ^{\alpha} & H_3 \end{array} \right),$$

where $H_{\frac{1}{2}}$ is a submatrix of H.

3) Refine the output from step 2 to get 1/4 of the number of eigenpairs (i.e., $n/4$) of the matrix H.

We are currently developing another implementation where blocks of columns of the matrix are stored on different processors.

8 Numerical Results and Performance

In this section we present the results of the implementation of the algorithm on a number of machines. The serial version of the code is available through NETLIB where it's called "nonsymdc".

The same algorithm has been run on the IBM RS/6000-550, the Alliant FX/8, the Intel iPSC/2 and the Intel iPSC/860. We compared our results to those of HQR2 from the EISPACK collection. We have used randomly generated matrices in these tests. In these implementations the linear solver used for the computation of the correction at each Newton step is column oriented. The storage requirement for our algorithm in a serial implementation is $4n^2 + O(n)$. The following are the results from the IBM RS/6000/500 implementation. The IBM RS/6000/500 is a single processor computer with a RISC-based Architecture.

Order	HQR2	D&C	Ratio HQR2/D&C	Distinct λ
100	1.04	1.12	.93	99
200	9.31	9.18	1.01	196
300	34.1	28.1	1.2	293
400	94.0	65.3	1.4	395
500	196	136	1.4	490
1000	1741	992	1.7	1000

In an implementation on a shared memory machine, the storage allocated to the Jacobian (in serial mode) is multiplied by the number of processors used; this is meant to prevent concurrent write to the same memory locations. The following are some results from the Alliant FX/8 implementation. The Alliant FX/8 is a parallel machine with 8 vector processors.

Order	No. of procs.	Levels	Ratio HQR2/D&C
100	2	1	1.7
	4	2	2.4
	8	3	4.0

The results on the Alliant were, in general, disappointing. The storage scheme for the Jacobian that we used on that machine seems to have inhibited the compiler performed vector optimizations. HQR2, running on a single processor *was* vector-optimized.

Finally we present results from the runs on the hypercubes. The following are some results from the Intel iPSC/2 implementation. The largest size used in each case was dictated by the memory capacity of a single node.

Order	No. of procs.	Levels	Ratio HQR2/D&C
100	2	1	2.2
	4	2	3.7
	8	3	6.0
	16	4	8.2
200	2	1	2.2
	4	2	3.5
	8	3	5.2
	16	4	9.1
300	2	1	2.2
	4	2	3.2
	8	3	6.3
	16	4	9.8
	32	5	13.2
	64	6	21.3

The following are some results from the Intel iPSC/860 implementation.

Order	No. of procs.	Levels	Ratio HQR2/D&C
100	2	1	1.9
	4	2	3.3
	8	3	5.1
400	2	1	2.4
	4	2	3.2
	8	3	6.0
	16	4	8.4
600	8	3	7.5
	16	4	13.5
	32	5	23
	64	6	32

We observe here that the speedups realized by our algorithm over the QR algorithm, did not remain linear for a large number of processors. This is due to the fact that our algorithm is much less efficient on small matrices and we had to work with this kind of matrices when the number of processors became large. For example, with a matrix of order 600 and using 64 processors, matrices of average size 20 had to be solved on each node at level 5.

References

[1] P. A. Businger. Numerically stable deflation of Hessenberg and symmetric tridiagonal matrices. *BIT*, 11:262–270, 1971.

[2] J. Cuppen. A divide and conquer method for the symmetric tridiagonal eigenproblem. *Numer. Math.*, 36:177–195, 1981.

[3] J. Dongarra. Improving the accuracy of computed matrix eigenvalues. Technical Report ANL-80-84, Argonne National Lab., August 1980.

[4] J. Dongarra and M. Sidani. A parallel algorithm for the non-symmetric eigenvalue problem. Technical Report CS-91-137, Univ. of Tennessee, Comp. Sci. Dept., 1991.

[5] J. Dongarra and D. Sorensen. A fully parallel algorithm for the symmetric eigenvalue problem. *SIAM J. Sci. Statist. Comput.*, 8:s139–s154, 1987.

[6] J. Dongarra, D. Sorensen, K. Connolly, and J. Patterson. Programming methodology and performance issues for advanced computer architectures. *Parallel Computing*, 8:41–58, 1988.

[7] G. Golub and J.H. Wilkinson. Ill-conditioned eigensystems and the computation of the jordan canonical form. *SIAM Review*, 18:578–619, 1976.

[8] Gene Golub and Charles Van Loan. *Matrix Computations*. Johns Hopkins University Press, Baltimore, MD, second edition, 1989.

[9] Konrad Knopp. *Theory of Functions*. Dover Publications, New York, NY, 1945.

[10] T.Y. Li, Zhonggang Zeng, and Luan Cong. Solving eigenvalue problems of real nonsymmetric matrices with real homotopies. Preprint, Michigan State University, E. Lansing, MI 48824-1027, 1990.

[11] G. Peters and J.H. Wilkinson. Inverse iteration, ill-conditioned equations and Newton's method. *SIAM Review*, 21:339–360, 1979.

[12] B. T. Smith, J. M. Boyle, J. J. Dongarra, B. S. Garbow, Y. Ikebe, V. C. Klema, and C. B. Moler. *Matrix Eigensystem Routines – EISPACK Guide*, volume 6 of *Lecture Notes in Computer Science*. Springer-Verlag, Berlin, 1976.

[13] J. H. Wilkinson. *The Algebraic Eigenvalue Problem*. Oxford University Press, 1965.

Jack J. Dongarra, Mathematical Sciences Section, Oak Ridge National Laboratory, Oak Ridge, TN 37831 and Computer Science Department, University of Tennessee, Knoxville, TN 37996-1301, dongarra@cs.utk.edu,

Majed Sidani, Mathematical Sciences Section, Oak Ridge National Laboratory, Oak Ridge, TN 37831 and Computer Science Department, University of Tennessee, Knoxville, TN 37996-1301, sidani@cs.utk.edu.

J A GREGORY

An introduction to bivariate uniform subdivision

1 Introduction

The generation of curves and surfaces by recursive subdivision is a well known technique in Approximation Theory and CAGD (Computer Aided Geometric Design). Our purpose is not to provide a review of such techniques but rather to provide an introduction to the theory of uniform subdivision which has been developed in recent years. In particular, we will concentrate attention on some of the practical tools which can be used in the study of continuity and differentiability of the limits of bivariate, uniform, binary subdivision schemes.

The work presented here is but a small part of uniform subdivision theory. A much more extensive review of uniform subdivision is given in Dyn[3], where a full bibliography of the subject can be found, and the paper by Dahmen, Cavaretta and Micchelli[1] is a major contribution to the subject. The shorter review article of Caveratta and Micchelli[2] provides another introduction to this area.

Ideally, an introduction to uniform subdivision would begin with a study of the univariate case but, for brevity, we consider only the theory for the bivariate case. The multivariate case is then an immediate obvious generalization and the univariate case is but a simplification. The discussion is also restricted to the case of binary (diadic) subdivision schemes, although the generalization to p-adic schemes is immediately apparent.

Uniform subdivision schemes generate sets of 'control points' according to some fixed subdivision rule and, in the bivariate case, we are concerned with whether the points become dense on some continous, and possibly differentiable, limit surface. This concern will be resolved in a very simple way, namely, that for a continuously differentiable limit, divided differences will be converging to a continuous limit, and, for a continuous limit, differences will be converging to zero. After introducing some preliminary notation and theory in Section 2, the theory of differentiability is considered in Section 3 and the analysis for continuous limits is considered in Section 4. The theory is illustrated for the case of box splines in Section 3 and for the case of an interpolatory 'butterfly' subdivision scheme in Section 5.

2 Preliminaries

2.1 Binary subdivision scheme

A bivariate, uniform, 'binary subdivision scheme' generates sets of 'control points'

$$\mathbf{f}^k := \{f_\alpha^k \in I\!\!R^m : \alpha \in \mathbb{Z}^2\}, \ k = 0, 1, 2, \dots , \tag{2.1}$$

according to the rule

$$f_\alpha^{k+1} := \sum_{\beta \in \mathbb{Z}^2} a_{\alpha-2\beta} f_\beta^k, \quad \alpha \in \mathbb{Z}^2. \tag{2.2}$$

Here there are really four different rules which can be exhibited as

$$f_{2\alpha+\gamma}^{k+1} := \sum_{\beta \in \mathbb{Z}^2} a_{2(\alpha-\beta)+\gamma} f_\beta^k = \sum_{\beta \in \mathbb{Z}^2} a_{\gamma-2\beta} f_{\alpha+\beta}^k, \quad \alpha \in \mathbb{Z}^2, \tag{2.3}$$

where

$$\gamma \in E := \{(0,0), (1,0), (0,1), (1,1)\}. \tag{2.4}$$

The set

$$\mathbf{a} := \{a_\alpha \in \mathbb{R} : \alpha \in \mathbb{Z}^2\} \tag{2.5}$$

is called the 'mask' of the scheme, where the 'support'

$$\operatorname{supp}(\mathbf{a}) := \{\alpha \in \mathbb{Z}^2 : a_\alpha \neq 0\} \tag{2.6}$$

is assumed finite.

As a simple example, consider the mask with

$$\operatorname{supp}(\mathbf{a}) = \{-1, 0, 1\}^2. \tag{2.7}$$

Then the binary subdivision scheme is described, for $(i,j) \in \mathbb{Z}^2$, by

$$\left.\begin{aligned}
f_{2i,2j}^{k+1} &:= a_{0,0} f_{i,j}^k, \\
f_{2i+1,2j}^{k+1} &:= a_{1,0} f_{i,j}^k + a_{-1,0} f_{i+1,j}^k, \\
f_{2i,2j+1}^{k+1} &:= a_{0,1} f_{i,j}^k + a_{0,-1} f_{i,j+1}^k, \\
f_{2i+1,2j+1}^{k+1} &:= a_{1,1} f_{i,j}^k + a_{-1,1} f_{i+1,j}^k + a_{1,-1} f_{i,j+1}^k + a_{-1,-1} f_{i+1,j+1}^k.
\end{aligned}\right\} \tag{2.8}$$

The subdivision scheme (2.2) defines a bounded linear operator $S_\mathbf{a}$ on $\ell_\infty(\mathbb{Z}^2)$, namely, for $\mathbf{f} \in \ell_\infty(\mathbb{Z}^2)$, $S_\mathbf{a}\mathbf{f} \in \ell_\infty(\mathbb{Z}^2)$ is defined by

$$(S_\mathbf{a}\mathbf{f})_\alpha := \sum_{\beta \in \mathbb{Z}^2} a_{\alpha-2\beta} f_\beta. \tag{2.9}$$

The norm of this operator is

$$\|S_\mathbf{a}\| := \sup_{\|\mathbf{f}\|_{\ell_\infty}=1} \|S_\mathbf{a}\mathbf{f}\|_{\ell_\infty} = \max_{\alpha \in E} \sum_{\beta \in \mathbb{Z}^2} |a_{\alpha-2\beta}|. \tag{2.10}$$

The subdivision scheme can now be written as

$$\mathbf{f}^{k+1} := S_\mathbf{a}\mathbf{f}^k = S_\mathbf{a}^{k+1}\mathbf{f}^0, \tag{2.11}$$

where

$$\mathbf{f}^0 := \{f_\alpha^0 \in \mathbb{R}^m : \alpha \in \mathbb{Z}^2\}. \tag{2.12}$$

denotes the set of given initial control points. Finally, we associate with each subdivision operator $S_\mathbf{a}$ the bivariate Laurent polynomial

$$a(z) := \sum_{\alpha \in \mathbb{Z}^2} a_\alpha z^\alpha, \quad \alpha := (i,j), \quad z^\alpha := z_1^i z_2^j, \quad z_1, z_2 \in \mathbb{C}. \tag{2.13}$$

104

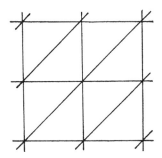

Figure 1: The $(1,1)$ triangulation

For example, the subdivision scheme (2.8) has the Laurent polynomial

$$a(z) = \sum_{i=-1}^{1} \sum_{j=-1}^{1} a_{i,j} z_1^i z_2^j. \tag{2.14}$$

This is called the *'generating polynomial'* for the subdivision scheme and provides an extremely useful tool for the analysis.

2.2   The control polygon

The control points $f_\alpha^k \in I\!\!R^m$, at level k, are associated with the diadic rectangular grid *'domain points'*

$$2^{-k}\alpha = (2^{-k}i, 2^{-k}j) \in I\!\!R^2, \ \alpha = (i,j) \in Z\!\!\!Z^2. \tag{2.15}$$

Hence the control points $f_{2\alpha+\gamma}^{k+1}$, $\gamma \in E$, at level $k+1$ are associated with the domain points

$$2^{-k}\alpha + 2^{-k-1}\gamma, \ \alpha \in Z\!\!\!Z^2, \ \gamma \in E, \tag{2.16}$$

given on the finer diadic grid obtained by binary subdivision. We now consider a particular definition of a *'control polygon'* whose vertices are the control points f_α^k, $\alpha \in Z\!\!\!Z^2$.

Suppose the grid at level k is triangulated by subdivision along the $(1,1)$ direction, giving triangles T_α^1 and T_α^2, $\alpha \in Z\!\!\!Z^2$, with vertices $2^{-k}\{\alpha, \alpha + (1,0), \alpha + (1,1)\}$ and $2^{-k}\{\alpha, \alpha + (1,1), \alpha + (0,1)\}$ respectively, see Figure 1. The piecewise linear interpolant on the $(1,1)$ triangulation is now defined by

$$L_k[\mathbf{f}^k](s,t) := \begin{cases} (1-\theta_i)f_{i,j}^k + (\theta_i - \theta_j)f_{i+1,j}^k + \theta_j f_{i+1,j+1}^k, & (s,t) \in T_{i,j}^1, \\ (1-\theta_j)f_{i,j}^k + \theta_i f_{i+1,j+1}^k + (\theta_j - \theta_i)f_{i,j+1}^k, & (s,t) \in T_{i,j}^2, \end{cases} \tag{2.17}$$

where

$$\theta_i := 2^k s - i, \ \theta_j := 2^k t - j. \tag{2.18}$$

Thus

$$L_k[\mathbf{f}^k](2^{-k}\alpha) = f_\alpha^k, \ \alpha \in Z\!\!\!Z^2, \tag{2.19}$$

and we define $L_k[\mathbf{f}^k]$ as the control polygon of $\mathbf{f}^k$ with respect to the $(1,1)$ triangulation.

The control polygon with respect to the $(-1,1)$ triangulation can be similarly defined. More generally, consider a rectilinear partition of the diadic points $2^{-k}\alpha$, $\alpha \in \mathbb{Z}^2$, along skew directions, which is then triangulated along either of the diagonals. A control polygon can then be defined with respect to this skew triangulation. Finally, a control polygon can be defined with respect to any rectilinear partition as the piecewise bilinear interpolant on that partition. The choice of the definition of an appropriate control polygon is usually determined a priori in the construction of a particular subdivision scheme.

2.3   The fundamental solution and convergence

Let

$$\varphi^k := S_{\mathbf{a}}^k \varphi^0 \tag{2.20}$$

denote the subdivision scheme applied to the 'cardinal set' of initial scalar data

$$\varphi^0 := \{\varphi_\alpha^0 := \delta_{\alpha,(0,0)} : \alpha \in \mathbb{Z}^2\}. \tag{2.21}$$

Thus

$$\varphi_\alpha^0 := \begin{cases} 1, & \alpha = (0,0), \\ 0, & \alpha \in \mathbb{Z}^2 \backslash (0,0). \end{cases} \tag{2.22}$$

We then have:

Definition 1 (Uniform convergence.)   *The subdivision scheme is said to be uniformly convergent (with respect to the diadic point parameterization (2.15)) if there exists $\varphi \in C(\mathbb{R}^2)$ such that*

$$\lim_{k\to\infty} \sup_{\alpha\in\mathbb{Z}^2} |\varphi_\alpha^k - \varphi(2^{-k}\alpha)| = 0. \tag{2.23}$$

Equivalently, in terms of the behaviour of the control polygon sequence,

$$\lim_{k\to\infty} \|L_k[\varphi^k] - \varphi\|_\infty = 0. \tag{2.24}$$

If there exists such a continuous function φ, we call it the 'fundamental solution' of the subdivision scheme and write

$$\varphi = S_{\mathbf{a}}^\infty \varphi^0. \tag{2.25}$$

This function has the important property of having 'local support', since it can be shown that

$$\operatorname{supp}(\varphi) := \{(s,t) \in \mathbb{R}^2 : \varphi(s,t) \neq 0\} \subset [\operatorname{supp}(\mathbf{a})], \tag{2.26}$$

where $[\operatorname{supp}(\mathbf{a})]$ denotes the convex hull in $\mathbb{R}^2$ of $\operatorname{supp}(\mathbf{a}) \subset \mathbb{Z}^2$. In fact,

$$\operatorname{supp}(L_k[\varphi^k]) \subset \operatorname{supp}(L_{k+1}[\varphi^{k+1}]) \subset [\operatorname{supp}(\mathbf{a})]. \tag{2.27}$$

The limit for bounded initial data $\mathbf{f}^0$ can now be defined in terms of translates of the fundamental solution as

$$f(s,t) := \sum_{(i,j)\in\mathbb{Z}^2} f_{i,j}^0 \, \varphi(s-i, t-j). \tag{2.28}$$

The fundamental solution can be characterized in the following way: Observe that

$$\varphi(s,t) = (S_{\mathbf{a}}^{\infty}\varphi^0)(s,t) = (S_{\mathbf{a}}^{\infty}\varphi^1)(2s,2t) = (S_{\mathbf{a}}^{\infty}\mathbf{a})(2s,2t), \tag{2.29}$$

where it has been observed that, for cardinal initial data,

$$\varphi^1 = S_{\mathbf{a}}\varphi^0 = \mathbf{a}. \tag{2.30}$$

Thus, from (2.28),

$$\varphi(s,t) = \sum_{(i,j)\in\mathbb{Z}^2} a_{i,j}\,\varphi(2s-i,2t-j). \tag{2.31}$$

This is called the *'functional equation'* of the subdivision scheme and plays an important role in the study of uniform subdivision, see [1] and [3], although we will not pursue its study here.

A simple consequence of the definition of the binary subdivision scheme as in (2.3) is:

Lemma 2 *A necessary condition for uniform convergence is that*

$$\sum_{\beta\in\mathbb{Z}^2} a_{\gamma-2\beta} = 1, \ \gamma \in E. \tag{2.32}$$

This condition can be characterized in terms of the generating polynomial as

$$a(1,1) = 4, \ a(-1,-1) = a(1,-1) = a(-1,1) = 0. \tag{2.33}$$

and implies that the subdivision scheme is invariant under affine transformations of the initial data in $\mathbb{R}^m$.

2.4   Examples

We conclude this preliminary section by considering two simple examples of convergent binary subdivision schemes, the first of which will be used as a building block for the theory of Sections 3 and 4.

2.4.1   Piecewise linear scheme

Consider the scheme (2.8) with

$$a_{0,0} = 1, \ a_{-1,1} = a_{1,-1} = 0, \ \text{and} \ a_{1,0} = a_{-1,0} = a_{0,1} = a_{0,-1} = a_{1,1} = a_{-1,-1} = \tfrac{1}{2}. \tag{2.34}$$

This scheme is symmetric with respect to the $(1,1)$ triangulation and the limit of the scheme is the initial control polygon with respect to the $(1,1)$ triangulation, since

$$L_{k+1}[\mathbf{f}^{k+1}] = L_k[\mathbf{f}^k] = L_0[\mathbf{f}^0]. \tag{2.35}$$

The scheme will be required in the later analysis and hence we distinguish its generating polynomial as

$$\begin{aligned}
\ell(z) &:= 1 + \tfrac{1}{2}(z_1 + z_1^{-1} + z_2 + z_2^{-1} + z_1z_2 + z_1^{-1}z_2^{-1}) & (2.36)\\
&= \tfrac{1}{2}(1 + z_1^{-1})(1 + z_2^{-1})(1 + z_1z_2). & (2.37)
\end{aligned}$$

The fundamental solution φ for this case is the well known *Courant hat function*, namely the piecewise linear interpolant on the $(1,1)$ triangulation having the value 1 at $(0,0)$ and zero at all the other vertices of $\mathbb{Z}^2$.

One can similarly define a piecewise linear binary subdivision scheme with respect to the (-1,1) triangulation which we leave as an exercise for the reader.

2.4.2 Piecewise bilinear scheme

Consider the scheme (2.8) with

$$a_{0,0} = 1, \; a_{1,0} = a_{-1,0} = a_{0,1} = a_{0,-1} = \tfrac{1}{2}, \text{ and } a_{1,1} = a_{-1,1} = a_{1,-1} = a_{-1,-1} = \tfrac{1}{4}. \quad (2.38)$$

This scheme is symmetric with respect to the rectangular diadic grid and it is easily seen that the limit of the scheme is the initial bilinear control polygon with respect to the rectilinear partition of $\mathbb{Z}^2$. Thus the scheme has a tensor product structure which is reflected in its generating polynomial factorization

$$a(z) := \left(\tfrac{1}{2}z_1^{-1} + 1 + \tfrac{1}{2}z_1\right)\left(\tfrac{1}{2}z_2^{-1} + 1 + \tfrac{1}{2}z_2\right). \quad (2.39)$$

Here, each factor represents the generating polynomial of a univariate (piecewise linear) binary subdivision scheme and further factorization gives

$$a(z) := \tfrac{1}{4}(1 + z_1^{-1})(1 + z_1)(1 + z_2^{-1})(1 + z_2). \quad (2.40)$$

(A significance of a factorization of the generating polynomial will become apparent in Subsection 3.2.)

3 Differentiable limits and Box splines

3.1 Differentiability

Given the set of control points $\mathbf{f}^k$ at level k, let

$$\Delta_\gamma \mathbf{f}^k := \{\Delta_\gamma f_\alpha^k := f_{\alpha+\gamma}^k - f_\alpha^k : \alpha \in \mathbb{Z}^2\} \quad (3.1)$$

define the set of *'differences'* and

$$D_\gamma \mathbf{f}^k := \{D_\gamma f_\alpha^k := 2^k \Delta_\gamma f_\alpha^k : \alpha \in \mathbb{Z}^2\} \quad (3.2)$$

define the set of *'divided differences'* along the direction $\gamma = (m,n) \in \mathbb{Z}^2 \setminus (0,0)$. Also, let

$$\partial_\gamma := m\partial/\partial s + n\partial/\partial t \quad (3.3)$$

define the derivative operator along the direction γ with respect to differentiable functions of (s,t). We now consider the divided difference sequence $\{D_\gamma \varphi^k\}_{k=0}^\infty$ and have the following:

Theorem 3 (Differentiability.) *Suppose there exists $g \in C(\mathbb{R}^2)$, with* $\mathrm{supp}(g) \subset [\mathrm{supp}(\mathbf{a})]$, *such that*

$$\lim_{k \to \infty} \sup_{\alpha \in \mathbb{Z}^2} |D^\gamma \varphi_\alpha^k - g(2^{-k}\alpha)| = 0. \tag{3.4}$$

Thus the divided differences of the binary subdivision scheme $S_{\mathbf{a}}$, with cardinal initial data, converge uniformly to a continuous, compactly supported function g (see Definition 1). Then the subdivision scheme $S_{\mathbf{a}}$ is uniformly convergent with fundamental solution

$$\varphi = I_\gamma[g], \tag{3.5}$$

where

$$I_\gamma[g](s,t) := \int_{-\infty}^{0} g((s,t) + \theta\gamma)d\theta \tag{3.6}$$

defines the indefinite integral of g along the direction γ. Thus

$$\partial_\gamma \varphi = g. \tag{3.7}$$

Proof. Consider, in particular, $\gamma \in G := \{(1,0), (0,1), (1,1)\}$ and let $L_k[\varphi^k]$ denote the piecewise linear interpolant of φ^k with respect to the $(1,1)$ triangulation, see (2.17). We will show that $\{L_k[\varphi^k]$ converges uniformly to $I_\gamma[g]$. Since $L_k[\varphi^k]$ is a continuous, piecewise linear function, it can be written as the indefinite integral of its piecewise constant derivative along the γ direction, that is

$$L_k[\varphi^k] = I_\gamma[\partial_\gamma L_k[\varphi^k]]. \tag{3.8}$$

Also, observe that for all bounded, compactly supported functions f (with $\mathrm{supp}(f) \subset [\mathrm{supp}(\mathbf{a})]$)

$$\|I_\gamma[f]\|_\infty \leq C\|f\|_\infty, \tag{3.9}$$

where C is a constant dependent only on the support. We then have

$$
\begin{aligned}
\|L_k[\varphi^k] - I_\gamma[g]\|_\infty &= \|I_\gamma[\partial_\gamma L_k[\varphi^k]] - I_\gamma[g]\|_\infty \\
&\leq \|I_\gamma[\partial_\gamma L_k[\varphi^k]] - I_\gamma[L_k[D_\gamma\varphi^k]]\|_\infty + \|I_\gamma[L_k[D_\gamma\varphi^k]] - I_\gamma[g]\|_\infty \\
&\leq C\|\partial_\gamma L_k[\varphi^k] - L_k[D_\gamma\varphi^k]\|_\infty + C\|L_k[D_\gamma\varphi^k] - g\|_\infty \\
&\leq C\max_{\mu \in G}\|\Delta_\mu D_\gamma\varphi^k\|_{\ell_\infty} + C\|L_k[D_\gamma\varphi^k] - g\|_\infty.
\end{aligned} \tag{3.10}
$$

Here, the first term on the the right hand side of the last inequality follows from the definition of L_k. For example, with $\gamma = (1,0)$, (2.17) gives

$$\partial_{1,0}L_k[\varphi^k](s,t) - L_k[D_{1,0}\varphi^k](s,t) = \begin{cases} -\theta_i\Delta_{1,0}D_{1,0}f_{i,j}^k - \theta_j\Delta_{1,0}f_{i,j}^k, & (s,t) \in T_{i,j}^1, \\ (1-\theta_j)\Delta_{0,1}D_{1,0}f_{i,j}^k - \theta_i\Delta_{1,0}f_{i,j}^k, & (s,t) \in T_{i,j}^2, \end{cases} \tag{3.11}$$

and, by symmetry on the $(1,1)$ triangulation, similar relations hold for $\gamma = (0,1)$ and $\gamma = (1,1)$. Both terms on the right hand side of the last inequality of (3.10) tend to zero as $k \to \infty$, by the hypothesis (3.4). This completes the proof for the particular choice of $\gamma \in G$ and for general $\gamma \in \mathbb{Z}^2 \setminus (0,0)$ the above proof can be generalized by defining L_k with respect to a skew triangulation. $\square$

3.2 Divided difference schemes and Box splines

The previous theorem indicates that differentiability of the limits of uniform subdivision schemes is related to the behaviour of their divided differences. We now consider a special case where the divided differences themselves satisfy binary subdivision schemes. An illustration of this case for box spline subdivision schemes is then given.

Proposition 4 (Difference and divided difference schemes.) *Suppose that there exist Laurent polynomials $b(z)$ and $\hat{a}(z) := 2b(z)$ such that*

$$a(z) = (1 + z^{-\gamma})b(z) = \tfrac{1}{2}(1 + z^{-\gamma})\hat{a}(z), \tag{3.12}$$

where $\gamma \in \mathbb{Z}^2 \setminus (0,0)$. Then

$$\Delta_\gamma \mathbf{f}^{k+1} = S_\mathbf{b} \Delta_\gamma \mathbf{f}^k \text{ and } D_\gamma \mathbf{f}^{k+1} = S_{\hat{\mathbf{a}}} D_\gamma \mathbf{f}^k, \tag{3.13}$$

that is, the differences and divided differences satisfy binary subdivision schemes with generating polynomials $b(z)$ and $\hat{a}(z)$ respectively.

Proof. From (3.12), $a_\alpha = b_\alpha + b_{\alpha+\gamma}$. Hence, from (2.2),

$$
\begin{aligned}
f_{\alpha+\gamma}^{k+1} - f_\alpha^{k+1} &= \sum_{\beta \in \mathbb{Z}^2} (a_{\alpha+\gamma-2\beta} - a_{\alpha-2\beta}) f_\beta^k, \\
&= \sum_{\beta \in \mathbb{Z}^2} (b_{\alpha+2\gamma-2\beta} - b_{\alpha-2\beta}) f_\beta^k, \\
&= \sum_{\beta \in \mathbb{Z}^2} b_{\alpha-2\beta} (f_{\beta+\gamma}^k - f_\beta^k).
\end{aligned}
\tag{3.14}
$$

This is the subdivision scheme for the differences and multiplying both sides by 2^{-k-1} gives the divided difference scheme. $\square$

Remark. In the case of a univariate uniform subdivision scheme, the existence of a difference and a divided difference scheme follows from the univariate form of Lemma 2. In this case $a(-1) = 0$, which implies that $(1 + z^{-1})$ is a factor of the univariate polynomial $a(z)$. In the bivariate case, however, factorization of the generating polynomial does not necessarily follow from (2.33).

The function g of Theorem 3 can be considered as the limit of the divided difference scheme applied to the initial data $D_\gamma \varphi^0$. Thus

$$g(s,t) = -\hat{\varphi}(s,t) + \hat{\varphi}((s,t) + \gamma), \tag{3.15}$$

cf. (2.28), where $\hat{\varphi}$ is the fundamental solution of the divided difference scheme. Thus application of (3.6) of Theorem 3 gives:

Corollary 5 *Suppose that there exists a uniformly convergent divided difference scheme, with generating polynomial $\hat{a}(z)$ satisfying (3.12) and with fundamental solution $\hat{\varphi} \in C(\mathbb{R})$. Then the basic scheme $S_\mathbf{a}$ is uniformly convergent with fundamental solution*

$$\varphi(s,t) = \int_0^1 \hat{\varphi}((s,t) + \theta\gamma)d\theta. \tag{3.16}$$

More generally, we have:

Corollary 6 *Suppose that*

$$a(z) = 2^{-n} \prod_{i=1}^{n} (1 + z^{-\gamma_i}) \hat{a}(z), \quad \gamma_i \in \mathbb{Z}^2 \setminus (0,0), \tag{3.17}$$

where $\hat{a}(z)$ is the generating polynomial of a uniformly convergent subdivision scheme with fundamental solution $\hat{\varphi}$. Then the subdivision scheme $S_{\mathbf{a}}$ is uniformly convergent with fundamental solution

$$\varphi(s,t) = \int_0^1 \cdots \int_0^1 \hat{\varphi}((s,t) + \theta_1 \gamma_1 + \ldots + \theta_n \gamma_n) d\theta_1 \ldots d\theta_n. \tag{3.18}$$

Box splines. A simple consequence of Corollary 6 is that it gives a binary subdivision development for the theory of box splines. For example, let $\hat{a}(z) = \ell(z)$ in (3.17), where $\ell(z)$ is the generating polynomial of the piecewise linear scheme on the $(1,1)$ triangulation. Then $\hat{\varphi}$ is the piecewise linear Courant hat function on the triangulation with centre the origin. Equation (3.17) then gives the generating polynomial of a bivariate spline subdivision scheme with fundamental solution defined by (3.18). Each integral along a direction γ_i in (3.18) corresponds to an increase by one of the polynomial degree and continuity of the fundamental spline along that direction. The survey paper [2] gives more details of such subdivision schemes. It can also be observed that the factorizations (2.37) and (2.40) reflect the simple fact that piecewise linear and bilinear schemes can be considered as the 'integrals' of piecewise constant schemes, although we have chosen here to define convergence of subdivision schemes with respect to their having continuous limits.

4 A C^0 convergence analysis

We now consider how to determine if a subdivision scheme $S_{\hat{a}}$ is uniformly convergent, in the case where the fundamental solution limit is not known explicitly. Here, $\hat{a}(z)$ may be the generating polynomial of a basic scheme, or may correspond to the special case of schemes having divided difference polynomial factors as in (3.12) or (3.17). In a later subsection, we briefly consider the need to generalize the theory to *'matricial schemes'*, for the case where such special factorizations are not available.

4.1 A preliminary result

Proposition 7 *Let $S_{\mathbf{c}}$ be a binary subdivision operator, with finite mask $\mathbf{c}$, such that*

$$\sum_{\beta \in \mathbb{Z}^2} c_{\gamma - 2\beta} = 0 \quad \text{for} \quad \gamma \in E, \tag{4.1}$$

cf. (2.32). Then given any two directions $\lambda, \mu \in \mathbb{Z}^2 \setminus (0,0)$, $\lambda \neq \mu$, which generate a rectilinear partition of $\mathbb{Z}^2$, there exist (non-unique) finite masks $\mathbf{b}^{\lambda}$ and $\mathbf{b}^{\mu}$ such that

$$S_{\mathbf{c}} = S_{\mathbf{b}^{\lambda}} \Delta_{\lambda} + S_{\mathbf{b}^{\mu}} \Delta_{\mu}. \tag{4.2}$$

Proof. The subdivision operator $S_{\mathbf{c}}$ is defined for $\mathbf{f} \in \ell_\infty(\mathbf{Z}^2)$ by

$$(S_{\mathbf{c}}\mathbf{f})_{2\alpha+\gamma} = \sum_{\beta \in \mathbf{Z}^2} c_{\gamma-2\beta} f_{\alpha+\beta}, \quad \alpha \in \mathbf{Z}^2, \ \gamma \in E. \tag{4.3}$$

The proof of the lemma is then based on the observation that, for each $\gamma \in E$, there exists a (non-unique) finite path through the mask $\mathbf{c}$, covering all the non-zero coefficients, each step of which is taken along either the λ or μ direction. The fact that the sum of coefficients is zero then means that the linear combination can be written as a sum of differences along the path, that is

$$(S_{\mathbf{c}}\mathbf{f})_{2\alpha+\gamma} = \sum_{\beta \in \mathbf{Z}^2} b^\lambda_{\gamma-2\beta} \Delta_\lambda f_{\alpha+\beta} + \sum_{\beta \in \mathbf{Z}^2} b^\mu_{\gamma-2\beta} \Delta_\mu f_{\alpha+\beta}, \quad \alpha \in \mathbf{Z}^2, \ \gamma \in E, \tag{4.4}$$

for some finite masks $\mathbf{b}^\lambda$ and $\mathbf{b}^\mu$. $\square$

Remark. The proof of Proposition 7 can be argued in terms of the generating polynomial $c(z)$ as follows: The hypothesis (4.1) is equivalent to the condition

$$c(1,1) = c(-1,-1) = c(1,-1) = c(-1,1) = 0. \tag{4.5}$$

It can then be shown that this condition gives the generating polynomial decomposition

$$c(z) = (-1 + z^{-2\lambda})b^\lambda(z) + (-1 + z^{-2\mu})b^\mu(z), \tag{4.6}$$

for some non-unique Laurent polynomials $b^\lambda(z)$ and $b^\mu(z)$. The result (4.2) now follows by applying the following lemma to each term of (4.6):

Lemma 8 *Suppose that*

$$c(z) = (-1 + z^{-2\gamma})b(z), \quad \gamma \in \mathbf{Z}^2 \setminus (0,0), \tag{4.7}$$

for some Laurent polynomials $c(z)$ and $b(z)$. Then

$$S_{\mathbf{c}} = S_{\mathbf{b}} \Delta_\gamma. \tag{4.8}$$

4.2 Uniform convergence

We wish to find conditions for which the scheme $S_{\hat{\mathbf{a}}}$ is uniformly convergent. Consider the control polygon sequence $\{L_k[\hat{\varphi}^k]\}_{k=0}^\infty$ where, for example, L_k is the piecewise linear interpolation operator defined by (2.17), and $\hat{\varphi}^k$ denotes the values at level k produced by the subdivision scheme applied to cardinal initial data. Then we seek conditions for which $\{L_k[\hat{\varphi}^k]\}_{k=0}^\infty$ is a Cauchy sequence. Proposition 7 leads to:

Lemma 9 *Suppose that $\hat{a}(z)$ satisfies the necessary convergence condition*

$$\hat{a}(1,1) = 4, \quad \hat{a}(-1,-1) = \hat{a}(1,-1) = \hat{a}(-1,1) = 0, \tag{4.9}$$

see Lemma 2. Then

$$\|L_{k+1}[\hat{\varphi}^{k+1}] - L_k[\hat{\varphi}^k]\|_\infty \leq C \max\{\|\Delta_\lambda \hat{\varphi}^k\|_{\ell_\infty}, \|\Delta_\mu \hat{\varphi}^k\|_{\ell_\infty}\}, \tag{4.10}$$

for $\lambda, \mu \in \{(1,0), (0,1), (1,1)\}$, $\lambda \neq \mu$. More generally, defining L_k with respect to a skew triangulation, then (4.10) holds for the λ, μ directions defining any rectilinear partition of the points $\mathbf{Z}^2$.

Proof. Observe that, for any $\mathbf{f} \in \ell_\infty(\mathbb{Z}^2)$,

$$\|L_{k+1}[\mathbf{f}]\|_\infty = \|\mathbf{f}\|_{\ell_\infty}, \qquad (4.11)$$

since any piecewise linear interpolant achieves its extreme values at the vertices. We thus have that

$$
\begin{aligned}
\|L_{k+1}[\hat{\varphi}^{k+1}] - L_k[\hat{\varphi}^k]\|_\infty &= \|L_{k+1}[(S_{\hat{\mathbf{a}}} - S_{\boldsymbol{\ell}})\hat{\varphi}^k]\|_\infty, \\
&= \|(S_{\hat{\mathbf{a}}} - S_{\boldsymbol{\ell}})\hat{\varphi}^k\|_{\ell_\infty}, \\
&= \|S_{\mathbf{c}}\hat{\varphi}^k\|_{\ell_\infty}, \qquad (4.12)
\end{aligned}
$$

where

$$c(z) := \hat{a}(z) - \ell(z) \qquad (4.13)$$

is a generating polynomial satisfying conditions (4.5). Thus Proposition 7 can be applied and (4.10) follows by expressing $S_{\mathbf{c}}$ in the form (4.2). $\square$

We now make the simplifying assumption that there exist difference schemes for $\Delta_\lambda \hat{\varphi}^k$ and $\Delta_\mu \hat{\varphi}^k$. Thus

$$\hat{a}(z) = (1 + z^{-\lambda})\hat{b}^\lambda(z) \quad \text{and} \quad \hat{a}(z) = (1 + z^{-\mu})\hat{b}^\mu(z), \qquad (4.14)$$

where $\hat{b}^\lambda(z)$ and $\hat{b}^\mu(z)$ are the generating polynomials for the difference schemes (see Proposition 4). Lemma 9 now leads to the following convergence result:

Theorem 10 (Convergence.) *Let $S_{\hat{\mathbf{a}}}$ define a binary subdivision scheme having difference schemes $S_{\hat{\mathbf{b}}^\lambda}$ and $S_{\hat{\mathbf{b}}^\mu}$, where the directions λ and μ define a rectilinear partition of $\mathbb{Z}^2$. Furthermore, suppose that there exists a positive integer L such that the the Lth iterated difference operators have the 'contractive property' that*

$$\|S_{\hat{\mathbf{b}}^\lambda}^L\| < 1 \quad \text{and} \quad \|S_{\hat{\mathbf{b}}^\mu}^L\| < 1. \qquad (4.15)$$

Then $S_{\hat{\mathbf{a}}}$ is uniformly convergent.

The proof of Theorem 10 follows from the fact that the differences along the λ and μ directions will be contracting over L steps. This condition, together with Lemma 9, can then be used to show that $\{L_k[\hat{\varphi}^k]\}_{k=0}^\infty$ is a Cauchy sequence and hence that the scheme is uniformly convergent.

To apply Theorem 10 we require the Lth iterated operators of the difference schemes, together with their norms. These are given by the following proposition:

Proposition 11 *Let $\hat{b}(z)$ be the generating polynomial of a bivariate binary subdivision scheme $S_{\hat{\mathbf{b}}}$. Then $S_{\hat{\mathbf{b}}}^L$ is defined by*

$$(S_{\hat{\mathbf{b}}}^L \mathbf{f})_\alpha := \sum_{\beta \in \mathbb{Z}^2} \hat{b}^{[L]}_{\alpha - 2^L \beta} f_\beta, \quad \mathbf{f} \in \ell_\infty(\mathbb{Z}^2), \qquad (4.16)$$

with generating polynomial

$$\hat{b}^{[L]}(z) := \hat{b}(z)\hat{b}(z^2) \ldots \hat{b}(z^{2^{L-1}}) \qquad (4.17)$$

113

and norm

$$\|S_{\mathbf{b}}^{L}\| := \max_{\alpha \in \{0,\ldots,2^L-1\}^2} \left\{ \sum_{\beta \in \mathbb{Z}^2} |\hat{b}_{\alpha-2^L\beta}^{[L]}| \right\}. \tag{4.18}$$

Proof. (Levin[6]) Define the *z-transform*

$$G_k(z) := \sum_{\alpha \in \mathbb{Z}^2} (S_{\mathbf{b}}^k \mathbf{f})_\alpha z^\alpha. \tag{4.19}$$

Then it is easily shown that

$$G_{k+1}(z) = \hat{b}(z) G_k(z^2) \tag{4.20}$$

and hence that

$$G_L(z) = \hat{b}(z)\hat{b}(z^2) \ldots \hat{b}(z^{2^{L-1}}) G_0(z^{2^L}) = \hat{b}^{[L]}(z) G_0(z^{2^L}). \tag{4.21}$$

Equating coefficients then gives the Lth iterated subdivision operator defined by (4.16). The norm of this operator then immediately follows (cf. (2.10)). $\square$

4.3   Matricial schemes

To prove convergence to differentiable limits using the theory of the previous subsection, we must assume that $a(z)$ has divided difference generating polynomial factors $\hat{a}(z)$. Also, in Theorem 10, the simplifying assumption has been made that $\hat{a}(z)$ can be factored appropriately to give difference schemes along directions λ and μ. A generalization of the theory to cover *'matricial schemes'* avoids these simplifying assumptions. We thus conclude by briefly showing how matricial schemes arise in the study of differentiable limits by considering a generalization of Proposition 4:

Proposition 12 *Let $S_{\mathbf{a}}$ be a binary subdivision operator with generating polynomial $a(z)$ satisfying the necessary uniform convergence condition (2.33). Also, let λ and μ be two directions defining a rectilinear partition of $\mathbb{Z}^2$. Then given $\gamma \in \mathbb{Z}^2 \backslash (0,0)$, there exist (non-unique) Laurent polynomials $b^{\gamma,\lambda}(z)$, $b^{\gamma,\mu}(z)$ and $\hat{a}^{\gamma,\lambda}(z) := 2b^{\gamma,\lambda}(z)$, $\hat{a}^{\gamma,\mu}(z) := 2b^{\gamma,\mu}(z)$ such that*

$$\Delta_\gamma S_{\mathbf{a}} = S_{\mathbf{b}^{\gamma,\lambda}} \Delta_\lambda + S_{\mathbf{b}^{\gamma,\lambda}} \Delta_\mu, \tag{4.22}$$

$$D_\gamma S_{\mathbf{a}} = S_{\hat{\mathbf{a}}^{\gamma,\lambda}} D_\lambda + S_{\hat{\mathbf{a}}^{\gamma,\mu}} D_\mu. \tag{4.23}$$

Proof. The operator

$$S_{\mathbf{c}^\gamma} := \Delta_\gamma S_{\mathbf{a}} \tag{4.24}$$

has generating polynomial coefficients $c_\alpha^\gamma := a_{\alpha+\gamma} - a_\alpha$. Thus

$$c^\gamma(z) = (z^{-\gamma} - 1)a(z) \tag{4.25}$$

and hence $c(1,1) = 0$. Thus, using (2.33), it follows that $c(z)$ satisfies condition (4.5) and hence

$$c^\gamma(z) = (-1 + z^{-2\lambda})b^{\gamma,\lambda}(z) + (-1 + z^{-2\mu})b^{\gamma,\mu}(z) \tag{4.26}$$

for some non-unique Laurent polynomials $b^{\gamma,\lambda}(z)$ and $b^{\gamma,\mu}(z)$. Proposition 7 then gives (4.22) and multiplying by 2^{-k-1} gives (4.23). $\square$

In the special case where

$$a(z) = (1 + z^{-\gamma})b(z) \tag{4.27}$$

we obtain

$$c^\gamma(z) = (-1 + z^{-2\gamma})b(z) \tag{4.28}$$

in the above proof. Hence Proposition 12 gives

$$\Delta_\gamma S_{\mathbf{a}} = S_{\mathbf{b}}\Delta_\gamma \text{ and } D_\gamma S_{\mathbf{a}} = S_{\hat{\mathbf{a}}}D_\gamma, \tag{4.29}$$

where $\hat{a}(z) = 2b(z)$. This is the case of Proposition 4, when there exist difference and hence divided difference schemes. When such divided difference schemes are not available, we can take $\gamma = \lambda$ and $\gamma = \mu$ in Proposition 12 to give the matricial divided difference scheme

$$\begin{bmatrix} D_\lambda \mathbf{f}^{k+1} \\ D_\mu \mathbf{f}^{k+1} \end{bmatrix} = \begin{bmatrix} S_{\hat{\mathbf{a}}\lambda,\lambda} & S_{\hat{\mathbf{a}}\lambda,\mu} \\ S_{\hat{\mathbf{a}}\mu,\lambda} & S_{\hat{\mathbf{a}}\mu,\mu} \end{bmatrix} \begin{bmatrix} D_\lambda \mathbf{f}^k \\ D_\mu \mathbf{f}^k \end{bmatrix}. \tag{4.30}$$

This suggests the analysis of matricial schemes per se.

5 Example of the Butterfly subdivision scheme

We conclude this introduction to uniform subdivision by applying the theory to the interpolatory 'butterfly' subdivision scheme described in [4]. This scheme has been analysed by Dyn, Levin and Micchelli [5], who show that there exists an interval for a shape parameter ω for which the scheme converges to a C^1 limit. Here, we give more precise details of the calculation of the norm of the 2nd iterate of the appropriate subdivision operator. This calculation is equivalent to that of Qu [7], who uses a matrix norm approach.

The butterfly scheme is defined with respect to the $(1,1)$ triangulation by

$$\left.\begin{aligned}
f_{2i,2j}^{k+1} &:= f_{i,j}^k, \\
f_{2i+1,2j}^{k+1} &:= \tfrac{1}{2}(f_{i,j}^k + f_{i+1,j}^k) + 2\omega(f_{i,j-1}^k + f_{i+1,j+1}^k) \\
&\quad -\omega(f_{i-1,j-1}^k + f_{i+1,j-1}^k + f_{i,j+1}^k + f_{i+2,j+1}^k), \\
f_{2i,2j+1}^{k+1} &:= \tfrac{1}{2}(f_{i,j}^k + f_{i,j+1}^k) + 2\omega(f_{i-1,j}^k + f_{i+1,j+1}^k) \\
&\quad -\omega(f_{i-1,j-1}^k + f_{i-1,j+1}^k + f_{i+1,j}^k + f_{i+1,j+2}^k), \\
f_{2i+1,2j+1}^{k+1} &:= \tfrac{1}{2}(f_{i,j}^k + f_{i+1,j+1}^k) + 2\omega(f_{i+1,j}^k + f_{i,j+1}^k) \\
&\quad -\omega(f_{i,j-1}^k + f_{i-1,j}^k + f_{i+2,j+1}^k + f_{i+1,j+2}^k).
\end{aligned}\right\} \tag{5.1}$$

This scheme is symmetric with respect to the $(1,1)$ triangulation and is interpolatory by definition of the first rule in (5.1). The description of the scheme is derived from the butterfly appearance of the individual masks for the second, third and fourth rules. The parameter ω can be used to control the shape of the limit surface. The case $\omega = 0$ gives the piecewise linear scheme of subsection 2.4.1 and the case $\omega = \frac{1}{16}$ gives a scheme which reproduces cubic polynomials. Here we will indicate that $0 < \omega < \frac{1}{12}$ is a sufficient condition for the scheme to have a C^1 limit.

The generating poynomial $a(z)$, for the subdivision scheme defined by (5.1), has factors $\frac{1}{2}(1+z_1^{-1})$, $\frac{1}{2}(1+z_2^{-1})$, and $\frac{1}{2}(1+z_1^{-1}z_2^{-1})$. Thus there exist uniform subdivision schemes for the divided difference sets $D^\gamma f^k$, for $\gamma = (1,0),(0,1),(1,1)$. We now seek conditions for which these divided difference schemes have C^0 limits where by symmetry, it is sufficient to consider only $\gamma = (1,0)$. It will then follow, by Theorem 3, that the butterfly scheme converges to a C^1 limit.

Writing

$$a(z) = \tfrac{1}{2}(1 + z_1^{-1})\hat{a}(z), \tag{5.2}$$

then, with $\lambda = (-1,0)$ and $\mu = (-1,-1)$,

$$\hat{a}(z) = (1 + z_2^{-1})\hat{b}^\lambda(z) \text{ and } \hat{a}(z) = (1 + z_1^{-1}z_2^{-1})\hat{b}^\mu(z), \tag{5.3}$$

where

$$
\begin{aligned}
\hat{b}^\lambda(z) \; := \; & (1 - 8w)(1 + z_1 z_2) + 4w(z_1^{-1}z_2^{-1} + z_1^2 z_2^2 + 2w(z_2^{-1} + z_1^{-1} + z_1^2 z_2 + z_1 z_2^2) \\
& -2w(z_1^{-1}z_2^{-2} + z_1^{-2}z_2^{-1} + z_1 z_2^{-1} + z_1^{-1}z_2 + z_1^2 + z_2^2 + z_1^3 z_2^2 + z_1^2 z_2^3) \quad (5.4)
\end{aligned}
$$

with a dual expression for $\hat{b}^\mu(z)$. The subdivision operator $S_{\hat{b}^\lambda}$ has norm $\|S_{\hat{b}^\lambda}\| \geq 1$. However, calculation of the generating polynomial $\hat{b}^\lambda(z)\hat{b}^\lambda(z^2)$ for the iterated operator $S^2_{\hat{b}^\lambda}$ leads to

$$\|S^2_{\hat{b}^\lambda}\| = max\{A_1(w), A_2(w), A_3(w), A_4(w), A_5(w), A_6(w)\}, \tag{5.5}$$

where

$$
\left.
\begin{aligned}
A_1(w) & := 16w^2 + 4|4w^2 + w| + 8|9w^2 - w| + |72w^2 - 16w + 1|, \\
A_2(w) & := 44w^2 + 6|6w^2 - w| + 2|24w^2 - w|, \\
A_3(w) & := 52w^2 + 4|8w^2 - w| + 2|16w^2 - w| + 6|6w^2 - w|, \\
A_4(w) & := 104w^2 + 4|2w^2 - w|, \\
A_5(w) & := 48w^2 + 8|6w^2 - w| + |12w - 1|, \\
A_6(w) & := 40w^2 + 4|6w^2 + w|.
\end{aligned}
\right\} \tag{5.6}
$$

Here, the expected sixteen terms in (5.5), see Proposition 11, reduce to six because of symmetries and repetitions. Also, by symmetry, the expression for $\|S^2_{\hat{b}^\mu}\|$ is identical to (5.5). A careful analysis of the terms (5.6) now leads to

$$\|S^2_{\hat{b}^\lambda}\| = \|S^2_{\hat{b}^\mu}\| < 1 \text{ for } 0 < w < \tfrac{1}{12}. \tag{5.7}$$

In particular, $w < \frac{1}{12}$ is obtained from the condition $A_4(w) < 1$, where $A_4(w)$ is the dominant term in (5.5) in the neighbourhood of $w = \frac{1}{12}$. It now follows, from Theorem 10, that the subdivision scheme for the divided difference converges to a C^0 limit, and hence the butterfly scheme converges to a C^1 limit, for $0 < w < \frac{1}{12}$.

References

[1] Cavaretta, A.S., W. Dahmen and C.A. Micchelli, Stationary subdivision, to appear in Memoirs of AMS.

[2] Cavaretta, A.S., and C.A. Micchelli, The design of curves and surfaces by subdivision algorithms, in *Mathematical Methods in Computer Aided Geometric Design*, T. Lyche and L.L. Schumaker (eds.), Academic Press, New York, 1989, 115-153.

[3] Dyn, N., Subdivision schemes in CAGD, to appear in the proceedings of the 1990 Lancaster Summer School on Multivariate Approximation.

[4] Dyn, N., J.A. Gregory and D. Levin, A butterfly subdivision scheme for surface interpolation with tension control, ACM Trans. on Graphics **9**, 1990, 160-169.

[5] Dyn, N., D. Levin and C.A. Micchelli, Using parameters to increase smoothness of curves and surfaces generated by subdivision, Computer Aided Geometric Design **7**, 1990, 129-140.

[6] Levin, D., Generating function techniques in the analysis of subdivision schemes, private communication.

[7] Qu, R., Recursive subdivision algorithms for curve and surface design, Ph.D. Thesis, Department of Mathematics and Statistics, Brunel University, 1990.

John A. Gregory
Department of Mathematics and Statistics
Brunel University
Uxbridge UB8 3PH
England, UK.

J LORENZ
Computation of invariant manifolds

1   Invariant Manifolds

For illustration, we start with a simple example of a single nonlinear oscillator. Let

$$(\phi, r) \in S^1 \times [0, \infty]$$

denote polar coordinates in the plane, where $S^1 = \mathbf{R} \, mod \, 2\pi$ denotes the circle. Consider the dynamical system

$$\dot{\phi} = 1, \quad \dot{r} = r(1 - r^2) . \tag{1.1}$$

Clearly, if $r(0) > 0$ then $r(t) \to 1$ as $t \to \infty$. The unit circle

$$M = \{(\phi, 1) : \phi \in S^1\}$$

is an attracting invariant manifold for the system (1.1).

Here a manifold $M \subset W$ is called invariant for a dynamical system $\dot{w} = f(w), w(t) \in W$, if for all initial conditions $w(0) = w_0 \in M$ the solution $w(t)$ exists for all $t \in \mathbf{R}$ and satisfies $w(t) \in M$ for all $t \in \mathbf{R}$.

We now couple two oscillators of the type described above and consider a system

$$\begin{bmatrix} \dot{\phi}_1 \\ \dot{\phi}_2 \\ \dot{r}_1 \\ \dot{r}_2 \end{bmatrix} = \begin{bmatrix} 1 \\ 1 \\ r_1(1 - r_1^2) \\ r_2(1 - r_2^2) \end{bmatrix} + \lambda F(\phi_1, \phi_2, r_1, r_2, \lambda) . \tag{1.2}$$

Here $\lambda \in \mathbf{R}$ is the coupling constant and $F = F(\phi_1, \phi_2, r_1, r_2, \lambda)$ is a C^∞ function,

$$F : S^1 \times S^1 \times \mathbf{R}^2 \times (-\infty, \infty) \to \mathbf{R}^4 .$$

(For specific examples, see [1].) Clearly, for zero coupling, $\lambda = 0$, the system has the invariant manifold

$$M(\lambda = 0) = \{(\phi_1, \phi_2, 1, 1) : \phi_1, \phi_2 \in S^1\} ,$$

which is just the product of two invariant circles. In other words, $M(\lambda = 0)$ is diffeomorphic to the standard 2-torus $T^2 = S^1 \times S^1$. What happens to this invariant manifold for $\lambda \neq 0$? Analytic perturbation results cover the case of small coupling, $|\lambda| < \epsilon$. If $|\lambda|$ increases, one must expect a breakdown of the invariant torus $M(\lambda)$. We first describe a perturbation result, which can be deduced from more general theorems in [4], [9], [11].

THEOREM 1.1 *Consider the dynamical (1.2) with a C^∞-function F. For all $k = 1, 2...$ there is $\lambda_k > 0$ and a locally unique C^k-function*

$$R : T^2 \times (-\lambda_k, \lambda_k) \to \mathbf{R}^2$$

such that the manifold

$$M(\lambda) = \{(\phi, R(\phi, \lambda)) : \phi = (\phi_1, \phi_2) \in T^2\}$$

is invariant under (1.2) for $-\lambda_k < \lambda < \lambda_k$.

To obtain this result, one has to observe the following two facts about the unperturbed ($\lambda = 0$) case: First, the flow on $M(\lambda = 0)$ is parallel (there is no attractivity *within* $M(\lambda = 0)$); second, the flow *towards* $M(\lambda = 0)$ is exponentially attracting. Perturbation results for cases where there is attractivity *within* the unperturbed manifold are also covered in [4], [9], [11]. Theorem 1 indicates that the disappearance of $M(\lambda)$ for increasing $|\lambda|$ is connected with a loss of smoothness; the bifurcation theory of invariant manifolds is not governed by operators of Fredholm type. Indeed, the discussion in [6] suggest that in the region of final disappearance of an invariant torus, there might be a Cantor set Λ of λ values so that the torus exists for $\lambda \in \Lambda$, but does not exist for $\lambda \notin \Lambda$.

2   Relation to Weierstrass' Functions

In dealing with elliptic equations, one is used to solutions which are at least as smooth as the coefficients that define the problem. This is not the case if one determines invariant manifolds. For illustration, we consider a system

$$
\begin{aligned}
\dot{\phi}_1 &= f(\phi) \\
\dot{\phi}_2 &= 1 \\
\dot{r} &= -r + g(\phi)
\end{aligned}
\tag{2.1}
$$

where $\phi = (\phi_1, \phi_2) \in T^2$ and where $f = f(\phi)$ and $g = g(\phi)$ are C^∞-functions. We ask for a manifold of the form

$$M = \{(\phi, R(\phi)) : \phi \in T^2\} \tag{2.2}$$

which is invariant under this system, and try to determine an appropriate function $R : T^2 \to \mathbf{R}$. The following considerations do not require any smoothness of $\mathbf{R}$. Let $\phi(t; \alpha)$ denote the solution of

$$\dot{\phi}_1 = f(\phi), \quad \dot{\phi}_2 = 1; \quad \phi_1(0) = \alpha, \quad \phi_2(0) = 0 .$$

Then the solution of

$$\dot{r} = -r + g(\phi(t; \alpha)), \quad r(0) = r_0 ,$$

is

$$r(t) = e^{-t}r_0 + \int_0^t e^{-(t-s)}g(\phi(s;\alpha))ds .$$

Now suppose that (2.2) is invariant, and let

$$R_0(\alpha) := R(\alpha, 0), \quad \alpha \in S^1 .$$

Choosing $r_0 = R_0(\alpha)$, one obtains after time $t = 2\pi$,

$$R_0(\gamma(\alpha)) = e^{-2\pi} R_0(\alpha) + q(\alpha), \quad \alpha \in S^1 , \tag{2.3}$$

with

$$\gamma(\alpha) = \phi(2\pi; \alpha)$$
$$q(\alpha) = \int_0^{2\pi} e^{-(2\pi-s)}g(\phi(s;\alpha))ds .$$

Conversely, it is easily shown that any solution $R_0 = R_0(\alpha), \alpha \in S^1$, of the functional equation (2.3) generates a set M (see (2.2)) which is invariant under (2.1).

The functional equation (2.3) is linear, and we can discuss it. The circle map $\alpha \to \gamma(\alpha) = \phi(2\pi; \alpha), \alpha \in S^1$, is C^∞-smooth and has a C^∞-inverse, $\kappa := \gamma^{-1}$. We set $Q(\alpha) = q(\kappa(\alpha))$ and obtain

$$R_0(\alpha) = e^{-2\pi} R_0(\kappa(\alpha)) + Q(\alpha), \quad \alpha \in S^1 . \tag{2.4}$$

This equation is equivalent to (2.3). Let $\mathcal{C}$ denote the Banach space of all continuous functions

$$R_0 : S^1 \to \mathbf{R}$$

with norm

$$|R_0|_\infty = \max\{|R_0(\alpha)| : \alpha \in S^1\}$$

and let $L : \mathcal{C} \to \mathcal{C}$ denote the linear operator defined by

$$(LR_0)(\alpha) = e^{-2\pi} R_0(\kappa(\alpha)), \quad \alpha \in S^1 .$$

Then (2.4) is nothing but the fixed point equation

$$R_0 = LR_0 + Q$$

with the unique solution

$$R_0 = \sum_{n=0}^\infty L^n Q ,$$

i.e.,

$$R_0(\alpha) = \sum_{n=0}^\infty e^{-2\pi n} Q(\kappa(...(\kappa(\alpha)))) .$$

As shown in [8], the formally differentiated series converges (and represents $R_0'(\alpha)$) if

$$e^{-2\pi}|\kappa'(\alpha)| < 1 \quad \forall \alpha \in S^1 . \tag{2.5}$$

If (2.5) is violated, then one can, in general, not expect that R_0 is everywhere differentiable. Indeed, in 1875, Weierstrass has considered the functions

$$R_0(\alpha) = \sum_{n=0}^{\infty} a^n \cos(b^n \alpha), \quad \alpha \in \mathbf{R}, \tag{2.6}$$

which are continuous for $|a| < 1$. If $|ab| > 1$, then (2.6) is nowhere differentiable [5]. Numerical approximations to smooth and non-smooth solutions of the functional equation (2.4) are given in [8].

3 P.D.E. Formulation

Let

$$T^p = S^1 \times ... \times S^1 \qquad \text{(p times)}$$

denote the standard p-torus. We consider partitioned systems

$$\dot{\phi} = f(\phi, r) \quad , \quad \phi(t) \in T^p ,$$

$$\dot{r} = g(\phi, r) \quad , \quad r(t) \in \mathbf{R}^q , \tag{3.1}$$

where f and g are smooth functions. We want to determine a manifold M of the form

$$M = \{(\phi, R(\phi)) : \phi \in T^p\} \subset T^p \times \dot{\mathbf{R}}^q , \tag{3.2}$$

which is invariant under (3.1). Here $R : T^p \to \mathbf{R}^q$ needs to be determined; we will assume that $R \in C^1$, at least.

It is not difficult to show that the manifold (3.2) is invariant under (3.1) if and only if for each point $Q \in M$ the vector

$$\begin{pmatrix} f(Q) \\ g(Q) \end{pmatrix} \in \mathbf{R}^p \times \mathbf{R}^q$$

lies in the tangent space $T_Q(M)$ of M at Q. Since $T_Q(M)$ is spanned by the p vectors

$$\begin{pmatrix} 0 \\ \vdots \\ 1 \\ \vdots \\ 0 \\ \dfrac{\partial R}{\partial \phi_\nu} \end{pmatrix} \qquad (1 \quad \text{in component } \nu), \quad \nu = 1, ..., p ,$$

the invariance condition reads

$$\sum_{\nu=1}^{p} f_\nu(\phi, R(\phi)) \frac{\partial R}{\partial \phi_\nu}(\phi) = g(\phi, R(\phi)), \quad \phi \in T^p. \qquad (3.3)$$

Any C^1-solution $R : T^p \to \mathbf{R}^q$ of (3.3) leads to a C^1-manifold (3.2) invariant under (3.1); and conversely, if (3.2) is a C^1-manifold which is invariant under (3.1), then R solves (3.3).

4   Leap-Frog Scheme

We restrict ourselves to the case $p = 2$ and to meshes which are equidistant in each coordinate direction. Let $N_1, N_2 \in \mathbf{N}$; let $h_1 = 2\pi/N_1, h_2 = 2\pi/N_2, h = (h_1, h_2)$. With

$$T_h^2 = \{(ih_1, jh_2) : i \in \mathbf{Z} \, mod N_1, j \in \mathbf{Z} \, mod N_2\}$$

we denote the grid replacing the torus T^2. If

$$R_h : T_h^2 \to \mathbf{R}^q$$

is a grid function, then

$$(E_1 R_h)(ih_1, jh_2) = R_h((i+1)h_1, jh_2).$$

Here $i + 1$ is computed modulo N_1. Similarly, E_2 is the shift operator for the 2nd coordinate direction. As usual, the forward, backward, and centered divided difference operators are defined by

$$D_{+\nu} = \frac{1}{h_\nu}(E_\nu - I), \quad D_{-\nu} = \frac{1}{h_\nu}(I - E_\nu^{-1}),$$

$$D_{0\nu} = \frac{1}{2}(D_{+\nu} + D_{-\nu}), \quad \nu = 1, 2.$$

In the **leap-frog scheme**, one discretizes $\partial/\partial\phi_\nu$ by $D_{0\nu}$ to obtain the difference equations

$$\sum_{\nu=1}^{2} f_\nu(\phi_{ij}, R_h(\phi_{ij})) D_{0\nu} R_h(\phi_{ij}) = g(\phi_{ij}, R_h(\phi_{ij})), \quad \phi_{ij} = (ih_1, jh_2) \in T_h^2. \qquad (4.1)$$

Open Problem: Suppose that (3.3) (with $p = 2$) has a unique smooth solution R and that the formal linearization of (3.3) at R is well-behaved. Also, let $\max\{h_1, h_2\}$ be sufficiently small. Show that (4.1) has a solution R_h which is locally unique in a small neighborhood of $R|_h$, where $R|_h$ denotes the restriction of R to T_h^2. Show that $\|R_h - R|_h\| = 0(h^2)$ in an appropriate norm.

So far, we have analyzed (4.1) only for the very special case where

$$f_\nu(\phi, r) = f_\nu = const, \quad \nu = 1, 2,$$
$$g(\phi, r) = -Cr + \psi(\phi),$$

with a constant $q \times q$ matrix C; see [3]. In this case one can employ discrete Fourier analysis to discuss the problem. It is only required that the matrix C has no eigenvalues on the imaginary axis in order to derive suitable estimates for the inverse of the global discretization matrix; see [3].

5 Block M-Functions

To make further progress, we restrict ourselves to dynamical systems of the form

$$\dot{\phi} = f(\phi) \quad , \quad \phi(t) \in T^p$$

$$\dot{r} = g(\phi, r) \quad , \quad r(t) \in \mathbf{R}^q \tag{5.1}$$

To discretize the corresponding pde (3.3), we choose a grid T_h^p and replace (3.3) by

$$F(R_h) \equiv \sum_{\nu=1}^{p} [f_\nu(\phi) D_{0\nu} R_h - \frac{1}{2} h_\nu \sigma(f_\nu(\phi)) D_{+\nu} D_{-\nu} R_h] - g(\phi, R_h) = 0, \quad \phi \in T_h^p. \tag{5.2}$$

Here $\sigma = \sigma(\alpha)$ is a real function satisfying $\sigma(\alpha) \geq |\alpha|$ for all $\alpha \in \mathbf{R}$. For example, one may take $\sigma(\alpha) = |\alpha|$ corresponding to upwinding, or $\sigma(\alpha) = (1 + \alpha^2)^{\frac{1}{2}}$; see [12]. (To justify Richardson extrapolation, it seems to be necessary to choose a smooth function σ; see [2] for details.)

Here we want to address the question of unique solvability of (5.2). It will be convenient to generalize the concepts of an M-matrix and an M-function.
Notation: If $B, C \in \mathbf{R}^{q \times q}$ we write

$$B \geq_Q C \Leftrightarrow x^T B x \geq x^T C x \quad \forall x \in \mathbf{R}^q.$$

The index Q in the symbol $\geq_Q$ shall remind of quadratic form.
Definition 5.1: Let $A \in \mathbf{R}^{Nq \times Nq}$ denote a real square matrix consisting of N^2 blocks where each block has dimensions $q \times q$:

$$A = (A_{ij})_{i,j=1,\ldots,N}, \quad A_{ij} \in \mathbf{R}^{q \times q}.$$

We call A a block M-matrix of type (β, q) if
1) $A_{ij} = \sigma_{ij} I$ with $\sigma_{ij} \in \mathbf{R}$, $\sigma_{ij} \leq 0$, for all $i \neq j$;
2) $\beta > 0$ and

$$\sum_{j=1}^{N} A_{ij} \geq_Q \beta I, \quad i = 1, \ldots, N.$$

We denote the Euclidian norm in $\mathbf{R}^q$ by $|\cdot|$ and set

$$||y|| = \max\{|y_i| : i = 1, \ldots, N\}$$

for vectors

$$y = (y_1, ..., y_N) \in \mathbf{R}^{Nq}, \quad y_i \in \mathbf{R}^q .$$

The same notation is used for the corresponding matrix norm.

The following result is proved in [2]:

LEMMA 5.1 *Let* $A \in \mathbf{R}^{Nq \times Nq}$ *denote a block M-matrix of type* (β, q)*. Then* A *is non-singular and*

$$||A^{-1}|| \leq \frac{1}{\beta} .$$

We now generalize to nonlinear functions and show the following result.

THEOREM 5.1 *Let* F *denote a* C^1*-function from* $\mathbf{R}^{Nq}$ *into itself and assume that for all* $y \in \mathbf{R}^{Nq}$ *the matrix* $F'(y)$ *is a block M-matrix of type* (β, q)*. Here* $\beta > 0$ *and* q *are independent of* y*. Then* F *is a bijection of* $\mathbf{R}^{Nq}$ *onto itself and*

$$||F^{-1}(\widetilde{z}) - F^{-1}(z)|| \leq \frac{1}{\beta}||\widetilde{z} - z|| \quad \forall z, \widetilde{z} \in \mathbf{R}^{Nq} .$$

Proof: Fix $y, \widetilde{y} \in \mathbf{R}^{Nq}$ and set

$$\psi(s) = y + s(\widetilde{y} - y), \quad 0 \leq s \leq 1 .$$

We have

$$
\begin{aligned}
F(\widetilde{y}) - F(y) &= F(\psi(1)) - F(\psi(0)) \\
&= \int_0^1 \{\frac{d}{ds}F(\psi(s))\}ds \\
&= \{\int_0^1 F'(\psi(s))ds\}(\widetilde{y} - y) .
\end{aligned}
$$

Here

$$\Delta F := \int_0^1 F'(\psi(s))ds$$

is a block M-matrix of type (β, q), and therefore

$$||(\Delta F)^{-1}|| \leq \frac{1}{\beta}$$

by Lemma 5.1. This implies

$$||\widetilde{y} - y|| \leq \frac{1}{\beta}||F(\widetilde{y}) - F(y)|| \quad \forall y, \widetilde{y} \in \mathbf{R}^{Nq} . \tag{5.3}$$

2)   The remainder of the proof follows from Hadamard's Theorem; see [10]. We give a short argument for completeness. It remains to show that the set $\Omega := F(\mathbf{R}^{Nq})$ equals $\mathbf{R}^{Nq}$. If $y_n \in \mathbf{R}^{Nq}$ and $F(y_n) =: z_n \to z$ as $n \to \infty$, then

$$||y_n - y_m|| \leq \frac{1}{\beta}||z_n - z_m|| ,$$

and therefore $y_n \to y$. One obtains

$$z_n = F(y_n) \to F(y) = z \,.$$

This proves that Ω is closed. Also, the sets $\mathbf{R}^{Nq}$ and Ω are homeomorphic since F and F^{-1} are continuous. Brower's Theorem of the invariance of domain implies that Ω is open. Being both open and closed, the set Ω must equal $\mathbf{R}^{Nq}$.

Application: Linearization of (5.2) at R_h yields

$$F'(R_h)\eta_h = \sum_{\nu=1}^{p}[f_\nu(\phi)D_{0\nu}\eta_h - \frac{1}{2}h_\nu\sigma(f_\nu(\phi))D_{+\nu}D_{-\nu}\eta_h - g_r(\phi, R_h(\phi))\eta_h \,.$$

In a standard matrix representation, the outer-diagonal $q \times q$ blocks are

$$\left\{\pm\frac{f_\nu(\phi)}{2h_\nu} - \frac{1}{2h_\nu}\sigma(f_\nu(\phi))\right\}I \,.$$

Since we have assumed $\sigma(\alpha) \geq |\alpha|$, the condition 1) of Definition 5.1 is fulfilled. If we require

$$g_r(\phi, r) \leq_Q -\beta I \quad \forall \phi \in T^p, \forall r \in \mathbf{R}^q \tag{5.4}$$

for some $\beta > 0$, then $F'(R_h)$ is a block M-matrix of type (β, q). We obtain

THEOREM 5.2 *Let $g \in C^2$ and assume (5.4). For all grids T_h^p the system (5.2) of difference equations has a unique solution R_h. Newton's method to compute R_h converges locally quadratic. If the system of pdes (3.3) has a C^2-solution R, then*

$$\|R_h - R|_h\| = O(h) \,.$$

Proof: Unique solvability of (5.2) follows from Theorem 5.1. Since $F'(R_h)$ is a block M-matrix – and therefore non-singular – the quadratic convergence of Newton's method follows from standard arguments; one has to observe that F' is locally lipschitz. Also, consistency arguments yield

$$\|F(R|_h)\| = \mathcal{O}(h) \,.$$

Then (5.3) implies $\|R_h - R|_h\| = \mathcal{O}(h) \,.$

6 Additional Remarks

After discretization of the pde (3.3) one obtains a system of difference equations such as (4.1) or (5.2). To solve these systems, we have used Newton's method and – in case of a parameter dependent equation – arclength continuation. To solve the sparse linear systems, we have gained experience with implementations on distributed memory machines [7], [13] and on serial computers [3].

An interesting aspect of continuation procedures on distributed memory machines is the following: Once the Newton matrix is distributed over the process grid, the execution time for computing an LU-factorization depends critically on the location of the pivots. If the pivot locations are known in advance, then an ideal data distribution can be chosen. In a continuation process, acceptable pivot locations have to be changed only occasionally. Therefore, almost ideal speed-ups can be obtained. There are still some open question about determining 'acceptability' of pivots from the point of view of numerical stability; also, a proper treatment of sparsity and of fill is not obvious. For more details, see [7], [13].

In [3] we have investigate another approach to treat the linear equations to be solved in each Newton or continuation step: The linearized difference equations can be used as marching schemes to generate a compactified system of smaller dimension. The process is analogous to shooting for ordinary boundary value problems or – in the language of dynamical systems – to applying the Poincaré map. The compactified system can be viewed as a discretization of a functional equation like (2.3). As in shooting, the conditioning of the compactified system is a critical issue. Unless one obeys the CFL – condition – with time corresponding to the compactification direction – the compactified system will be ill-conditioned. Also, instead of using leap-frog, in the form (3.3), it is better to split up the zero-under term properly. E.g., an equation

$$\frac{\partial R}{\partial \phi_1} + \frac{\partial R}{\partial \phi_2} + bR = \beta(x,t)$$

is discretized by leap-frog as

$$\frac{1}{2h_1}\left\{R_{i+1,j} - R_{i-1,j}\right\} + \frac{1}{2h_2}\left\{R_{i,j+1} - R_{i,j-1}\right\} + bR_{ij} = \beta_{ij} \, .$$

If $b > 0$ and if one wants to compactify in increasing ϕ_1- direction, it is better to use

$$\frac{1}{2}b\left\{R_{i+1,j} + R_{i-1,j}\right\}$$

instead of bR_{ij}. Under restrictive assumptions, the conditioning question has been analyzed in [3].

References

[1] D. G. Aronson, E. J. Doedel, and H. G. Othmer, "An analytical and numerical study of the bifurcations in a system of linearly-coupled oscillators," Phys. D **25**, pp. 20-104 (1987).

[2] L. Dieci, J. Lorenz, "Block M-matrices and computation of invariant tori." To appear in SIAM J. Sci. Stat. Comput.

[3] L. Dieci, J. Lorenz, and R. D. Russell, "Numerical calculation of invariant tori," SIAM J. Sci. Stat. Comput. **12**, pp. 607-647 (1991).

[4] N. Fenichel, "Persistence and smoothness of invariant manifolds for flows," Indiana Univ. Math. J. **21**, pp. 193-226 (1971).

[5] G. H. Hardy, "Weierstrass's Non-differentiable Function." Trans. Amer. Math. Soc. **17**, pp. 301-325 (1916).

[6] R. A. Johnson, "Hopf bifurcation from non-periodic solutions of differential equations I: Linear theory." Preprint, 1989.

[7] J. Lorenz, E. van de Velde, "Concurrent computations of invariant manifolds." March 1989. In: The proceedings of the Fourth Conference on Hypercubes, Concurrent Computers, and Applications. Golden Gate Enterprise, Los Altos, 1990.

[8] A. C. Morlet, J. Lorenz, "Numerical solution of a functional equation on a circle." Submitted for publication.

[9] J. Moser, "A rapidly convergent iteration method and non-linear partial differential equations – I," Ann Scuola Normale Superiore Pisa Serie 3 XX, pp. 265-315 (1966).

[10] J. M. Ortega, W. C. Rheinboldt, "Iterative solution of nonlinear equations in several variables". Academic Press, New York, 1970.

[11] R. Sacker, "A new approach to the perturbation theory of invariant surfaces," Comm. Pure and Applied Mathematics **18**, pp. 717-732 (1965).

[12] G. Stoyan, "Monotone difference schemes for diffusion-convection problems," ZAMM **59**, pp. 361-372 (1979).

[13] E. van de Velde, J. Lorenz, "Adaptive data distribution for concurrent continuation. Technical report, Center for Research on Parallel Computation, Caltech (1989). Submitted for publication.

Dr. Jens Lorenz
Dept. of Mathematics and Statistics
University of New Mexico
Albuquerque, New Mexico 87131
USA

K W MORTON

Upwinded test functions for finite element and finite volume methods

1 Introduction

Partial differential equations which model simultaneous convection and diffusion, but with the convection dominant, are among the most widely occurring in science and engineering; and they present some of the most severe problems to the numerical analyst. Dispersal of pollutants or other material by fluids, transport of vorticity or turbulence parameters in models of fluid flow, flow through porous media, the spread of disease, are just some of the examples that spring to mind.

We consider here only linear problems of the following form,

$$\frac{\partial u}{\partial t} + \boldsymbol{a} \cdot \nabla u - \epsilon \nabla^2 u = f \quad \text{in } \Omega \subset I\!R^d \tag{1.1a}$$

$$u = g \quad \text{on } \partial\Omega_D, \quad \frac{\partial u}{\partial n} = 0 \quad \text{on } \partial\Omega_N, \tag{1.1b}$$

where $\partial\Omega_D, \partial\Omega_N$ constitute a partition of the boundary $\partial\Omega$ of Ω such that $\text{meas}(\partial\Omega_D) \neq 0$ and $\partial\Omega_D$ includes all inflow points (i.e. those such that the outward normal $\boldsymbol{n}$ satisfies $\boldsymbol{n} \cdot \boldsymbol{a} < 0$). We also assume that the convective velocity field $\boldsymbol{a}$ is incompressible, $\nabla \cdot \boldsymbol{a} = 0$, and that the diffusion coefficient ϵ is a positive constant. We shall either be concerned with the solution of the steady problem, where f and g are independent of the time t and the $\partial u/\partial t$ term is omitted, or the full unsteady problem where initial data $u(\boldsymbol{x}, 0)$ is given at $t = 0$ and the solution sought for $t > 0$. We shall show that close links exist between the theoretical foundations of upwinding in the two cases.

In these two cases, upwinding is by far the most widely used technique to overcome the disadvantages of central discretisation methods when convection becomes dominant. Looking across finite difference, finite element and finite volume methods one may distinguish three fairly distinct themes and objectives:

(i) upwinded modifications of the discrete equations for steady convection-diffusion problems, in order to obtain both better accuracy and a better conditioned system of equations to solve;

(ii) upwinded discretisations for unsteady hyperbolic problems in order to ensure desired properties of the approximation, such as monotonicity preservation, the TVD property etc.;

(iii) upwinding to design the iteration procedures and boundary conditions for a given discrete system approximating the steady problem.

We shall address each of these in the following.

2 Steady convection-diffusion problems

Let us consider the steady problem first. Finite element methods are based on the weak form obtained from multiplying (1.1a) by a test function $v \in H^1(\Omega)$ satisfying the homogeneous Dirichlet boundary conditions: find $u \in H^1_E$ such that

$$B(u, v) = \langle f, v \rangle \quad \forall v \in H^1_{E_0} \tag{2.1}$$

where

$$B(u, v) := \int_\Omega [\epsilon \nabla u \cdot \nabla v + (\boldsymbol{a} \cdot \nabla u)v] d\Omega \tag{2.2a}$$

$$H^1_E := \{w \in H^1(\Omega) | w = g \text{ on } \partial\Omega_D\} \tag{2.2b}$$

$$H^1_{E_0} := \{w \in H^1(\Omega) | w = 0 \text{ on } \partial\Omega_D\}, \tag{2.2c}$$

and $\langle \cdot, \cdot \rangle$ denotes the L_2 spatial inner product. Our assumptions on $\boldsymbol{a}$ and $\partial\Omega_D$ ensure that $B(\cdot, \cdot)$ is a coercive form on $H^1_{E_0}(\Omega) \times H^1_{E_0}(\Omega)$ and hence, by the Lax-Milgram Lemma, (2.1) has a unique solution for all $f \in H^{-1}(\Omega)$ and $g \in H^{\frac{1}{2}}(\partial\Omega_D)$.

2.1 Petrov-Galerkin methods

For a conforming Petrov-Galerkin approximation we need a trial space $S^h_E \subset H^1_E$ and a test space $T^h_0 \subset H^1_{E_0}$; then the approximation $U \in S^h_E$ is given by

$$B(U, V) = \langle f, V \rangle \quad \forall V \in T^h_0. \tag{2.3}$$

Exploiting the conforming property, $T^h_0 \subset H^1_{E_0}$, enables (2.1) to be combined with (2.3) to give the error projection property

$$B(u - U, V) = 0 \quad \forall V \in T^h_0. \tag{2.4}$$

The numerical problem is to choose these trial and test spaces, especially the latter, so as to give a well-conditioned system of discrete equations and an accurate numerical approximation with a sharp error bound.

For the self-adjoint, pure diffusion problem it is appropriate to take the test space identical to the trial space to get the Galerkin approximation, $U^G \in S^h_E$ such that

$$B(U^G, V) = \langle f, V \rangle \quad \forall V \in S^h_0; \tag{2.5}$$

for then the bilinear form reduces to the symmetric form

$$B_1(u, v) := \int_\Omega (\nabla u \cdot \nabla v) d\Omega \tag{2.6}$$

and (2.4) yields the optimal approximation property

$$\|u - U^G\|_{B_1} \le \|u - W\|_{B_1} \quad \forall W \in S_E^h. \tag{2.7}$$

However, when convection is introduced the optimality of the Galerkin approximation is lost. This loss can be characterised by the introduction of a constant multiplying the right-hand side of (2.7), which from the general Lax-Milgram argument can be taken as the ratio of the constants in the upper and lower bounds to the bilinear form. By more careful argument, and using the norm $\|\cdot\|_{B_1}$, one can show [12, 14] that it is of the form $1 + K\beta$, where β is the mesh Péclet number

$$\beta := \frac{|a|h}{\epsilon}. \tag{2.8}$$

Thus in many practical cases the bound becomes completely worthless and, in fact, the approximation may be worthless.

It is worth noting that for pure diffusion in one dimension the optimality of the Galerkin approximation implies that a piecewise linear approximation is exact at the mesh points. Indeed, it follows from (2.4) that in this case any Petrov-Galerkin approximation for which the test space includes the continuous piecewise linears has this property. Now consider the opposite extreme of pure convection $au' = f$: from

$$\int a(u' - U')V \, \mathrm{d}x = 0 \quad \forall V \in T_0^h \tag{2.9}$$

it is clear that any test space which includes the characteristic functions for each mesh interval yields exact values at the mesh points. This is an important motivation for finite volume methods. In particular, suppose $a > 0$ so that on the interval $(0, 1)$ the solution value $u(0) = g_0$ is given. Then on the mesh

$$0 = x_0 < x_1 < x_2 \ldots < x_J = 1, \tag{2.10}$$

application of (2.9) gives the values of an approximation $U(x)$ at successive mesh points through the equations

$$a[U(x_j) - U(x_{j-1})] = \int_{x_{j-1}}^{x_j} f(x) \, \mathrm{d}x, \quad j = 1, 2 \ldots, J. \tag{2.11}$$

These are clearly exact; and in the sense that $U(x_j)$ is determined by the test function $\chi_{(x_{j-1}, x_j)}$, the characteristic function of the mesh interval (x_{j-1}, x_j), it uses a fully upwinded test function to achieve this property.

2.2   Optimal and near-optimal test functions

In this one dimensional case these are just the two extremes of the one-parameter family of Hemker test functions which give exact mesh point values for all mesh Péclet numbers

[7]. More generally, as introduced by Barrett and Morton [2], suppose that $B_m(\cdot, \cdot)$ is any symmetric form, continuous and coercive on $H^1_{E_0}(\Omega) \times H^1_{E_0}(\Omega)$. Then we can introduce, by the Riesz Representation Theorem, an invertible operator R_m for which

$$B(u, v) = B_m(u, R_m v) \quad \forall u, v \in H^1_{E_0} \tag{2.12a}$$

and

$$B(u, R_m^{-1} v) = B_m(u, v) \quad \forall u, v \in H^1_{E_0}; \tag{2.12b}$$

the inverse R_m^{-1} can be regarded as the symmetrizer of $B(\cdot, \cdot)$. In terms of $B_m(\cdot, \cdot)$ the error approximation property (2.4) of a conforming Petrov-Galerkin method can be written

$$B_m(u - U, R_m V) = 0 \quad \forall V \in T^h_0. \tag{2.13}$$

Hence if $R_m T^h_0 = S^h_0$ the solution U is optimal in the norm $\|\cdot\|_{B_m}$ induced by $B_m(\cdot, \cdot)$. More generally we have a theorem, first given by Morton [12], which relates the optimality of U to the approximation power of $R_m T^h_0$ in the space S^h_0. A similar theorem has been given by D'yakonov [6].

Theorem *Suppose T^h_0 and S^h_0 have the same dimension and $\exists$ a constant $\Delta_m \in [0, 1)$ such that*

$$\inf_{V \in T^h_0} \|W - R_m V\|_{B_m} \le \Delta_m \|W\|_{B_m} \quad \forall W \in S^h_0. \tag{2.14}$$

Then a unique solution U to (2.3) exists and satisfies

$$\|u - U\|_{B_m} \le (1 - \Delta_m^2)^{-\frac{1}{2}} \inf_{W \in S^h_E} \|u - W\|_{B_m}. \tag{2.15}$$

Proof See [2]. □

The best known case is that referred to already in which the symmetric part of $B(\cdot, \cdot)$, given by (2.6), is used as the bilinear form in one dimension. Then the test space in each element or mesh interval is spanned by the pair of functions, in the local co-ordinate $s := (x - x_{j-1})/(x_j - x_{j-1}) \in [0, 1]$,

$$N_1(s) := (1 - e^{-\beta_j s})/(1 - e^{-\beta_j}) \tag{2.16a}$$

$$N_2(s) := (e^{-\beta_j s} - e^{-\beta_j})/(1 - e^{-\beta_j}); \tag{2.16b}$$

here $\beta_j := a(x_j - x_{j-1})/\epsilon$ is the local mesh Péclet number. Clearly $N_1(s) + N_2(s) \equiv 1$, so that compared with (2.11) we see that the element integral is split into two parts

$$\int_0^1 \left\{ \frac{dU}{dx} \left[\epsilon \frac{dN_\alpha(s)}{ds} \cdot \frac{1}{x_j - x_{j-1}} + a N_\alpha(s) \right] - f(x) N_\alpha(s) \right\} (x_j - x_{j-1}) ds, \tag{2.17}$$

with $\alpha = 1$ contributing to the equation associated with node x_{j-1} and $\alpha = 2$ to that at x_j. The gradual increase of the proportion to x_j as β_j increases is the essence of upwinding. In Fig.1 we show these test functions for $\beta = 2$ and $\beta = 10$; for comparison we also show the conforming test functions used in [5], where the trial function is modified by an upwinded quadratic term; included too is the nonconforming test function given by the streamline diffusion technique of [8] and [9], which is of the form $\phi_j + \gamma\phi'_j$ where ϕ_j is the trial space basis function. In [2] and [17] the calculated values of the optimality constant in (2.15) are tabulated for the latter schemes. They remain bounded and close to unity as $\beta \to \infty$, showing how effective the upwinding technique is in one dimension. The streamline diffusion method, as its name implies, was designed for use in multi-dimensions, where it is widely used. Exponential fitting, which is closely related to the use of Hemker test functions, was also devised by Allen and Southwell [1] for two-dimensional fluid flow problems. The theoretical basis for the methods is less well established in multi-dimensions, however. In [14] it is shown how the optimality constant can be calculated for a conforming method without having an explicit representation of the operator R_1 or its inverse: but results for the quadratic modification of the trial space show that it is far from optimal in the $B_1(\cdot, \cdot)$ induced norm for a well-known two-dimensional problem. Numerical results in [17] when compared with those in [10] bear this out.

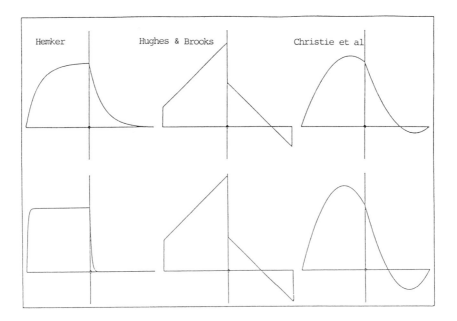

Figure 1: Test functions for convection-diffusion with $\beta = 2$ (top) and $\beta = 10$ (bottom).

2.3 Moore's upwinded control volumes

Let us turn now to finite volume methods. Suppose a piecewise bilinear trial space is used on a structured quadrilateral mesh. (Strictly speaking, this is bilinear in the local coordinates after an isoparametric coordinate transformation on each element.) For pure diffusion, the test space would normally consist of the characteristic functions for the dual quadrilateral mesh. That is, the equation generated for the (i, j) node would be obtained by integrating $\nabla^2 u - f$ over the dual quadrilateral $\Omega_{i,j}$ whose vertices are the centroids of the four primary quadrilaterals which meet at the (i, j) node. Gauss' theorem is used to convert the $\nabla^2 u$ integral into a line integral, giving the equations for the approximation U

$$\int_{\partial \Omega_{i,j}} \frac{\partial U}{\partial n} \mathrm{d}l = \int_{\Omega_{i,j}} f \mathrm{d}\Omega. \tag{2.18}$$

In practice and more precisely, the integrals are carried out element-by-element on the primary quadrilaterals in the local coordinate system, that is for $(\xi, \eta) \in [-1, 1] \times [-1, 1]$. In effect, then, each of these integrals is divided into four parts along the lines $\xi = 0$ and $\eta = 0$; and the contribution from each quarter is allocated to the equation associated with its vertex. The point $\xi = 0, \eta = 0$ is called the divide point for the element. Note that in the line integral form (2.18) only the fluxes along $\xi = 0$ and $\eta = 0$ need to be computed in each element.

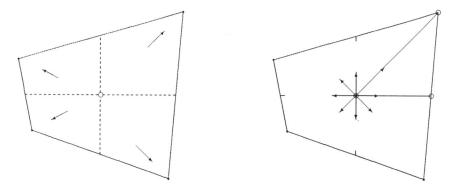

Figure 2: The divide point used in the Moores' method: on the left, allocation of element sub-integrals to vertices; on the right, possible divide point movements.

We are now in a position to introduce the upwinding used by Joan and John Moore [11] when convection is present, see also [3]. If the mesh Péclet number, averaged for the element, is greater than 2 the divide point is moved in one of eight upwind directions according to the direction of the convective velocity $\mathbf{a}$. The eight final positions correspond to the four vertices and four edge midpoints — see Fig.2; various algorithms have been used to decide between these. For pure convection which is uniform on a rectangular mesh, when the divide points move to a vertex one obtains the box difference scheme, or cell vertex finite volume scheme; and when they move to an edge midpoint one obtains a Crank-Nicolson scheme, with either a vertical or horizontal orientation. Care needs to be taken with the evaluation of the diffusion terms when the divide point moves onto the element boundary; but if the integrals are interpreted as the limit when $\beta \to \infty$ of those obtained using tensor product Hemker test functions, as in [3], the procedures can be properly justified.

This is a very useful upwinding idea that, like the Petrov-Galerkin methods described above, may be directed towards the first objective outlined in the Introduction. However, it is also used in [13] and [10] to derive appropriate equations and solution procedures for the cell vertex method. This latter method, as in the simple example (2.11), gives a fully centred discretisation scheme, with the centring based on the cell or element; but needs an upwinding to relate equations to unknowns in order to derive a well-conditioned system of equations for a well-behaved accurate solution. Thus the upwinding of control volumes can also have objectives of the type (iii) described in the Introduction.

3 Unsteady convection-diffusion problems

3.1 Petrov-Galerkin methods

Let us now consider unsteady problems. For pure diffusion there is no need to look beyond the Galerkin method: in the semi-discrete form we have

$$\langle \frac{\partial U}{\partial t}, V \rangle + \epsilon B_1(U, V) = \langle f, V \rangle \quad \forall V \in S_0^h, \tag{3.1}$$

yielding a system of ordinary differential equations for the parameters $\{U_j(t)\}$ in an expansion

$$U(\boldsymbol{x}, t) = \sum U_j(t) \phi_j(\boldsymbol{x}). \tag{3.2}$$

Such a method of lines approach, with the ODEs solved by a library package, is very popular and effective; and the stability and error analysis using the elliptic projection $B_1(\cdot, \cdot)$ is quite straightforward.

However, once convection becomes important this approach runs into difficulties which are similar to those occurring in the steady case. It is better then to discretise in time first and as in the steady case seek improvement through the Petrov-Galerkin formulation. Using the simplest forward difference we obtain, writing U_j^n for $U_j(t_n)$,

$$\langle \frac{U^{n+1} - U^n}{\Delta t}, V \rangle + B(U^n, V) = \langle f, V \rangle \quad \forall V \in T_0^h. \tag{3.3}$$

For the pure convection case in one dimension, Morton and Parrott [15] devised test functions for a number of time stepping schemes that restored the "unit CFL condition", i.e. gave stability for $|a|\Delta t = \Delta x$, and much improved the accuracy. For example, with a piecewise linear trial space the Euler scheme (3.3) is stable only for $\Delta t = O((\Delta x)^2)$ if the Galerkin method is used; but discontinuous, upwind test functions restore stability for $|a|\Delta t \le \Delta x$; and give second order or even third order accuracy. Similarly, for the leap-frog difference scheme, discontinuous but centred test functions extend the stability range from $3(a\Delta t)^2 \le (\Delta x)^2$ to $|a|\Delta t \le \Delta x$ and fourth order accuracy can be achieved.

3.2 Evolution-Galerkin methods

However, experience with two dimensional problems suggested that more explicit use should be made of the characteristics of a hyperbolic problem in carrying out the upwinding. Suppose in general we have an equation

$$\frac{\partial u}{\partial t} + Lu = 0, \tag{3.4}$$

where the operator L contains all the spatial derivatives, and that this generates an evolution operator

$$E(\Delta t)u(t) \mapsto u(t + \Delta t). \tag{3.5}$$

Let P be a projector onto the trial space S_E^h and E_Δ an approximation to $E(\Delta t)$. Then we can generate a sequence $\{U^n \in S_E^h, \; n = 1, 2, \ldots\}$ from initial data U^0 by

$$U^{n+1} = PE_\Delta U^n. \tag{3.6}$$

For the pure convection problem, given by (1.1) with $\epsilon = 0 = f$, many authors working in different fields of application have made use of a characteristic Galerkin method. Introducing characteristics given by

$$\frac{\mathrm{d}X(x, s; t)}{\mathrm{d}t} = a(X(x, s; t)) \tag{3.7}$$

and using the notation

$$X(x, t_n; t_n) = x, \quad X(x, t_n; t_{n+1}) = y, \tag{3.8}$$

then the evolution of the exact solution through one time step is given by

$$u(y, t_{n+1}) = u(x, t_n). \tag{3.9}$$

Then a direct formulation of the characteristic Galerkin method, using the L_2 projection, gives

$$\langle U^{n+1}, \phi_i \rangle = \int U^n(x)\phi_i(y)\mathrm{d}y. \tag{3.10}$$

Note that an ODE-solver will be needed to approximate (3.7) and a quadrature for evaluating the right-hand side of (3.10). As shown in [16] care must be taken with the latter choice and a variety of schemes have resulted. For our theme of upwinded test functions, it is most appropriate to consider the ECG formulation described in [4] and earlier papers. Using the Euler approximation for (3.7) in one dimension

$$y = x + a(x, t_n)\Delta t, \tag{3.11}$$

an update form of (3.10) can be written

$$\langle U^{n+1} - U^n, \phi_i \rangle = \int U^n(x) \left[\phi_i(y)dy - \phi_i(x)dx \right]. \tag{3.12}$$

Define the test function

$$\Phi_i^n(x) := \frac{1}{a(x, t_n)\Delta t} \int_x^y \phi_i(z)dz, \tag{3.13}$$

with y given by (3.11), so that $d[a\Phi_i^n] = \phi_i(y)dy - \phi_i(x)dx$. Then, after integration by parts and ignoring possible boundary terms, one obtains

$$\langle \frac{U^{n+1} - U^n}{\Delta t}, \phi_i \rangle + \langle a\frac{dU^n}{dx}, \Phi_i^n \rangle = 0. \tag{3.14}$$

Clearly Φ_i^n is an upwind-averaged trial basis function and with piecewise linears this scheme is third order accurate and unconditionally stable. Fig.3 shows Φ_i^n for piecewise constants and piecewise linears.

The question is prompted — is there a relation between this upwinded test function and that given by the $R_m^{-1}\phi_i$ used in the steady case? We resolve this question by comparing the defining identities in the two cases.

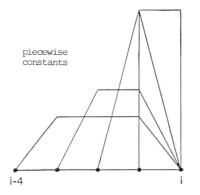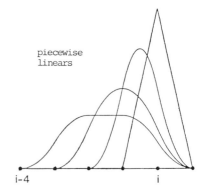

Figure 3: Characteristic Galerkin test functions Φ_i^n for CFL numbers 0,1,2 and 3.

3.3 Comparison with steady case

To compare with (2.12) we need to convert (3.14) into an identity which relates the evolution operator E_Δ or the differential operator L to the mapping from ϕ_i to Φ_i^n. Thus for the homogeneous convection-diffusion equation written in the form (3.4) on the whole real line, we generalise (3.14) to

$$\langle \frac{1}{\Delta t}(E_\Delta - I)u, \phi_i \rangle_\gamma + \langle Lu, \Phi_i \rangle_\gamma = 0 \quad \forall \phi_i \in S_h, u \in H^2, \tag{3.15}$$

where we have also generalised the projection to the mixed norm

$$\langle u, v \rangle_\gamma := \int [uv + \gamma^2 u'v']dx. \tag{3.16}$$

Now let E_Δ^* be the γ-norm adjoint of E_Δ and introduce the function χ_i and the operator Q such that

$$\chi_i := \frac{1}{\Delta t}(E_\Delta^* - I)\phi_i , \quad \Phi_i =: -Q\chi_i. \tag{3.17}$$

Substitution into (3.15) gives

$$\langle u, \chi_i \rangle_\gamma = \langle Lu, Q\chi_i \rangle_\gamma \quad \forall \phi_i \in S_h, u \in H^2. \tag{3.18}$$

For the pure convection of (3.14) and the L_2 projection, it is clear that

$$(Q\chi_i)(x) = \frac{1}{a} \int_x^\infty \chi_i(\eta)d\eta = -\frac{1}{a\Delta t} \int_x^{x+a\Delta t} \phi_i(\eta)d\eta \tag{3.19}$$

so that (3.18) specialises to (3.14) when v is taken as U^n. For the general convection-diffusion case we generalise (3.19) to

$$a(Q\chi_i) + \epsilon(Q\chi_i)' = \int_x^\infty \chi_i(\eta)d\eta. \tag{3.20}$$

Then, with the special choice $\gamma = \epsilon/a$, factorisation of the operator $(a^2 - \epsilon^2 d^2/dx^2)$ gives

$$\begin{aligned}
\langle Lu, Q\chi_i \rangle_{\epsilon/a} &= \langle Lu, \frac{1}{a^2}(a - \epsilon\frac{d}{dx}) \int_.^\infty \chi_i d\eta \rangle \\
&= \langle Lu, \frac{1}{a}\left[\frac{\epsilon}{a}\chi_i + \int_.^\infty \chi_i d\eta\right] \rangle,
\end{aligned} \tag{3.21}$$

which defines the operator Q in the identity (3.18).

Now for the steady problem, the choice $\gamma = \epsilon/a$ corresponds to the inner product $B_2(\cdot, \cdot)$ used by Barrett and Morton [2]. On the interval $(0,1)$ with Dirichlet boundary conditions at each end, the identity (2.12b) therefore becomes

$$\begin{aligned}
\langle u, v \rangle_{\epsilon/a} &= B(u, R_2^{-1}v) \\
&= \langle Lu, R_2^{-1}v \rangle \quad \forall u \in H_{E_0}^1 \cap H^2, v \in H_{E_0}^1
\end{aligned} \tag{3.22}$$

written in the form of (3.18). In [2] R_2^{-1} is given as

$$R_2^{-1}v = \frac{1}{a}\left[\frac{\epsilon}{a}v + \int_x^1 (v - \frac{\bar{v}e^{-a/\epsilon}}{(\epsilon/a)(1 - e^{-a/\epsilon})})\mathrm{d}\eta\right] \tag{3.23}$$

which, apart from the term in the average $\bar{v}$ of v which is required to satisfy the boundary conditions, is identical to the operator on the right-hand side of (3.21).

Thus for this particular choice of the norm used in the projections, the upwinded test functions used in these two cases are the same, bearing in mind the differences in the domain and boundary conditions.

3.4   Use of a convolution integral

The approximation scheme which is implicitly defined by (3.15)–(3.20) has been given explicit form by Morton and Sobey [18], as follows. In one dimension and on the whole real line the evolution operator for the convection-diffusion equation can be written as a convolution integral:

$$u(x, t + \Delta t) = \int_{-\infty}^\infty u(\eta, t)G(x - \eta; \Delta t)\mathrm{d}\eta \tag{3.24}$$

where

$$G(z; \tau) = \frac{1}{\sqrt{(\epsilon\tau\pi)}}e^{-(z-a\tau)^2/4\epsilon\tau}. \tag{3.25}$$

Finite difference approximations, such as Lax-Wendroff and Leonard's QUICKEST, can be derived from this by substituting a local interpolating polynomial in the integral.

More to the point for the present paper, finite element approximations of the evolution-Galerkin type which generalise (3.10) can be written as

$$\langle U^{n+1}, \phi_i\rangle_\gamma = \int_{-\infty}^\infty \int_{-\infty}^\infty [\phi_i(x)U^n(\eta) + \gamma^2\phi_i'(x)U^{n'}(\eta)]G(x - \eta; \Delta t)\mathrm{d}\eta\mathrm{d}x, \tag{3.26}$$

where we have used the mixed norm for the projection. On a uniform mesh and with a piecewise linear basis $\{\phi_i\}$, one of these integrals can be readily carried out; with $x = s\Delta x$ and writing $\phi_j(x)$ as the second order B-spline $\chi^{(2)}(s - j)$, we can use the convolution property of splines to get

$$\langle E(\Delta t)\phi_j, \phi_i\rangle_\gamma = \int_{-\infty}^\infty [(\Delta x)^2\chi^{(4)}(s + j - i) - \gamma^2\delta^2\chi^{(2)}(s + j - i)]G(s\Delta x; \Delta t)\mathrm{d}s \tag{3.27}$$

where δ^2 is the usual second order central difference operator. Multiplying this by U_j^n and summing over j gives (3.26). Unfortunately, the remaining integrals cannot be expressed in closed form. However, suppose that the characteristic $\mathrm{d}x/\mathrm{d}t = a$ drawn down from (x_i, t_{n+1}) to level t_n falls in the $(p+1)^{th}$ interval to the left of x_i; that is, the CFL number has the form

$$\frac{a\Delta t}{\Delta x} = p + \hat{\nu} \tag{3.28}$$

where $p \in \mathbb{Z}$ and $\hat{\nu} \in [0, 1)$. Then the sum over j of the square bracket in (3.27) multiplied by U_j^n has a cubic form in this interval, and let us suppose that this form holds on the whole real line. Then the corresponding integral can be evaluated and the result yields an approximation which is valid for small values of $\epsilon \Delta t / (\Delta x)^2 =: \mu$. Written as a difference scheme it has the form

$$[1 + (\tfrac{1}{6} - (\tfrac{\gamma}{\Delta x})^2)\delta^2](U_i^{n+1} - U_{i-p}^n)$$

$$= [-\hat{\nu}\Delta_0 + (\tfrac{1}{2}\hat{\nu}^2 + \mu)\delta^2 + \hat{\nu}((\tfrac{\gamma}{\Delta x})^2 - \tfrac{1}{6}\hat{\nu}^2 - \mu)\delta^2\Delta_-]U_{i-p}^n. \quad (3.29)$$

With $p = 0$ and $\gamma^2 = \tfrac{1}{6}(\Delta x)^2$ it reduces to the QUICKEST scheme, having the identity as its mass matrix. When $\mu = 0$ and $\gamma = 0$ it reduces to the ECG scheme obtained from (3.14). The presence of the backward difference Δ_- in the last term shows the upwind nature of the approximation even for $p = 0$.

4   Conclusions

From the foregoing it is difficult to avoid the conclusion that some form of upwinding is needed to handle convection well.

For unsteady problems this arises naturally by making use of the Lagrangian derivative, or the characteristic of the hyperbolic problem.

For steady problems we have shown a direct connection between this upwinding in the unsteady case and the use of Petrov-Galerkin methods generated by symmetrising the operator, thus giving further credence to these methods.

Even when highly accurate, well behaved, centre schemes are devised for the steady problem, such as the cell vertex scheme, we have seen that some form of upwinding comes into play in setting up and solving the discrete equations.

References

[1] D. Allen and R. Southwell. Relaxation methods applied to determining the motion, in two dimensions, of a viscous fluid past a fixed cylinder. *Quarterly Journal of Mechanics and Applied Mathematics*, 8:129–145, 1955.

[2] J.W. Barrett and K.W. Morton. Approximate symmetrization and Petrov-Galerkin methods for diffusion-convection problems. *Computer Methods in Applied Mechanics and Engineering*, 45:97–122, 1984.

[3] D.J.A. Benson, K.W. Morton, and T. Murdoch. On Moore's upwinded control volume method for convection dominated flows. In preparation, 1991.

[4] P.N. Childs and K.W. Morton. Characteristic Galerkin methods for scalar conservation laws in one dimension. *SIAM Journal of Numerical Analysis*, 27:553–594, 1990.

[5] I. Christie, D.F. Griffiths, A.R. Mitchell, and O.C. Zienkiewicz. Finite element methods for second order differential equations with significant first derivatives. *International Journal for Numerical Methods in Engineering*, 10:1389–1396, 1976.

[6] E.G. D'yakonov. *Minimization of Computational Work. Asymptotically Optimal Algorithms for Elliptic Problems*. Hayka, Moscow, 1989.

[7] P.W. Hemker. *A numerical study of stiff two-point boundary problems*. PhD thesis, Mathematisch Centrum, Amsterdam, 1977.

[8] T.J.R. Hughes and A. Brooks. A theoretical framework for Petrov-Galerkin methods with discontinuous weighting functions: application to the streamline-upwind procedure. In R.H. Gallagher, D.H. Norrie, J.T. Oden, and O.C. Zienkiewicz, editors, *Finite Elements in Fluids*, volume 4, pages 47–65. Wiley, New York, 1982.

[9] C. Johnson and U. Nävert. An analysis of some finite element methods for advection-diffusion problems. In O. Axelsson, L.S. Frank, and A. van der Sluis, editors, *Conference on Analytical and Numerical Approaches to Asymptotic Problems in Analysis*. North-Holland, Amsterdam, 1981.

[10] J.A. Mackenzie and K.W. Morton. Finite volume solutions of convection-diffusion test problems. Technical Report NA90/1, Oxford University Computing Laboratory, 11 Keble Road, Oxford, OX1 3QD, 1990. Submitted for publication.

[11] J. Moore and J. Moore. Calculation of horseshoe vortex flow without numerical mixing. Technical Report JM/83-11, Virginia Polytechnic Inst. and State University, Blacksburg, Virginia 24061, 1983. Prepared for presentation at the 1984 Gas Turbine Conference, Amsterdam.

[12] K.W. Morton. Finite element methods for non-self-adjoint problems. In P.R. Turner, editor, *Topics in Numerical Analysis Proceedings of the SERC Summer School, Lancaster, 1981*, volume 965 of *Lecture Notes in Mathematics*, pages 113–148. Springer-Verlag, 1982.

[13] K.W. Morton. Finite volume methods and their analysis. Technical Report NA90/11, Oxford University Computing Laboratory, 11 Keble Road, Oxford, OX1 3QD, 1990. To appear in the Proceedings of the Conference of The Mathematics of Finite Elements and Applications VII MAFELAP 1989.

[14] K.W. Morton, T. Murdoch, and E. Süli. Optimal error estimation for Petrov-Galerkin methods in two dimensions. Technical Report NA90/22, Oxford University Computing Laboratory, 11 Keble Road, Oxford, OX1 3QD, 1990. Submitted for publication.

[15] K.W. Morton and A.K. Parrott. Generalised Galerkin methods for first-order hyperbolic equations. *Journal of Computational Physics*, 36:249–270, 1980.

[16] K.W. Morton, A. Priestley, and E. Süli. Stability analysis of the Lagrange-Galerkin method with non-exact integration. *Mathematical Modelling and Numerical Analysis*, 22:625–653, 1988.

[17] K.W. Morton and B.W. Scotney. Petrov-Galerkin methods and diffusion-convection problems in 2D. In J.R. Whiteman, editor, *Proceedings of the Conference of The Mathematics of Finite Elements and Applications V MAFELAP 1984*, pages 343–366. Academic Press, 1985.

[18] K.W. Morton and I.J. Sobey. Discretisation of a convection-diffusion equation. Technical Report NA91/4, Oxford University Computing Laboratory, 11 Keble Road, Oxford, OX1 3QD, 1991.

M J D POWELL

The complexity of Karmarkar's algorithm for linear programming

Abstract   Karmarkar's original algorithm for linear programming is described in the case when the inequality constraints on the variables are general. We analyse the number of iterations that are required to achieve a prescribed accuracy in the final value of the objective function without introducing slack variables that convert the inequality constraints to nonnegativity conditions on the variables. A similar analysis is presented for a log barrier version of Karmarkar's algorithm. Thus, under some assumptions that include boundedness of the feasible region and the existence of a suitable starting point, we find that the numbers of iterations of these methods are bounded above by constant multiples of m and $\sqrt{m} \log m$ respectively, where m is the number of inequality constraints. Although these bounds are known already, our theory includes some new features such as the use of a vector norm that removes the need for projective transformations of the variables. A numerical example is considered that suggests that the $\mathcal{O}(m)$ bound on the complexity of the original algorithm cannot be improved.

1. Introduction

Karmarkar's algorithm for linear programming [4] has received much attention which was due initially to the excellent efficiency that was claimed. Further, the algorithm includes several brilliant techniques that provide some highly interesting theoretical properties. Usually these features are presented and analysed in the case when the only inequality constraints on the variables are that they shall be nonnegative and general linear equality constraints are allowed too. However, we will work with general inequality constraints directly, which has the advantage of avoiding the need for projection operators whose purpose is to preserve equality conditions. Thus we derive the main complexity property of the algorithm in a way that seems to be simpler than the derivations that are given elsewhere.

The most general form of a linear programming problem is to calculate the least value of a linear function of several variables subject to linear constraints on the values of the variables, where these constraints can be a mixture of equalities and inequalities. We let the objective function be the expression

$$c^T x, \quad x \in \mathcal{R}^n, \tag{1.1}$$

where c is a nonzero constant vector in $\mathcal{R}^n$ and where x is the vector of variables. Further, in our analysis of the original algorithm we assume that the constraints have the form

$$\left.\begin{aligned} \mathbf{a}_0^T\mathbf{x} &= 1 \\ \mathbf{a}_j^T\mathbf{x} &\geq 0, \quad j=1,2,\ldots,m \end{aligned}\right\}, \tag{1.2}$$

the vectors $\{\mathbf{a}_j : j=0,1,\ldots,m\}$ being data. This assumption does not lose generality because any given equality conditions can be used to eliminate variables, so the remaining constraints are the inequalities

$$\hat{\mathbf{a}}_j^T\hat{\mathbf{x}} \geq b_j, \quad j=1,2,\ldots,m, \tag{1.3}$$

where $\hat{\mathbf{x}}$ is the vector of variables after the eliminations. Further, expression (1.3) is equivalent to the conditions

$$\left.\begin{aligned} \xi &= 1 \\ \hat{\mathbf{a}}_j^T\hat{\mathbf{x}} - b_j\xi &\geq 0, \quad j=1,2,\ldots,m \end{aligned}\right\}. \tag{1.4}$$

Therefore, by letting n be one more than the number of components of $\hat{\mathbf{x}}$, by regarding ξ as a new variable, by using the notation

$$\mathbf{x} = \begin{pmatrix} \hat{\mathbf{x}} \\ \cdots \\ \xi \end{pmatrix} \quad \text{and} \quad \mathbf{a}_j = \begin{pmatrix} \hat{\mathbf{a}}_j \\ \cdots \\ -b_j \end{pmatrix}, \quad j=1,2,\ldots,m, \tag{1.5}$$

and by defining $\mathbf{a}_0$ to be the last coordinate vector in $\mathcal{R}^n$, we can express the constraints (1.3) in the form (1.2). An advantage of this direct treatment of inequalities is that n is at most one more than the number of variables of the original general problem. And if the original problem includes any nonnegativity conditions on variables that have not been eliminated, then the corresponding vectors $\mathbf{a}_j$ of expression (1.2) are also coordinate vectors.

We say that a vector $\mathbf{x}$ is "feasible" if it satisfies the constraints (1.2) and we let $\mathcal{S}$ be the open convex set

$$\mathcal{S} = \{\mathbf{x} : \mathbf{a}_j^T\mathbf{x} > 0, \ j=1,2,\ldots,m\}. \tag{1.6}$$

We require three assumptions that lose a little generality, namely that feasible vectors are bounded, that the constraint gradients $\{\mathbf{a}_j : j=1,2,\ldots,m\}$ span $\mathcal{R}^n$, and that we are given a starting point $\mathbf{x}_1$ for our calculation that is not only feasible but also in $\mathcal{S}$. It is important to note that these assumptions imply that $\mathbf{a}_0^T\mathbf{x}$ is positive for every nonzero $\mathbf{x}$ in the closure of $\mathcal{S}$, namely $\bar{\mathcal{S}}$. Indeed, if $\mathbf{a}_0^T\mathbf{x}$ were zero for such an $\mathbf{x}$, then $\mathbf{x}_1 + \lambda\mathbf{x}$ would be feasible for all positive values of λ, which would contradict the boundedness assumption. Alternatively, if $\mathbf{a}_0^T\mathbf{x}$ were negative for such an $\mathbf{x}$, then the point

$$\mathbf{y} = \mathbf{x} - (\mathbf{a}_0^T\mathbf{x})\mathbf{x}_1 \tag{1.7}$$

would satisfy $\mathbf{a}_0^T\mathbf{y}=0$ and would be nonzero because $\mathbf{x}\in\bar{\mathcal{S}}$ and $\mathbf{x}_1\in\mathcal{S}$ imply $\mathbf{y}\in\mathcal{S}$, so we would have a contradiction as before. Therefore the assertion is true.

143

The final requirement for our analysis of the original algorithm is that at a solution to the linear programming problem the objective function has the value

$$\mathbf{c}^T\mathbf{x}^* = 0, \tag{1.8}$$

where $\mathbf{x}^*$ is any optimal vector of variables. This condition is called the "restrictive assumption" but it is not intolerable. Indeed, it can always be achieved without altering the other features of the calculation by adding a suitable multiple of $\mathbf{a}_0$ to $\mathbf{c}$. Further, Todd and Burrell [7] present a way of determining the multiplying factor automatically, but we do not study their valuable technique.

We use the restrictive assumption in two ways. Firstly, because $\mathbf{c}^T\mathbf{x}^* = 0$, $\mathbf{a}_0^T\mathbf{x}^* = 1$ and $\mathbf{c} \neq 0$ imply that $\mathbf{c}$ is not a multiple of $\mathbf{a}_0$, we deduce from the Kuhn–Tucker conditions at $\mathbf{x}^*$ that every optimal vector of variables is on the boundary of at least one inequality constraint. Therefore no feasible point of $\mathcal{S}$ is optimal, which gives the condition

$$\mathbf{c}^T\mathbf{x} = (\mathbf{a}_0^T\mathbf{x})\,(\mathbf{c}^T\{\mathbf{x}/\mathbf{a}_0^T\mathbf{x}\}) > 0, \quad \mathbf{x} \in \mathcal{S}, \tag{1.9}$$

the vector inside the braces being feasible. Secondly, for every $\mathbf{x}$ in $\mathcal{S}$ at least one of the numbers

$$\mathbf{a}_\ell^T [\sum_{j=1}^{m} \frac{\mathbf{a}_j\mathbf{a}_j^T}{(\mathbf{a}_j^T\mathbf{x})^2}]^{-1}\mathbf{c}, \quad \ell = 1, 2, \ldots, m, \tag{1.10}$$

is nonpositive, the inverse matrix being well-defined because the vectors $\{\mathbf{a}_j : j = 1, 2, \ldots, m\}$ span $\mathcal{R}^n$. Indeed, if all of these numbers were positive we would have the following contradiction. Let $\mathbf{x}^*$ be optimal. Because it is feasible and nonzero the sum

$$\sum_{\ell=1}^{m} \frac{\mathbf{a}_\ell^T\mathbf{x}^*}{(\mathbf{a}_\ell^T\mathbf{x})^2} \mathbf{a}_\ell^T [\sum_{j=1}^{m} \frac{\mathbf{a}_j\mathbf{a}_j^T}{(\mathbf{a}_j^T\mathbf{x})^2}]^{-1}\mathbf{c} \tag{1.11}$$

would be positive. This sum, however, can be expressed in the form

$$\mathbf{x}^{*T} [\sum_{\ell=1}^{m} \frac{\mathbf{a}_\ell\mathbf{a}_\ell^T}{(\mathbf{a}_\ell^T\mathbf{x})^2}]\,[\sum_{j=1}^{m} \frac{\mathbf{a}_j\mathbf{a}_j^T}{(\mathbf{a}_j^T\mathbf{x})^2}]^{-1}\mathbf{c} = \mathbf{x}^{*T}\mathbf{c}, \tag{1.12}$$

which is zero due to the restrictive assumption.

Karmarkar's algorithm proceeds steadily towards a solution of the linear programming problem by making changes to the variables that reduce the "potential function"

$$V(\mathbf{x}) = m \log \mathbf{c}^T\mathbf{x} - \sum_{j=1}^{m} \log \mathbf{a}_j^T\mathbf{x}, \quad \mathbf{x} \in \mathcal{S}. \tag{1.13}$$

This expression is suitable for the minimization of $\mathbf{c}^T\mathbf{x}$ subject to the constraints (1.2) when the given assumptions hold, and it is analogous to the potential function that occurs in the "standard form" of the calculation, which means that general linear equality constraints are allowed but the only inequalities are that the components of $\mathbf{x}$ shall be

144

nonnegative. The equivalence of the two potential functions is shown by Bayer and Lagarias [1] and by Powell [5]. We note four important features of the form (1.13).

Firstly, it follows from expressions (1.6) and (1.9) that $V(\mathbf{x})$ is well-defined for every $\mathbf{x}$ in $\mathcal{S}$. Secondly, for all $\mathbf{x} \in \mathcal{S}$ the homogeneity condition

$$V(\lambda \mathbf{x}) = V(\mathbf{x}), \quad \lambda > 0, \tag{1.14}$$

is satisfied and we shall see that this property avoids the need for $V(\cdot)$ to depend on $\mathbf{a}_0$. Thirdly, if $\mathbf{x}$ tends to a point on the boundary of $\mathcal{S}$ at which $\mathbf{c}^T\mathbf{x}$ is positive, then $V(\mathbf{x})$ tends to $+\infty$. Fourthly, if a sequence of values of $\mathbf{x}$ is confined to a straight line segment in $\mathcal{S}$ and if it tends to a solution $\mathbf{x}^*$ of the linear programming problem, then $V(\mathbf{x})$ tends to $-\infty$. This assertion can be deduced from equation (1.8) and from the fact that $\mathbf{x}^* \neq 0$ implies that $\mathbf{x}^*$ is on the boundaries of at most $m-1$ inequality constraints.

The algorithm is iterative, the k-th iteration calculating $\mathbf{x}_{k+1}$ from $\mathbf{x}_k$. The iterations continue until a vector of variables is found that satisfies the inequality

$$V(\mathbf{x}) \leq -\Delta, \tag{1.15}$$

where Δ is a suitably large constant. Each new vector of variables has the form

$$\mathbf{x}_{k+1} = \mathbf{x}_k + \alpha_k \mathbf{d}_k, \tag{1.16}$$

where $\mathbf{x}_k$ is in $\mathcal{S}$, where $\mathbf{d}_k$ is a search direction that has the descent property

$$\mathbf{d}_k^T \nabla V(\mathbf{x}_k) < 0, \tag{1.17}$$

and where α_k is a positive steplength that provides the reduction

$$V(\mathbf{x}_{k+1}) < V(\mathbf{x}_k) \tag{1.18}$$

in the potential function. It follows from the third and fourth properties of the previous paragraph that, if condition (1.15) does not hold at $\mathbf{x}_{k+1}$, then $\mathbf{x}_{k+1}$ is in $\mathcal{S}$, even if α_k is set to the steplength that minimizes the resultant value of $V(\mathbf{x}_{k+1})$. The details of suitable choices of $\mathbf{d}_k$ and α_k are given in Section 2 with the important result that they cause the difference $V(\mathbf{x}_k) - V(\mathbf{x}_{k+1})$ to be bounded below by a positive constant, ω say. Therefore, if at least one iteration is performed, then the final iteration number satisfies the condition

$$k \leq [V(\mathbf{x}_1) + \Delta]/\omega. \tag{1.19}$$

We let $\mathbf{x}_{k+1}$ be the final vector of variables that is given by this calculation. Because it is a nonzero element of the closure of $\mathcal{S}$, we know that $\mathbf{a}_0^T\mathbf{x}_{k+1}$ is positive. Hence the vector $\mathbf{x} = \mathbf{x}_{k+1}/\mathbf{a}_0^T\mathbf{x}_{k+1}$ is feasible and, due to equation (1.14), it also satisfies inequality (1.15). We now consider the choice of Δ.

We take the point of view that, if $\mathbf{x}$ is feasible and if $\mathbf{c}^T\mathbf{x}$ is very close to its optimal value $\mathbf{c}^T\mathbf{x}^* = 0$, then either $\mathbf{x}$ is acceptable as an approximation to $\mathbf{x}^*$ or a solution of

145

the linear programming problem can be found by a refinement procedure that moves $\mathbf{x}$ to the nearest vertex of the feasible region. Therefore we presume that it is sufficient to calculate a feasible point $\mathbf{x}$ that satisfies the condition

$$\mathbf{c}^T\mathbf{x} \leq e^{-L} \tag{1.20}$$

for some constant L. It is elementary that this condition holds if we achieve the inequality

$$m \log \mathbf{c}^T\mathbf{x} - \sum_{j=1}^{m} \log \mathbf{a}_j^T\mathbf{x} \leq -mL - \bar{\Delta}, \tag{1.21}$$

where $\bar{\Delta}$ is any upper bound on the number

$$\max \{ \sum_{j=1}^{m} \log \mathbf{a}_j^T\mathbf{x} : \mathbf{x} \text{ feasible}\}, \tag{1.22}$$

which is finite because feasible vectors are bounded. Therefore, because the left hand side of expression (1.21) is $V(\mathbf{x})$, the choice

$$\Delta = mL + \bar{\Delta} \tag{1.23}$$

is suitable. Hence, assuming also that $V(\mathbf{x}_1)$ is at most a constant multiple of m, it follows from expression (1.19) that the number of iterations of Karmarkar's original algorithm is $\mathcal{O}(m)$. The principal question that we address is whether there are stronger bounds on the number of iterations.

A good reason for believing that it may be possible to bound the number of iterations by a multiple of $\sqrt{m} \log m$ is that this complexity is achieved by an extension of Karmarkar's original algorithm that employs the "log barrier method" (see [3], for instance). A proof of this assertion is given in Section 3, in order to demonstrate the details of an analysis that works with general linear inequality constraints instead of the "standard form". Specifically, we retain the notation (1.1) for the linear objective function and, corresponding to the derivation of expression (1.3), we assume without loss of generality that the only constraints on the variables are the inequalities

$$\mathbf{a}_j^T\mathbf{x} \geq b_j, \quad j=1,2,\dots,m. \tag{1.24}$$

We continue to assume that feasible vectors are bounded, which now implies that the constraint gradients $\{\mathbf{a}_j : j=1,2,\dots,m\}$ span $\mathcal{R}^n$. Further, we require a starting point, $\mathbf{x}_0$ say, that satisfies the strict inequalities

$$\mathbf{a}_j^T\mathbf{x}_0 > b_j, \quad j=1,2,\dots,m. \tag{1.25}$$

There is no longer a "restrictive assumption" because the log barrier potential function has the form

$$V_r(\mathbf{x}) = r\,\mathbf{c}^T\mathbf{x} - \sum_{j=1}^{m} \log(\mathbf{a}_j^T\mathbf{x} - b_j), \quad \mathbf{x} \in \mathcal{S}, \tag{1.26}$$

146

which, in contrast to expression (1.13), can be well-defined when $\mathbf{c}^T\mathbf{x}$ is negative. Here r is a nonnegative parameter that is going to be adjusted automatically. Further, instead of the definition (1.6), we now let $\mathcal{S}$ be the set

$$\mathcal{S} = \{\mathbf{x} : \mathbf{a}_j^T\mathbf{x} > b_j,\ j=1,2,\ldots,m\}. \tag{1.27}$$

In order to explain the main idea of a log barrier method, we deduce that there is a unique vector of variables, $\mathbf{x}(r)$ say, that minimizes the function (1.26) for each choice of r. Because the condition $V_r(\mathbf{x}(r)) \leq V_r(\mathbf{x_0})$ would hold if $\mathbf{x}(r)$ existed, we can restrict attention to the values of $\mathbf{x}$ in the set

$$\mathcal{S}_0 = \mathcal{S} \cap \{\mathbf{x} : V_r(\mathbf{x}) \leq V_r(\mathbf{x_0})\}. \tag{1.28}$$

Now the function (1.26) is continuous and becomes infinite if $\mathbf{x}$ tends to any point on the boundary of $\mathcal{S}$, so the definition (1.28) implies that $\mathcal{S}_0$ is closed. Further, $\mathcal{S}_0$ inherits boundedness from $\mathcal{S}$. Therefore $\mathcal{S}_0$ is compact. It follows from the continuity of the function (1.26) that there exist one or more vectors $\mathbf{x}(r)$ in $\mathcal{S}_0$ that minimize $\{V_r(\mathbf{x}), \mathbf{x} \in \mathcal{S}\}$. If uniqueness failed then there would be a line segment in $\mathcal{S}$ on which $V_r(\cdot)$ would not be strictly convex. The potential function, however, has the second derivative matrix

$$\nabla^2 V_r(\mathbf{x}) = \sum_{j=1}^{m} \frac{\mathbf{a}_j\mathbf{a}_j^T}{(\mathbf{a}_j^T\mathbf{x}-b_j)^2}, \tag{1.29}$$

which is positive definite for every $\mathbf{x}$ in $\mathcal{S}$, because the constraint gradients $\{\mathbf{a}_j : j = 1,2,\ldots,m\}$ span $\mathcal{R}^n$. Therefore $V_r(\cdot)$ is a strictly convex function. It follows that $\mathbf{x}(r)$ is well-defined for every finite value of the parameter r.

We consider the set of points $\{\mathbf{x}(r) : 0 \leq r < \infty\}$, which form a curve in the feasible region

$$\bar{\mathcal{S}} = \{\mathbf{x} : \mathbf{a}_j^T\mathbf{x} \geq b_j,\ j=1,2,\ldots,m\}, \tag{1.30}$$

this set being the closure of $\mathcal{S}$. The curve is useful because its limit points as $r \to \infty$ are solutions of the linear programming problem. In order to prove this assertion, we recall the notation $\mathbf{x}^*$ for an optimal vector of variables, we let r be positive and finite, and we make use of the fact that the definition of $\mathbf{x}(r)$ implies the equation $\nabla V_r(\mathbf{x}(r)) = 0$. Further, we deduce from the definition (1.26) that this zero gradient gives the identity

$$\mathbf{c} = r^{-1} \sum_{j=1}^{m} \frac{\mathbf{a}_j}{\mathbf{a}_j^T\mathbf{x}(r)-b_j}. \tag{1.31}$$

It follows from the feasibility of $\mathbf{x}^*$ that we have the inequality

$$\mathbf{c}^T\mathbf{x}^* = r^{-1}\sum_{j=1}^{m} \frac{\mathbf{a}_j^T\mathbf{x}^*}{\mathbf{a}_j^T\mathbf{x}(r)-b_j} \geq r^{-1}\sum_{j=1}^{m} \frac{b_j}{\mathbf{a}_j^T\mathbf{x}(r)-b_j}$$

$$= r^{-1}\sum_{j=1}^{m}\{\frac{\mathbf{a}_j^T\mathbf{x}(r)}{\mathbf{a}_j^T\mathbf{x}(r)-b_j} - 1\} = \mathbf{c}^T\mathbf{x}(r) - r^{-1}m, \tag{1.32}$$

147

the identity (1.31) being invoked not only at the beginning but also at the end of this manipulation. We write expression (1.32) in the form

$$\mathbf{c}^T\mathbf{x}(r) \leq \mathbf{c}^T\mathbf{x}^* + r^{-1}m, \quad r > 0. \tag{1.33}$$

Moreover, the feasibility of $\mathbf{x}(r)$ and the optimality of $\mathbf{x}^*$ imply $\mathbf{c}^T\mathbf{x}^* \leq \mathbf{c}^T\mathbf{x}(r)$. Therefore $\mathbf{c}^T\mathbf{x}(r)$ tends to $\mathbf{c}^T\mathbf{x}^*$ as $r \to \infty$, which completes the proof.

Roughly speaking, the log barrier method finds a solution $\mathbf{x}^*$ by following the trajectory $\{\mathbf{x}(r) : 0 \leq r < \infty\}$ to its end. We presumed in the paragraph that includes expressions (1.20)–(1.23), however, that it is sufficient for practical purposes to obtain a feasible point $\mathbf{x}$ such that the difference $\mathbf{c}^T\mathbf{x} - \mathbf{c}^T\mathbf{x}^*$ is at most e^{-L}, where L is a constant. Therefore condition (1.33) tells us that, if $\mathbf{x}(r)$ can be calculated exactly, then the value $r = me^L$ would yield an acceptable approximation to an optimal vector of variables. It is expedient, however, to work with estimates of the relevant points on the trajectory, and to let the final value of r be a little larger than me^L, in order to compensate for the error of the final estimate.

The $\mathcal{O}(\sqrt{m} \log m)$ complexity bound that has been mentioned is obtained by moving gradually along the log barrier trajectory. We let $\mathbf{x}_1$ be the initial point of this procedure, this point being derived from $\mathbf{x}_0$ in a way that ensures that $\mathbf{x}_1$ is sufficiently close to $\mathbf{x}(r)$, it being usual for the initial r to be a moderate value of the trajectory parameter. For $k = 1, 2, 3, \ldots$, a Newton–Raphson procedure generates a new vector of variables $\mathbf{x}_{k+1}$ that provides the reduction

$$V_r(\mathbf{x}_{k+1}) < V_r(\mathbf{x}_k) \tag{1.34}$$

in the current potential function. On some iterations it is known that $\mathbf{x}_k$ is so close to $\mathbf{x}(r)$ that the value of r can be increased before $\mathbf{x}_{k+1}$ is calculated. A technique that applies this knowledge is included in our algorithm. The details of all these operations and the analysis that gives the $\mathcal{O}(\sqrt{m} \log m)$ bound on the number of iterations are presented in Section 3.

In Section 4 we address the question whether the number of iterations of the original Karmarkar algorithm can also be bounded by a multiple of $\sqrt{m} \log m$. We give particular attention to a numerical example that is taken from [5], where m is the number of sides of a regular polygonal approximation to the unit circle. Thus it is easy to investigate numerically how many iterations are needed as $m \to \infty$. We find, unfortunately, that more than $m/8$ iterations occur for every value of m that was tested, which suggests that the given $\mathcal{O}(m)$ complexity bound of the original algorithm cannot be reduced. It has been proved recently [6] that this conjecture is true.

2. Complexity of the original algorithm

In order to complete the material that is presented in the first half of Section 1, we have to specify the values of α_k and $\mathbf{d}_k$ that occur in the formula (1.16) that adjusts the vector of variables. Further, we have to prove that these values provide the condition

$$V(\mathbf{x}_{k+1}) \leq V(\mathbf{x}_k) - \omega, \tag{2.1}$$

where $V(\cdot)$ is the potential function (1.13) and where ω is a positive constant. Of course we retain the assumptions that are made in the first half of Section 1. We recall that the bound (2.1) is well-known, because it can be deduced from published work on linear programming problems that are expressed in "standard form". But the novel feature of our analysis is that we treat the general linear inequality constraints of expression (1.2) explicitly.

If we calculated $\mathbf{x}_{k+1}$ by applying the Newton–Raphson iteration to the minimization of $V(\cdot)$, then the difference $\mathbf{x}_{k+1} - \mathbf{x}_k$ would satisfy the equation

$$\nabla^2 V(\mathbf{x}_k)(\mathbf{x}_{k+1} - \mathbf{x}_k) = -\nabla V(\mathbf{x}_k). \tag{2.2}$$

In general, however, Newton–Raphson is suitable only if $\nabla^2 V(\mathbf{x}_k)$ is positive definite, and the homogeneity condition (1.14) implies that this second derivative matrix has at least one nonpositive eigenvalue. Specifically, analytic differentiation gives the formula

$$\nabla^2 V(\mathbf{x}_k) = \sum_{j=1}^m \frac{\mathbf{a}_j \mathbf{a}_j^T}{(\mathbf{a}_j^T \mathbf{x}_k)^2} - m \frac{\mathbf{c}\,\mathbf{c}^T}{(\mathbf{c}^T \mathbf{x}_k)^2}, \tag{2.3}$$

which has the property

$$\mathbf{x}_k^T \nabla^2 V(\mathbf{x}_k)\,\mathbf{x}_k = 0. \tag{2.4}$$

Now we have noted already that the first of the two terms on the right hand side of expression (2.3) is a positive definite matrix. Therefore the direction

$$\mathbf{d}_k = -\left[\sum_{j=1}^m \frac{\mathbf{a}_j \mathbf{a}_j^T}{(\mathbf{a}_j^T \mathbf{x}_k)^2}\right]^{-1} \nabla V(\mathbf{x}_k) \tag{2.5}$$

is well-defined, it is not unlike the Newton–Raphson search direction, and it provides the descent property (1.17). More importantly, formula (2.5) corresponds to the search direction that occurs in Karmarkar's original algorithm, which was proved for the standard form of linear programming problem by Gill, Murray, Saunders, Tomlin and Wright [2]. Therefore we let $\mathbf{d}_k$ be the vector (2.5).

We introduce a definition of the "length" of $\mathbf{d}_k$ that is suggested by a change of variables that occurs in the usual analysis of the standard form, this "length" being the norm

$$\|\mathbf{d}_k\|_* = \left\{ \mathbf{d}_k^T \left[\sum_{j=1}^m \frac{\mathbf{a}_j \mathbf{a}_j^T}{(\mathbf{a}_j^T \mathbf{x}_k)^2}\right] \mathbf{d}_k \right\}^{1/2} = \left[\sum_{j=1}^m \frac{(\mathbf{a}_j^T \mathbf{d}_k)^2}{(\mathbf{a}_j^T \mathbf{x}_k)^2}\right]^{1/2}. \tag{2.6}$$

No difficulties arise from the dependence of this norm on the iteration number because we study a single iteration.

We are going to deduce the bound

$$\|d_k\|_* \geq 1 \tag{2.7}$$

from equation (2.5) and from the "restrictive assumption". Our argument begins with the remark that the gradient of the potential function (1.13) can be expressed in the form

$$\nabla V(x_k) = m \frac{c}{c^T x_k} - [\sum_{j=1}^{m} \frac{a_j a_j^T}{(a_j^T x_k)^2}] x_k. \tag{2.8}$$

Therefore equation (2.5) gives the vector

$$d_k = x_k - \frac{m}{c^T x_k} [\sum_{j=1}^{m} \frac{a_j a_j^T}{(a_j^T x_k)^2}]^{-1} c. \tag{2.9}$$

Now we have deduced from the restrictive assumption that $c^T x_k$ is positive and that at least one of the numbers (1.10) is nonpositive. It follows from equation (2.9) that we can pick ℓ so that the inequality

$$a_\ell^T d_k \geq a_\ell^T x_k \tag{2.10}$$

is satisfied. Further, $a_\ell^T x_k$ is positive because x_k is in the set (1.6). Thus the ℓ-th term under the summation sign on the right hand side of expression (2.6) is at least one, which provides the required inequality (2.7).

We identify a value of α_k in equation (1.16) that gives the reduction (2.1) in the potential function for some positive constant ω by considering the function of one variable

$$\phi(\alpha) = V(x_k + \alpha d_k), \quad 0 \leq \alpha < \bar{\alpha}, \tag{2.11}$$

where $\bar{\alpha}$ is the number

$$\bar{\alpha} = 1 / \|d_k\|_*. \tag{2.12}$$

This bound on α ensures that $x_k + \alpha d_k$ is in the set (1.6), because the definition (2.6) provides the inequalities

$$a_j^T (x_k + \alpha d_k) = a_j^T x_k (1 + \alpha \frac{a_j^T d_k}{a_j^T x_k})$$

$$\geq a_j^T x_k (1 - \alpha \|d_k\|_*), \quad j = 1, 2, \ldots, m. \tag{2.13}$$

We are going to derive an upper bound on $\phi(\alpha)$ from the elementary identity

$$\phi(\alpha) = \phi(0) + \alpha \phi'(0) + \int_0^\alpha (\alpha - \theta) \phi''(\theta) \, d\theta, \quad 0 \leq \alpha < \bar{\alpha}. \tag{2.14}$$

Equations (2.11), (2.5) and (2.6) imply the value

$$\phi'(0) = \mathbf{d}_k^T \nabla V(\mathbf{x}_k) = -\mathbf{d}_k^T \left[\sum_{j=1}^m \frac{\mathbf{a}_j \mathbf{a}_j^T}{(\mathbf{a}_j^T \mathbf{x}_k)^2} \right] \mathbf{d}_k = -\|\mathbf{d}_k\|_*^2, \tag{2.15}$$

while expressions (2.11), (2.3), (2.13) and (2.6) provide the inequality

$$\begin{aligned}
\phi''(\theta) &= \mathbf{d}_k^T \nabla^2 V(\mathbf{x}_k + \theta \, \mathbf{d}_k) \, \mathbf{d}_k \\
&= \mathbf{d}_k^T \left[\sum_{j=1}^m \frac{\mathbf{a}_j \mathbf{a}_j^T}{\{\mathbf{a}_j^T (\mathbf{x}_k + \theta \, \mathbf{d}_k)\}^2} - m \frac{\mathbf{c} \, \mathbf{c}^T}{\{\mathbf{c}^T (\mathbf{x}_k + \theta \, \mathbf{d}_k)\}^2} \right] \mathbf{d}_k \\
&\le \mathbf{d}_k^T \left[\sum_{j=1}^m \frac{\mathbf{a}_j \mathbf{a}_j^T}{(\mathbf{a}_j^T \mathbf{x}_k)^2 (1 - \theta \, \|\mathbf{d}_k\|_*)^2} \right] \mathbf{d}_k = \frac{\|\mathbf{d}_k\|_*^2}{(1 - \theta \, \|\mathbf{d}_k\|_*)^2}.
\end{aligned} \tag{2.16}$$

Further, analytic integration gives the value

$$\int_0^\alpha \frac{(\alpha - \theta) \, \|\mathbf{d}_k\|_*^2}{(1 - \theta \, \|\mathbf{d}_k\|_*)^2} \, d\theta = -\alpha \, \|\mathbf{d}_k\|_* - \log(1 - \alpha \, \|\mathbf{d}_k\|_*). \tag{2.17}$$

Therefore the identity (2.14) implies the bound

$$\begin{aligned}
\phi(\alpha) &\le \phi(0) - \alpha \, \|\mathbf{d}_k\|_*^2 - \alpha \, \|\mathbf{d}_k\|_* - \log(1 - \alpha \, \|\mathbf{d}_k\|_*) \\
&\le \phi(0) - 2\alpha \, \|\mathbf{d}_k\|_* - \log(1 - \alpha \, \|\mathbf{d}_k\|_*), \quad 0 \le \alpha < \bar{\alpha},
\end{aligned} \tag{2.18}$$

where the last line is a consequence of condition (2.7).

It follows from the definition (2.11) that the required inequality (2.1) is satisfied for some constant ω if we can pick α so that the right hand side of expression (2.18) is $\phi(0)$ minus a positive constant. Differentiation with respect to α shows that this right hand side is least when the equation $\alpha \|\mathbf{d}_k\|_* = \frac{1}{2}$ is obtained. Fortunately the resultant steplength preserves feasibility because α is half of the bound (2.12). Therefore we may make the choice

$$\alpha_k = 1 / (2 \|\mathbf{d}_k\|_*), \tag{2.19}$$

and it gives the reduction

$$V\left(\mathbf{x}_k + \frac{\mathbf{d}_k}{2 \|\mathbf{d}_k\|_*} \right) \le V(\mathbf{x}_k) - 1 + \log 2 \tag{2.20}$$

in the potential function. Thus condition (2.1) is achieved for the constant

$$\omega = 1 - \log 2 = 0.307. \tag{2.21}$$

It follows *a fortiori* that the potential function is reduced by at least 0.307 if α_k is chosen to minimize the value of $V(\mathbf{x}_{k+1})$, which completes our analysis of the complexity of the original algorithm.

151

3. Complexity of the log barrier method

Some of the main features of the algorithm of this section are mentioned in Section 1. We recall that it generates a sequence of points $\{x_k : k = 1, 2, 3, \ldots\}$ in the set (1.27) that follows approximately the trajectory $\{x(r) : 0 \leq r < \infty\}$ to a large value of r, where $x(r)$ is the vector of variables that minimizes the strictly convex function (1.26). Each change to the variables satisfies inequality (1.34) for the current value of r. We recall also that the value of r may be increased at the beginning of any iteration. We are going to study first the calculation of x_{k+1} from x_k when r is fixed. Secondly, we consider the adjustment of r. Then we identify a stopping condition that guarantees the required accuracy

$$c^T x_k - c^T x^* \leq e^{-L} \tag{3.1}$$

in the objective function. Finally we address the problem of determining a suitable initial point x_1 for the iterative procedure from a given point $x_0 \in \mathcal{S}$. We find under mild assumptions that the number of iterations is of magnitude $\sqrt{m} \log m$. Much of our theory is equivalent to the analysis in [3] of the "standard form" of linear programming problem, but we do not mention the dual form of the calculation.

The potential function (1.26) has the advantage that its second derivative matrix (1.29) is positive definite for every x in the set (1.27). Therefore, given x_k and r, the Newton–Raphson method

$$d_k = -\left[\sum_{j=1}^{m} \frac{a_j a_j^T}{(a_j^T x_k - b_j)^2} \right]^{-1} \nabla V_r(x_k) \tag{3.2}$$

is used to generate a search direction from $x_k \in \mathcal{S}$ that satisfies the descent condition

$$d_k^T \nabla V_r(x_k) < 0. \tag{3.3}$$

Again we let x_{k+1} have the form

$$x_{k+1} = x_k + \alpha_k d_k, \tag{3.4}$$

where the positive steplength α_k is chosen by considering the function (2.11) of one variable, assuming that $V(\cdot)$ is replaced by $V_r(\cdot)$. A comparison of equations (2.5) and (3.2) shows that it is also appropriate to change expression (2.6) to the definition

$$\| d_k \|_* = \left[\sum_{j=1}^{m} \frac{(a_j^T d_k)^2}{(a_j^T x_k - b_j)^2} \right]^{1/2}. \tag{3.5}$$

Thus the inequalities (2.13) become the conditions

$$a_j^T (x_k + \alpha d_k) - b_j = (a_j^T x_k - b_j)\left(1 + \alpha \frac{a_j^T d_k}{a_j^T x_k - b_j}\right)$$

$$\geq (a_j^T x_k - b_j)(1 - \alpha \| d_k \|_*), \quad j = 1, 2, \ldots, m, \tag{3.6}$$

so we preserve feasibility by imposing the upper limit (2.12) on α for the new definition of $\|\cdot\|_*$.

We derive an upper bound on the new expression (2.14) by following the method of the penultimate paragraph of Section 2. Specifically, by making obvious changes to the intermediate steps of equation (2.15) and inequality (2.16), we deduce the relations

$$\phi'(0) = -\|\mathbf{d}_k\|_*^2 \tag{3.7}$$

and

$$\phi''(\theta) \leq \|\mathbf{d}_k\|_*^2 / (1 - \theta \|\mathbf{d}_k\|_*^2). \tag{3.8}$$

Therefore condition (2.18) is still valid. We find that the right hand side of its first line is least when α has the value

$$\alpha_k = 1/(1 + \|\mathbf{d}_k\|_*) \tag{3.9}$$

and we restrict attention to this choice of steplength. Hence formula (3.4) provides the reduction

$$V_r(\mathbf{x}_{k+1}) \leq V_r(\mathbf{x}_k) - \|\mathbf{d}_k\|_* + \log(1 + \|\mathbf{d}_k\|_*) \tag{3.10}$$

in the potential function, and it follows from expressions (3.6) and (3.9) that $\mathbf{x}_{k+1}$ is in the interior of the feasible region as required. Inequality (3.10) is going to be useful to our analysis.

It would be futile to refine the vector of variables when $\mathbf{x}_k$ is very close to $\mathbf{x}(r)$ for the current value of r if larger values of r were needed. Therefore we seek a bound on the difference $V_r(\mathbf{x}_k) - V_r(\mathbf{x}(r))$ that can be computed easily from the number $\|\mathbf{d}_k\|_*$ that occurs for the current r. We let $\mathbf{d}$ be any vector such that $\mathbf{x}_k + \mathbf{d}$ is feasible and we consider the identity

$$V_r(\mathbf{x}_k + \mathbf{d}) = V_r(\mathbf{x}_k) + \mathbf{d}^T \nabla V_r(\mathbf{x}_k) + \int_0^1 (1 - \theta) \mathbf{d}^T \nabla^2 V_r(\mathbf{x}_k + \theta \mathbf{d}) \mathbf{d} \, d\theta. \tag{3.11}$$

The definitions (3.2) and (3.5) and the Cauchy–Schwarz inequality give the condition

$$\mathbf{d}^T \nabla V_r(\mathbf{x}_k) = -\mathbf{d}^T \left[\sum_{j=1}^m \frac{\mathbf{a}_j \mathbf{a}_j^T}{(\mathbf{a}_j^T \mathbf{x}_k - b_j)^2} \right] \mathbf{d}_k \geq -\|\mathbf{d}\|_* \|\mathbf{d}_k\|_*, \tag{3.12}$$

while our treatment of the second derivative term of equation (3.11) begins with the remark that the first line of expression (3.6) leads to the inequalities

$$\mathbf{a}_j^T (\mathbf{x}_k + \theta \, \mathbf{d}) - b_j \leq (\mathbf{a}_j^T \mathbf{x}_k - b_j)(1 + \theta \|\mathbf{d}\|_*), \quad j = 1, 2, \ldots, m. \tag{3.13}$$

Therefore we have the bound

$$
\begin{aligned}
\mathbf{d}^T \nabla^2 V_r(\mathbf{x}_k + \theta \, \mathbf{d}) \, \mathbf{d} &= \sum_{j=1}^m \frac{(\mathbf{a}_j^T \mathbf{d})^2}{\{\mathbf{a}_j^T (\mathbf{x}_k + \theta \, \mathbf{d}) - b_j\}^2} \\
&\geq \sum_{j=1}^m \frac{(\mathbf{a}_j^T \mathbf{d})^2}{(\mathbf{a}_j^T \mathbf{x}_k - b_j)^2 (1 + \theta \|\mathbf{d}\|_*)^2} = \frac{\|\mathbf{d}\|_*^2}{(1 + \theta \|\mathbf{d}\|_*)^2}.
\end{aligned} \tag{3.14}
$$

153

Hence analytic integration provides the relation

$$V_r(\mathbf{x}_k+\mathbf{d}) \geq V_r(\mathbf{x}_k) - \|\mathbf{d}\|_* \|\mathbf{d}_k\|_* + \int_0^1 \frac{(1-\theta)\,\|\mathbf{d}\|_*^2}{(1+\theta\,\|\mathbf{d}\|_*)^2}\,d\theta$$

$$= V_r(\mathbf{x}_k) - \|\mathbf{d}\|_* \|\mathbf{d}_k\|_* + \|\mathbf{d}\|_* - \log(1+\|\mathbf{d}\|_*) \qquad (3.15)$$

for general feasible $\mathbf{d}$. We see that the right hand side is bounded below if we have $\|\mathbf{d}_k\|_* < 1$. In this case it is elementary that the right hand side is least when $\|\mathbf{d}\|_*$ has the value

$$\|\mathbf{d}\|_* = \|\mathbf{d}_k\|_* /(1-\|\mathbf{d}_k\|_*), \qquad (3.16)$$

which provides a lower bound on $V_r(\mathbf{x}_k+\mathbf{d})$ that is independent of $\mathbf{d}$. Thus, because $V_r(\mathbf{x}(r))$ is admissible on the left hand side of expression (3.15), we find the inequality

$$V_r(\mathbf{x}(r)) \geq V_r(\mathbf{x}_k) + \|\mathbf{d}_k\|_* + \log(1-\|\mathbf{d}_k\|_*). \qquad (3.17)$$

Our rule for increasing r depends on two constants that are given the values 0.5 and 0.1 for definiteness. Specifically, we satisfy the condition

$$\|\mathbf{d}_k\|_* \leq 0.5 \qquad (3.18)$$

immediately after any increase in r, and an increase is made when r is not at its final value and the old r provides $\|\mathbf{d}_k\|_* \leq 0.1$. It follows from expression (3.17) that condition (3.18) implies the inequality

$$V_r(\mathbf{x}_k) \leq V_r(\mathbf{x}(r)) + 0.193, \qquad (3.19)$$

so the sequence $\{\mathbf{x}_k : k=1,2,3,\ldots\}$ stays fairly close to the trajectory $\{\mathbf{x}(r) : 0 \leq r < \infty\}$. Moreover, if $\|\mathbf{d}_k\|_* > 0.1$, then condition (3.10) provides the bound

$$V_r(\mathbf{x}_{k+1}) < V_r(\mathbf{x}_k) - 0.0047. \qquad (3.20)$$

Because the definition of $\mathbf{x}(r)$ gives $V_r(\mathbf{x}(r)) \leq V_r(\mathbf{x}_{k+1})$, we deduce from inequalities (3.19) and (3.20) that the reduction (3.20) occurs at most four times for each r. Therefore the trajectory parameter is revised frequently until it reaches its final value.

Next we identify a suitable increase in r when the old value gives $\|\mathbf{d}_k\|_* \leq 0.1$. We let $\hat{r}$ and $\hat{\mathbf{d}}_k$ be the new trajectory parameter and the new Newton–Raphson step respectively. Further, equations (1.26) and (3.2) imply that $\hat{\mathbf{d}}_k$ can be expressed in the form

$$\hat{\mathbf{d}}_k = -[\sum_{j=1}^m \frac{\mathbf{a}_j \mathbf{a}_j^T}{(\mathbf{a}_j^T \mathbf{x}_k - b_j)^2}]^{-1} [\hat{r}\,\mathbf{c} - \sum_{j=1}^m \frac{\mathbf{a}_j}{\mathbf{a}_j^T \mathbf{x}_k - b_j}] = \frac{\hat{r}}{r}(\mathbf{d}_k - \boldsymbol{\delta}) + \boldsymbol{\delta}, \qquad (3.21)$$

where $\boldsymbol{\delta}$ is the vector

$$\boldsymbol{\delta} = [\sum_{j=1}^m \frac{\mathbf{a}_j \mathbf{a}_j^T}{(\mathbf{a}_j^T \mathbf{x}_k - b_j)^2}]^{-1} \sum_{j=1}^m \frac{\mathbf{a}_j}{\mathbf{a}_j^T \mathbf{x}_k - b_j}. \qquad (3.22)$$

Therefore the new Newton–Raphson step satisfies condition (3.18) if the change to the trajectory parameter is constrained by the inequality

$$(\hat{r}/r)\,\|\mathbf{d}_k\|_* + [\,(\hat{r}/r) - 1\,]\,\|\boldsymbol{\delta}\|_* \leq 0.5, \tag{3.23}$$

the definition (3.5) of $\|\cdot\|_*$ being independent of r.

Fortunately, $\boldsymbol{\delta}$ is endowed with the bound

$$\|\boldsymbol{\delta}\|_* \leq \sqrt{m}, \tag{3.24}$$

which is proved as follows. The definitions (3.5) and (3.22) imply the value

$$\|\boldsymbol{\delta}\|_*^2 = \Big(\sum_{j=1}^m \frac{\mathbf{a}_j}{\mathbf{a}_j^T\mathbf{x}_k - b_j}\Big)^T \Big[\sum_{j=1}^m \frac{\mathbf{a}_j\mathbf{a}_j^T}{(\mathbf{a}_j^T\mathbf{x}_k - b_j)^2}\Big]^{-1} \Big(\sum_{j=1}^m \frac{\mathbf{a}_j}{\mathbf{a}_j^T\mathbf{x}_k - b_j}\Big). \tag{3.25}$$

Therefore, letting A be the $n \times m$ matrix that has the columns $\{\mathbf{a}_j/(\mathbf{a}_j^T\mathbf{x}_k - b_j) : j = 1, 2, \ldots, m\}$ and letting $\mathbf{e}$ be the vector in $\mathcal{R}^m$ whose components are all one, we find the identity

$$\|\boldsymbol{\delta}\|_*^2 = (A\mathbf{e})^T(AA^T)^{-1}(A\mathbf{e}). \tag{3.26}$$

Now $A^T(AA^T)^{-1}A$ is a symmetric projection matrix from $\mathcal{R}^m$ to $\mathcal{R}^m$, so the right hand side of equation (3.26) is at most the square of the Euclidean norm of $\mathbf{e}$. Therefore inequality (3.24) is true.

It follows from expression (3.23) that, if we are increasing r because we have $\|\mathbf{d}_k\|_* \leq 0.1$, then it is suitable to define $\hat{r}$ by the equation

$$0.1\,(\hat{r}/r) + [\,(\hat{r}/r) - 1\,]\,\sqrt{m} = 0.5, \tag{3.27}$$

which is equivalent to the formula

$$\hat{r} = [\,(\sqrt{m}+0.5)/(\sqrt{m}+0.1)\,]\,r. \tag{3.28}$$

Thus the changes to the trajectory parameter are very gradual when m is large, but it will be shown that this formula provides the required $\mathcal{O}(\sqrt{m}\log m)$ bound on the total number of iterations.

Next we seek a positive number $\bar{r}$ such that, if the current trajectory parameter satisfies $r \geq \bar{r}$ and if its Newton–Raphson step at $\mathbf{x}_k$ satisfies $\|\mathbf{d}_k\|_* \leq 0.1$, then the iterative procedure can be terminated because the condition (3.1) must hold. Guided by Gonzaga [3], our derivation of a suitable value of $\bar{r}$ depends on the following upper bound on the norm $\|\mathbf{x}_k - \mathbf{x}(r)\|_*$. If the numbers $\|\mathbf{d}\|_*$ and $\|\mathbf{d}_k\|_*$ in expression (3.15) obey $\|\mathbf{d}\|_* \geq 0.3$ and $\|\mathbf{d}_k\|_* \leq 0.1$, then we have the relation

$$\begin{aligned} V_r(\mathbf{x}_k+\mathbf{d}) &\geq V_r(\mathbf{x}_k) + 0.9\,\|\mathbf{d}\|_* - \log(1+\|\mathbf{d}\|_*) \\ &\geq V_r(\mathbf{x}_k) + 0.0076, \end{aligned} \tag{3.29}$$

155

but we know that the inequality $V_r(\mathbf{x}(r)) \leq V_r(\mathbf{x}_k)$ holds. It follows that the condition $\|\mathbf{d}_k\|_* \leq 0.1$ implies the bound

$$\|\mathbf{x}_k - \mathbf{x}(r)\|_* < 0.3. \tag{3.30}$$

We are going to apply this bound to the last term of the elementary inequality

$$\mathbf{c}^T\mathbf{x}_k - \mathbf{c}^T\mathbf{x}^* \leq [\mathbf{c}^T\mathbf{x}(r) - \mathbf{c}^T\mathbf{x}^*] + |\mathbf{c}^T[\mathbf{x}_k - \mathbf{x}(r)]|. \tag{3.31}$$

Specifically, Cauchy–Schwarz with respect to the norm $\|\cdot\|_*$ and expression (3.30) imply the relation

$$|\mathbf{c}^T[\mathbf{x}_k - \mathbf{x}(r)]| \leq 0.3 \,\|\, [\sum_{j=1}^{m} \frac{\mathbf{a}_j \mathbf{a}_j^T}{(\mathbf{a}_j^T\mathbf{x}_k - b_j)^2}]^{-1}\mathbf{c}\,\|_*. \tag{3.32}$$

Now a rearrangement of the definition

$$\mathbf{d}_k = -[\sum_{j=1}^{m} \frac{\mathbf{a}_j\mathbf{a}_j^T}{(\mathbf{a}_j^T\mathbf{x}_k - b_j)^2}]^{-1}[r\,\mathbf{c} - \sum_{j=1}^{m} \frac{\mathbf{a}_j}{\mathbf{a}_j^T\mathbf{x}_k - b_j}] \tag{3.33}$$

gives the equation

$$[\sum_{j=1}^{m} \frac{\mathbf{a}_j\mathbf{a}_j^T}{(\mathbf{a}_j^T\mathbf{x}_k - b_j)^2}]^{-1}\mathbf{c} = r^{-1}(\boldsymbol{\delta} - \mathbf{d}_k), \tag{3.34}$$

where $\boldsymbol{\delta}$ is still the vector (3.22). Therefore, in view of the bound (3.24), the relation (3.32) implies the inequality

$$|\mathbf{c}^T[\mathbf{x}_k - \mathbf{x}(r)]| \leq 0.3\,r^{-1}(\sqrt{m} + 0.1) \tag{3.35}$$

in the case $\|\mathbf{d}_k\|_* \leq 0.1$. It follows from expressions (1.33) and (3.31) that the condition

$$\mathbf{c}^T\mathbf{x}_k - \mathbf{c}^T\mathbf{x}^* \leq r^{-1}(m + 0.3\sqrt{m} + 0.03) \tag{3.36}$$

is satisfied. Therefore the current $\mathbf{x}_k$ provides the accuracy (3.1) when we have $\|\mathbf{d}_k\|_* \leq 0.1$ and $r \geq \bar{r}$, where $\bar{r}$ is the number

$$\bar{r} = e^L(m + 0.3\sqrt{m} + 0.03). \tag{3.37}$$

Alternatively, because the inequality

$$(\sqrt{m} + 0.5)/(\sqrt{m} + 0.1) > (m + 0.3\sqrt{m} + 0.03)/m \tag{3.38}$$

holds for all $m \geq 1$, it is sufficient to make one further increase in r by formula (3.28) after r reaches the value me^L that is mentioned in Section 1.

Let us assume for the moment that a starting point $\mathbf{x}_1$ and an initial trajectory parameter r_1 are available such that the resultant Newton–Raphson search direction satisfies the condition $\|\mathbf{d}_1\|_* \leq 0.5$. Then it has been noted that each value of r is employed by at most four consecutive iterations and that r is increased by formula (3.28). Further, we have just found that the required accuracy (3.1) is achievable when r reaches the value

(3.37), so the number of applications of formula (3.28) is at most $\max[0, \ell+2]$, where ℓ is the greatest integer that satisfies the condition

$$r_1 [(\sqrt{m}+0.5)/(\sqrt{m}+0.1)]^\ell < m e^L. \tag{3.39}$$

It follows from the elementary inequality

$$\log[(\sqrt{m}+0.5)/(\sqrt{m}+0.1)] > 1/(4\sqrt{m}), \quad m \geq 1, \tag{3.40}$$

that we have the relation

$$\ell < (L+\log m - \log r_1)(4\sqrt{m}). \tag{3.41}$$

Therefore the number of iterations is bounded above by the expression

$$4\max[1, (L+\log m - \log r_1)(4\sqrt{m}) + 3]. \tag{3.42}$$

It remains to consider the calculation of $\mathbf{x}_1$ and r_1, given a general point $\mathbf{x}_0$ in the set (1.27).

One way of generating values of $\mathbf{x}_1$ and r_1 that satisfy the conditions of the previous paragraph is to pick $r = 1$, for example, and, starting at the point $\mathbf{x}_0$, to apply the iteration (3.4), where $\mathbf{d}_k$ and α_k are expressions (3.2) and (3.9), until inequality (3.18) is satisfied. This procedure will converge because $V_1(\cdot)$ is bounded below and because each iteration gives the reduction (3.20) in the potential function. Unfortunately, however, if the difference $V_1(\mathbf{x}_0)-V_1(\mathbf{x}(1))$ is of magnitude m, where $\mathbf{x}(1)$ is still the vector of variables that minimizes $V_1(\cdot)$, then inequality (3.20) gives a bound on the number of iterations of this process that is also $\mathcal{O}(m)$, which is too large for the complexity result that we are trying to establish. Further, the definition (1.26) shows that it is usual for $V_1(\mathbf{x}(1))$ to be of magnitude $-m$ when there are feasible points $\mathbf{x}$ at which all the constraint residuals $\{\mathbf{a}_j^T\mathbf{x}-b_j : j=1,2,\ldots,m\}$ exceed one. Therefore, even if $V_1(\mathbf{x}_0)$ is uniformly bounded, the value of $V_1(\mathbf{x}(1))$ may be so small that the procedure of this paragraph is unsuitable.

It is possible that the calculation of an acceptable $\mathbf{x}_1$ from $\mathbf{x}_0$ is as difficult as the problem that has been addressed already in this section, namely the calculation of a feasible point $\mathbf{x}_k$ that satisfies the accuracy condition (3.1) after $\mathbf{x}_1$ is available. In any case the log barrier technique that changes r gradually can be helpful to the initialization procedure. Specifically, given the interior feasible point $\mathbf{x}_0$, we let $\hat{\mathbf{c}}$ be the vector

$$\hat{\mathbf{c}} = \sum_{j=1}^{m} \frac{\mathbf{a}_j}{\mathbf{a}_j^T\mathbf{x}_0 - b_j}, \tag{3.43}$$

and we let $\hat{V}_r(\cdot)$ be the potential function

$$\hat{V}_r(\mathbf{x}) = r\,\hat{\mathbf{c}}^T\mathbf{x} - \sum_{j=1}^{m}\log(\mathbf{a}_j^T\mathbf{x}-b_j), \quad \mathbf{x}\in\mathcal{S}, \tag{3.44}$$

157

where $\mathcal{S}$ is still the set (1.27). We see that this function is strictly convex and we let $\hat{\mathbf{x}}(r)$ be the vector of variables that minimizes it. The definitions (3.43) and (3.44) are useful because they imply $\nabla \hat{V}_1(\mathbf{x}_0) = 0$, which gives the equation

$$\hat{\mathbf{x}}(1) = \mathbf{x}_0. \tag{3.45}$$

Therefore $\mathbf{x}_0$ is on the log barrier trajectory $\{\hat{\mathbf{x}}(r) : 0 \le r < \infty\}$. Further, a slightly modified version of the iterative algorithm of this section can be applied to follow the trajectory backwards from $\mathbf{x}_0$. In fact we are going to require a point, $\hat{\mathbf{x}}_1$ say, that is close to $\hat{\mathbf{x}}(r)$ for a value of the trajectory parameter that satisfies the condition

$$r \le 1/\|\hat{\mathbf{c}} - \mathbf{c}\|_2, \tag{3.46}$$

where $\| \cdot \|_2$ is the Euclidean vector norm. Therefore, if we have $\|\hat{\mathbf{c}} - \mathbf{c}\|_2 \le 1$, we set $r = 1$ and $\hat{\mathbf{x}}_1 = \hat{\mathbf{x}}(1) = \mathbf{x}_0$. Otherwise this trajectory parameter and vector of variables are used to start the modified algorithm.

In order to describe the modifications, we suppose that $V_r(\cdot)$ is replaced by $\hat{V}_r(\cdot)$ throughout the first half of this section. Then the only changes to an iteration are in the technique that revises r, this revision being triggered by the same test as before, namely when the old value of the trajectory parameter gives $\|\mathbf{d}_k\|_* \le 0.1$. We continue to let $\hat{r}$ be the new value of r but now we require $\hat{r} < r$. Therefore, instead of inequality (3.23), we preserve the bound (3.18) by satisfying the condition

$$(\hat{r}/r) \|\mathbf{d}_k\|_* + [1 - (\hat{r}/r)] \|\boldsymbol{\delta}\|_* \le 0.5. \tag{3.47}$$

Hence, corresponding to expression (3.27), it is suitable to define $\hat{r}$ by the equation

$$0.1 (\hat{r}/r) + [1 - (\hat{r}/r)] \sqrt{m} = 0.5, \tag{3.48}$$

which gives the formula

$$\hat{r} = [(\sqrt{m} - 0.5)/(\sqrt{m} - 0.1)] r. \tag{3.49}$$

The iterative procedure is terminated if inequality (3.46) holds for the new trajectory parameter. In this case we let $\hat{\mathbf{x}}_1$ be the current vector of variables, so expression (3.19) implies the bound

$$\hat{V}_r(\hat{\mathbf{x}}_1) \le \hat{V}_r(\hat{\mathbf{x}}(r)) + 0.193, \tag{3.50}$$

which is obtained trivially when inequality (3.46) allows $r = 1$.

We derive the complexity of this calculation from the remark that formula (3.49) is applied $\ell + 1$ times, where ℓ is now the greatest integer that satisfies the condition

$$[(\sqrt{m} - 0.5)/(\sqrt{m} - 0.1)]^\ell > 1/\|\hat{\mathbf{c}} - \mathbf{c}\|_2. \tag{3.51}$$

Thus the number of iterations of this part of the initialization procedure is at most a constant multiple of $\sqrt{m} \log(\|\hat{\mathbf{c}} - \mathbf{c}\|_2)$. Therefore the required complexity is preserved if we have the bound

$$\|\hat{\mathbf{c}} - \mathbf{c}\|_2 = \mathcal{O}(m^p) \tag{3.52}$$

for some constant p, which is usually a very mild constraint on $\mathbf{x}_0$.

We let the initial trajectory parameter of the main calculation, namely r_1, be the value of r that has just been chosen. Therefore, because its value is approximately $\min[1, 1/\|\hat{\mathbf{c}} - \mathbf{c}\|_2]$, the assumption (3.52) makes the $-\log r_1$ term of the bound (3.42) acceptably small.

Further, we begin the main calculation not at a point $\mathbf{x}_1$ that gives $\|\mathbf{d}_1\|_* \leq 0.5$ but at the vector of variables $\hat{\mathbf{x}}_1$. Therefore there is some preliminary work that is equivalent to the procedure of the complete paragraph that is between expressions (3.42) and (3.43). It follows that several iterations may be needed to achieve the inequality $\|\mathbf{d}_k\|_* \leq 0.5$, and then we can take the view that the current $\mathbf{x}_k$ is the $\mathbf{x}_1$ that has been mentioned already. Because we have noted that each of these preliminary iterations provides the reduction (3.20) in the potential function, the total calculation is of the required complexity if we can establish that conditions (3.46) and (3.50) imply the property

$$V_r(\hat{\mathbf{x}}_1) - V_r(\mathbf{x}(r)) = \mathcal{O}(\sqrt{m}). \tag{3.53}$$

The definition of $\hat{\mathbf{x}}(r)$ and much easy cancellation give the relation

$$[\hat{V}_r(\hat{\mathbf{x}}(r)) - \hat{V}_r(\hat{\mathbf{x}}_1)] + [\hat{V}_r(\hat{\mathbf{x}}_1) + r(\mathbf{c} - \hat{\mathbf{c}})^T \hat{\mathbf{x}}_1]$$
$$\leq [\hat{V}_r(\mathbf{x}(r)) + r(\mathbf{c} - \hat{\mathbf{c}})^T \mathbf{x}(r)] + r(\mathbf{c} - \hat{\mathbf{c}})^T [\hat{\mathbf{x}}_1 - \mathbf{x}(r)]. \tag{3.54}$$

We see that the second and third square brackets contain $V_r(\hat{\mathbf{x}}_1)$ and $V_r(\mathbf{x}(r))$ respectively. It follows from expression (3.50) and the Cauchy–Schwarz inequality that we have the condition

$$V_r(\hat{\mathbf{x}}_1) \leq V_r(\mathbf{x}(r)) + 0.193 + r\|\hat{\mathbf{c}} - \mathbf{c}\|_2 \|\hat{\mathbf{x}}_1 - \mathbf{x}(r)\|_2. \tag{3.55}$$

Now we are assuming that the feasible region (1.30) is bounded and we let Λ be a constant upper bound on the infinity norms $\{\|\mathbf{x}\|_\infty : \mathbf{x} \text{ feasible}\}$. Thus the relations (3.46) and (3.55) provide the property

$$V_r(\hat{\mathbf{x}}_1) \leq V_r(\mathbf{x}(r)) + 0.193 + 2\sqrt{n}\,\Lambda. \tag{3.56}$$

We see that the required condition (3.53) is a consequence of $m \geq n$. Therefore the given procedure that calculates $\mathbf{x}_1$ from $\mathbf{x}_0$ does not impair the $\mathcal{O}(\sqrt{m}\log m)$ bound on the number of iterations of the log barrier method that now follows from expression (3.42).

4. Discussion and a numerical example

Subject to some mild conditions, we have found that the original and the log barrier versions of Karmarkar's algorithm both achieve the proximity (3.1) to the optimal value of the objective function in a finite number of iterations, where $\mathbf{x}_k$ is feasible and L is a given constant. Further, we have derived upper bounds on the number of iterations that are of magnitudes m and $\sqrt{m}\log m$ respectively, which suggests that the log barrier version may be more efficient than the original algorithm. A severe disadvantage of the given log barrier approach, however, is that formula (3.28) changes the trajectory parameter so gradually that the number of iterations is hardly ever much less than $\sqrt{m}$. Therefore it happens frequently that the original algorithm requires fewer iterations (see [3], for instance). It follows that, if the given $\mathcal{O}(m)$ complexity bound cannot be improved, then the efficiency of the original algorithm depends strongly on the details of the linear programming problem that is being solved.

This dependence seems to exist. Indeed, as mentioned in Section 1, a numerical example is presented below, showing that the number of iterations of the original algorithm exceeds $m/8$ for a sequence of values of m that becomes quite large. The example is derived from the easy nonlinear calculation

$$\left. \begin{array}{ll} \text{minimize} & \hat{x}_2, \quad \hat{x} \in \mathcal{R}^2 \\ \text{subject to} & \hat{x}_1^2 + \hat{x}_2^2 \leq 1 \end{array} \right\}. \tag{4.1}$$

Specifically, we replace the nonlinear constraint by the regular polygonal approximation

$$\hat{x}_1 \sin(2\pi j/m) + \hat{x}_2 \cos(2\pi j/m) \geq -1, \quad j = 1, 2, \ldots, m, \tag{4.2}$$

where the integer m is a parameter. This expression corresponds to the inequalities (1.3). Therefore we achieve homogeneity by applying formulae (1.4) and (1.5), which yields the linear programming problem

$$\left. \begin{array}{l} \text{minimize} \quad x_2 + x_3, \quad x \in \mathcal{R}^3 \\ \text{subject to} \quad x_3 = 1, \\ x_1 \sin(2\pi j/m) + x_2 \cos(2\pi j/m) + x_3 \geq 0, \quad j = 1, 2, \ldots, m \end{array} \right\}. \tag{4.3}$$

We have added the term x_3 into the objective function in order that the restrictive assumption (1.8) is satisfied at the solution $x^* = (0, -1, 1)^T$. Therefore the calculation (4.3) has the form that is studied in Sections 1 and 2.

The original algorithm was started at the point

$$x_1 = (99/\{100\sqrt{2}\}, 99/\{100\sqrt{2}\}, 1)^T \tag{4.4}$$

for several values of m. Each iteration employed the search direction (2.5) and each steplength α_k was calculated in high precision to minimize the new value

$$V(x_{k+1}) = V(x_k + \alpha_k d_k) \tag{4.5}$$

of the potential function (1.13). One can take the view that the constant Δ of the termination condition (1.15) is very large, because in this example an optimal vector of variables x^* is always found in a finite number of iterations, due to the fact that optimality is achieved at every feasible point x that is on the boundary of the m-th inequality constraint of expression (4.3). Table 1 shows the number of iterations that were required. We see that it has the value $\frac{1}{8}m + 7$ in the last three columns of the table.

160

m	20	40	100	200	400	800
Iterations	6	10	19	32	57	107

Table 1: Increasing m in the problem (4.3)

k	$(\hat{\mathbf{x}}_k)_1$	$(\hat{\mathbf{x}}_k)_2$	$(\hat{\mathbf{v}}_k)_1$	$(\hat{\mathbf{v}}_k)_2$
1	0.700036	0.700036	0.707456	0.707456
2	0.938254	0.243569	0.960768	0.279129
3	0.937396	-0.299708	0.960768	-0.279129
4	0.825641	-0.558190	0.827489	-0.562361
5	0.755146	-0.654780	0.750481	-0.661638
6	0.728001	-0.685364	0.707456	-0.707456
7	0.707788	-0.706790	0.707456	-0.707456
8	0.655001	-0.755682	0.661638	-0.750481
9	0.611085	-0.791825	0.613210	-0.790545
10	0.559550	-0.829049	0.562361	-0.827489
11	0.507027	-0.862222	0.509293	-0.861167
12	0.452154	-0.892256	0.454215	-0.891446
13	0.395555	-0.918784	0.397344	-0.918208
14	0.337356	-0.941742	0.338905	-0.941345
15	0.277806	-0.961024	0.279129	-0.960768
16	0.217137	-0.976550	0.218251	-0.976399
17	0.155589	-0.988252	0.156512	-0.988176
18	0.093412	-0.996080	0.094155	-0.996053
19	0.030866	-1.000000	0.031426	-1.000000

Table 2: Values of $\hat{\mathbf{x}}_k$ and $\hat{\mathbf{v}}_k$ when $m = 100$

A reason for the $m/8$ term of the iteration count is apparent in the two-dimensional analogue of the calculated sequence $\{\mathbf{x}_k : k = 1, 2, 3, \ldots\} \subset \mathcal{R}^3$. Specifically, for each k we let $\hat{\mathbf{x}}_k$ be the vector in $\mathcal{R}^2$ whose components are $(\mathbf{x}_k)_1/(\mathbf{x}_k)_3$ and $(\mathbf{x}_k)_2/(\mathbf{x}_k)_3$, where $(\mathbf{x}_k)_i$ denotes the i-th component of $\mathbf{x}_k$. These ratios are well-defined because a comparison of expressions (4.3) and (1.2) yields the identity $(\mathbf{x}_k)_3 = \mathbf{a}_0^T \mathbf{x}_k$, so the strict positivity of $(\mathbf{x}_k)_3$ follows from the argument of the paragraph that includes equations (1.6) and (1.7). We consider the sequence $\{\hat{\mathbf{x}}_k : k = 1, 2, 3, \ldots\}$ that is shown explicitly in the first three columns of Table 2 in the case when $m = 100$. The second components $\{(\hat{\mathbf{x}}_k)_2 : k = 1, 2, 3, \ldots\}$ converge to -1, because the problem (4.3) that is solved by Karmarkar's algorithm is equivalent to minimizing $\hat{x}_2$ subject to the constraints (4.2). Further, because every $\mathbf{x}_k$ satisfies the inequality conditions of expression (4.3), the sequence $\{\hat{\mathbf{x}}_k : k = 1, 2, 3, \ldots\} \subset \mathcal{R}^2$ is enclosed by the m-sided regular polygon that is defined by the inequalities (4.2).

Each vector $\hat{\mathbf{v}}_k \in \mathcal{R}^2$, whose components are given in the last two columns of Table 2, is defined to be the vertex of this polygon that is nearest to $\hat{\mathbf{x}}_k$. The Euclidean distance between adjacent vertices when $m = 100$ is the number $2\tan(\pi/100) \approx 0.063$, and we see that, after the iterations have reduced $(\hat{\mathbf{x}}_k)_2$ to below $-2^{-1/2}$, the value of $\|\hat{\mathbf{x}}_k - \hat{\mathbf{v}}_k\|_2$ is much smaller than this distance. We also find that every one of the 12 vertices of the polygon that satisfies $(\hat{\mathbf{v}})_1 > 0$ and $(\hat{\mathbf{v}})_2 < -2^{-1/2}$ occurs once in this part of the table. Therefore the behaviour of the last 12 iterations of Table 2 is analogous to 12 consecutive iterations of the simplex method that move from vertex to adjacent vertex of the feasible region. Further, in all the cases of Table 1, the sequence $\{\hat{\mathbf{x}}_k : k = 1, 2, 3, \ldots\}$ followed consecutive vertices of the m-sided regular polygon very closely after the condition $(\hat{\mathbf{x}}_k)_2 < -2^{-1/2}$ was obtained, which accounts for $m/8$ of the iterations that occur.

Another remarkable property of this example is that the number of iterations is about $m/8$ even when the constant Δ of the termination condition (1.15) is not very large. We consider, for example, the modest choice $L = 10$ in formula (1.23). Also we set $\bar{\Delta} = 0$ because expression (1.22) is zero, which can be deduced from the remark that $\{\sum_{j=1}^m \log \mathbf{a}_j^T \mathbf{x}, \ \mathbf{x} \text{ feasible}\}$ is a concave function of $\mathbf{x}$ that satisfies the Kuhn–Tucker conditions for optimality when $\mathbf{x}$ has the components $(0, 0, 1)$. Thus the value $\Delta = 10m$ occurs in condition (1.15). It follows that the final calculated point, $\mathbf{x}_\ell$ say, satisfies the inequality

$$-10\,m \geq V(\mathbf{x}_\ell) = m \log \mathbf{c}^T \mathbf{x}_\ell - \sum_{j=1}^m \log \mathbf{a}_j^T \mathbf{x}_\ell$$

$$\geq m \log \mathbf{c}^T \mathbf{x}_\ell = m \log[(\mathbf{x}_\ell)_2 + 1], \tag{4.6}$$

where both the assertions in the last line depend on the normalization $(\mathbf{x}_\ell)_3 = 1$. Thus we deduce the condition

$$(\hat{\mathbf{x}}_\ell)_2 = (\mathbf{x}_\ell)_2 \leq -1 + e^{-10} < -0.99995. \tag{4.7}$$

Therefore, even when $L = 10$, the sequence $\{\hat{\mathbf{x}}_k : k = 1, 2, \ldots, \ell\}$ has to pass most or all of the vertices of the polygon that cause the $m/8$ complexity that is reported in Table 1.

We conclude that there are some linear programming problems that are not solved efficiently by the original algorithm. We have noted, however, that usually the given log barrier method requires far more iterations than the earlier algorithm because formula (3.28) increases the trajectory parameter too slowly. Gonzaga [3] mentions some other ways of adjusting this parameter that are more successful in practice, but it is not known whether they enjoy the $\mathcal{O}(\sqrt{m}\log m)$ complexity bound that is the subject of Section 3.

References

[1] D.A. Bayer and J.C. Lagarias, "Karmarkar's linear programming algorithm and Newton's method", *Math. Programming*, Vol. 50 (1991), pp. 291–330.

[2] P.E. Gill, W. Murray, M.A. Saunders, J.A. Tomlin and M.H. Wright, "On projected Newton barrier methods for linear programming and an equivalence to Karmarkar's projective method", *Math. Programming*, Vol. 36 (1986), pp. 183–209.

[3] C.C. Gonzaga, "Large step path-following methods for linear programming, Part 1: Barrier function method", *SIAM J. Optimization*, Vol. 1 (1991), pp. 268–279.

[4] N. Karmarkar, "A new polynomial-time algorithm for linear programming", *Combinatorica*, Vol. 4 (1984), pp. 373–393.

[5] M.J.D. Powell, "Karmarkar's algorithm: a view from nonlinear programming", *IMA Bulletin*, Vol. 26 (1990), pp. 165–181.

[6] M.J.D. Powell, "A worst case example of the complexity of Karmarkar's algorithm", presented at the 14th International Symposium on Mathematical Programming, Amsterdam (1991).

[7] M.J. Todd and B.P. Burrell, "An extension of Karmarkar's algorithm for linear programming using dual variables", *Algorithmica*, Vol. 1 (1986), pp. 409–424.

Department of Applied Mathematics and Theoretical Physics,
University of Cambridge,
Silver Street,
Cambridge CB3 9EW,
England.

H SCHWETLICK

Nonlinear parameter estimation: models, criteria and algorithms

1 The Problem

Let us consider a real process which is mathematically described by a nonlinear *model equation*

$$f(u, t, c, p) = 0 \quad \text{where} \quad f : R^{dimf} \times R \times R^{dimc} \times R^{dimp} \to R^{dimf}. \tag{1}$$

Here t is the *time*, the vector u denotes the *state variables* that describe the state of the process at time t where $dimf = dimu$, the vector c denotes additional *control variables* that can be varied within a certain feasible region to control the process, and p is the vector of *parameters* that characterize the specific process considered. The function f that is supposed to be sufficiently smooth expresses the relations between the *process variables* $\{u, t, c\}$ and parameters p. In general, for given c and p the state $u = u(t; c, p)$ is at least locally uniquely defined by (1). Exceptions are "critical" states which belong to so called "critical" values of the control variables and correspond, e.g., *turning* or *bifurcation points* of (1).

When the process is stationary then the time t does not occure and u describes the stationary state. For simplicity we denote in both cases all the process variables by z, i.e., $z = (u, t, c)$ in the unstationary and $z = (u, c)$ in the stationary case, call them states and rewrite the model equation (1) as

$$f(z, p) = 0 \quad \text{where} \quad f : R^{dimz} \times R^{dimp} \to R^{dimf}. \tag{2}$$

In order to exclude trivial situations we suppose $dimf < dimz$, i.e., $dim(t, c) > 0$ such that, for fixed p, the so-called *state manifold*

$$S_p := \{z : f(z, p) = 0\}$$

is generically a smooth manifold of dimension $dimS = dimz - dimf > 0$ in R^{dimz}.

Let us point out here that finite dimensional model equations arise in different settings.

(i) In the simpler case when no or only a weak insight into the underlying process is available then the type of the function f is not known a priori, and an appropriate *ansatz* parametrized by p has to be made. Examples for this case that is strongly related to *data analysis* are growth curves built from rational functions or exponentials with unknown coefficients in biology or linear combinations of Gaussian density curves for approximating spectra in physics. It is, of course, desirable to have as few parameters as possible and, moreover, the parameters chosen should bear a physical meaning which in most cases leads to a more stable parametrization of the state manifold S_p.

(ii) In the more complex case the real process is modelled by a theoretically founded *operator equation* — commonly a *differential equation* —

$$\mathcal{F}(u, t, c, p) = 0 \quad \text{plus} \quad \mathcal{IC} \quad \text{plus} \quad \mathcal{BC} \tag{3}$$

where $\mathcal{IC}$ and $\mathcal{BC}$ denote initial and boundary conditions, resp., which yield a unique solution $u = u(t; c, p)$ for fixed c and p. Unlike in (1), in (3) both u and c may be functions not only of time, but also of space so that (3) is, in general, a partial differential equation. In the stationary case it is a boundary value problem that can be reduced to a finite dimensional system by appropriate discretization in space. In the unstationary case, the space discretization has to be combined with a time integration that defines the value of the discretized state vector s at given times $t = t_i$ implicitly by an ODE solver or an integration scheme as, e.g., Crank-Nicholson. Examples of such problems are systems of ODE's describing chemical reactions where the rate constants are the parameters, and diffusion-convection or other transport problems modelled by parabolic PDE's whith the diffusion and further coefficients as parameters. We don't discuss this problem of *parameter estimation in differential equations* in detail, see, e.g., [11, 12, 72, 73, 5, 105, 108]. When p depends on time and space, too, then the problem is commonly called an *inverse problem in differential equations* and is, in general, incorrectly posed; we refer to the huge literature devoted to this interesting and difficult task.

Whereas the *direct problem* belonging to (2) consists in computing the manifold S_p of feasible states for given p the *parameter estimation problem* is an *inverse* one, namely

---

Given *observations* $\{\tilde{z}^i\}$ of unknown true states $\{z^{*i}\}$ that belong to the unknown true parameter p^* and are subject to *errors* $\{\delta z^i\}$, i.e.

$$\tilde{z}^i = z^{*i} + \delta z^i \quad \text{where} \quad f(z^{*i}, p^*) = 0 \quad (i = 1, \ldots, m)$$

Find *estimated values* $p = \hat{p}$ to p^* and $z^i = \hat{z}^i$ to z^{*i} such that

$$z^i - \tilde{z}^i \quad \text{``small''} \quad \text{and} \quad f(z^i, p) = 0 \quad (i = 1, \ldots, m)$$

Some authors speak also about *fitting the model to the data*.

In the following we assume the errors $\{\delta z^i\}$ to be stochastically independent and equally distributed with

$$\mathbf{E}(\delta z^i) = 0, \quad \mathbf{D}^2(\delta z^i) = \sigma^2 \, diag(\varepsilon_j) =: V, \quad \varepsilon_j \in \{0, 1\}. \tag{4}$$

This means that there are two groups of state variables $\{z_j\}_{j=1}^{dimz}$. The components z_j from the first group characterized by the index set $I_{fe} \subset \{1, \ldots, dimz\}$ are observed free of errors, i.e. $\sigma_j = \varepsilon_j = 0 \; \forall j \in I_{fe}$. The variables from the complementary second group I_e are subject to errors where, without loss of generality, these variables are supposed to be scaled such that the variances are equal, i.e. $\varepsilon_j = 1$, $\sigma_j = \sigma \; \forall j \in I_e$.

By introducing the notation

$$Z := \begin{bmatrix} - & z^{1^T} & - \\ & \vdots & \\ - & z^{m^T} & - \end{bmatrix}, \quad \tilde{Z} := \begin{bmatrix} - & \tilde{z}^{1^T} & - \\ & \vdots & \\ - & z^{m^T} & - \end{bmatrix}, \quad G(Z,p) := \begin{bmatrix} f(z^1,p) \\ \vdots \\ f(z^m,p) \end{bmatrix}$$

we can write the estimation problem in the compact form

Determine (Z,p) such that $\tilde{Z} - Z$ is "small" subject to $G(Z,p) = 0$. $\qquad(5)$

When doing parameter estimation one has obviously to do three things, namely

(i) Choose the *estimation model*, i.e., choose a model equation (2) and declare which components of z are measured subject to errors, i.e. define the index set I_e.

(ii) Choose the *estimation criterion*, i.e., define what "$\tilde{Z} - Z$ small" means in (5).

When t_l denotes one individual error $t_l := \tilde{z}_j^i - z_j^i$ ($i \in \{1, \ldots, m\}$, $j \in I_e$), then this error is usually evaluated by a so called *criterion function* $\varrho(.) : R \to R^+$, and all these errors are summed up to yield the scalar objective function

$$\psi(Z) := \sum_l \varrho(t_l) = \sum_{i=1}^m \sum_{j \in I_e} \varrho(\tilde{z}_j^i - z_j^i). \qquad(6)$$

When doing so problem (5) becomes the equality constrained, highly structured nonlinear optimization problem

$$\min \{ \psi(Z) : (Z,p) \text{ subject to } G(Z,p) = 0 \}. \qquad(7)$$

Under assumptions (4) the in a certain statistical sense "natural" estimation criterion is the *least squares criterion* defined by

$$\varrho(t) := \varrho_2(t) := \tfrac{1}{2} t^2,$$

one speaks about *least squares* or ℓ_2 *estimation*. However, recently also so-called *robust criterion functions* $\varrho(.)$ have found a lot of interest, but there are also other types of objective functions as certain determinants arising in *multiresponse parameter estimation* that do not have such a simple additive structure as (6).

(iii) Once the nonlinear estimation problem has been fixed *numerical algorithms* for solving the optimization problem (7) are needed.

The aim of this paper is to give an overview on standard as well as nonstandard nonlinear parameter estimation problems used today and the numerical algorithms for solving them. For the statistical background see, e.g., [5, 30, 81, 8, 98].

2   Models and Related ℓ_2 Estimation Problems

In this section estimation problems are classified with respect to the type of the model function and the structure of the errors. Moreover, the corresponding least squares optimization problems are formulated.

2.1 Implicit Models

A model is called *implicit* when the model equations (2) can not be solved for $dimf$ of the state variables. A simple two-dimensional example for such a situation is an ellipse described in Cartesian coordinates by the implicit relation

$$z_1^2/p_1^2 + z_2^2/p_2^2 = 1.$$

If all components of z are subject to errors then (2) is called *implicit model with errors in the variables*, sometimes also a *functional relation*. In this case the ℓ_2 version of (7) yields the *equality constrained nonlinear least squares problem*

$$\min \left\{ \psi(Z) := \frac{1}{2} \sum_{i=1}^m \|\tilde{z}^i - z^i\|^2 = \frac{1}{2}\|\tilde{Z} - Z\|_F^2 : (Z,p) \text{ s.t. } G(Z,p) = 0 \right\} \qquad (8)$$

where $\|v\| := \sqrt{\sum_j v_j^2}$ and $\|M\|_F := \sqrt{\sum_i \sum_j m_{ij}^2}$ denote the the Euclidean norm of a vector $v = (v_i)$ and the Frobenius norm of a matrix $M = (m_{ij})$, resp.

Here and in what follows we always assume that m is so large that the number of scalar squares in the objective function ψ ($= m \times dimz$ in (8)) is greater than or equal the number of scalar variables to be estimated minus the number of scalar equality constraints ($= m \times dimz + dimp - m \times dimf$ in (8)).

Let us note that when (Z,p) solves (8) then due to the necessary extremality conditions the optimal deviations $\tilde{z}^i - z^i$ are orthogonal to S_p, i.e., orthogonal to the tangential manifold of S_p at $z^i \in S_p$.

If certain components z_j ($j \in I_{fe}$) are free of errors then there holds

$$\delta z_j^i = 0 \quad \text{and hence} \quad z_j^i = \tilde{z}_j^i = z_j^{*i} \quad \forall j \in I_{fe}.$$

Consequently, these components have to be dropped in the objective function ψ, cf. (6), and to be fixed to the exactly observed values $\tilde{z}_j^i$ in the constraints. For simplicity suppose $I_{fe} = \{1, \ldots, dimz_{fe}\}$. Then z can be written as $z = \binom{z_{fe}}{z_e} =: (z_{fe}, z_e)$ with z_{fe} and z_e containing the variables observed free of or subject to errors, resp., and the corresponding estimation problem is

$$\min \left\{ \psi(Z_e) := \frac{1}{2} \sum_{i=1}^m \|\tilde{z}_e^i - z_e^i\|^2 : (Z_e,p) \text{ s.t. } G(Z_e,p) := (f(\tilde{z}_{fe}^i, z_e^i, p)) = 0 \right\} \qquad (9)$$

where Z_e is defined analoguously to Z and $f(z_{fe}, z_e, p) := f(z,p)$.

Let us finish the treatment of implicit models by discussing two special cases.

2.1.1 Additional Observations of Critical Points

Suppose that the process considered is a stationary one described by the model equations

$$f(u,c,p) = 0 \quad \text{with} \quad f : R^n \times R^{dimc} \times R^{dimp} \to R^n, \qquad (10)$$

cf. (1). Let $\lambda := c_1$ be a distinguished control variable, usually called the *bifurcation variable*, and rewrite (10) as

$$f(y,w,p) = 0 \quad \text{where} \quad f : R^{n+1} \times R^{dimw} \times R^{dimp} \to R^n \qquad (11)$$

167

with $y := (u, \lambda) = (u, c_1)$ and $w := (c_2, \ldots, c_{dimc})^T$, $dimw = dimc - 1$. Suppose further that the Jacobian $f_y = (\partial f_i / \partial y_j)$ of size $n \times (n + 1)$ has full rank n such that, for fixed (w, p), the solution manifold $S_{w,p} := \{ y : f(y, w, p) = 0 \}$ of (11) is a one-dimensional smooth curve in R^{n+1} which does not have branchings and has at most so-called *turning points* as critical points. At such points the tangent direction $v = v(y, w, p) \in R^{n+1}$ is orthogonal to the λ-axis, i.e., they are characterized by the scalar *turning point condition*

$$\varphi(y, w, p) := e^{n+1^T} v(y, w, p) = e^{n+1^T} B(y, w, p)^{-1} e^{n+1} = 0 \tag{12}$$

where v is defined as solution of the nonsingular square linear system

$$Bv = e^{n+1} \quad \text{with} \quad B := B(y, w, p) := \left[\frac{f_y(y, w, p)}{r^T} \right]. \tag{13}$$

Here $r \in R^{n+1}$ with $\|r\| = 1$ denotes an appropriately choosen normalizing vector the choice of which may depend on y, see [77, 93, 82] for details.

Suppose now that we have m observations $\{\tilde{y}^i, w^i\}_{i=1}^m$ belonging to "ordinary" states $y^{*i} \in S_{w^i, p^*}$ and additional M observations $\{\tilde{y}^i, w^i\}_{i=m+1}^{m+M}$ belonging to "critical" states y^{*i}, i.e., any such y^{*i} is a turning point of S_{w^i, p^*}. These turning points characterize unstable states of the underlying process, i.e. critical values of the control variables $c = (\lambda, w)$. The corresponding estimation problem is then

$$\min \left\{ \psi(Y) := \tfrac{1}{2} \sum_{i=1}^{m+M} \|y^i - \tilde{y}^i\|^2 = \tfrac{1}{2}\|Y - \tilde{Y}\|_F^2 : (Y, p) \right\} \tag{14}$$

s. t. $f(y^i, w^i, p) = 0 \ (i = 1, \ldots, m)$, $\varphi(y^i, w^i, p) = 0 \ (i = m + 1, \ldots, m + M)$.

Note that the constraints $\varphi(y^i, w^i, p) = 0$ contain first derivatives of f.

2.1.2 Implicit Models Linear in the State Variables

Let us go back to the implicit model (2) and consider the special case that f depends linearly on the states z, i.e., that the model is given by

$$f(z, p) := D(p)^T z = 0 \quad \text{with a matrix} \quad D(p) \in R^{dimz \times dimf}. \tag{15}$$

Such problems arise, e.g., in signal processing in connection with *linear prediction models*

$$y_i = a_1 y_{i-1} + \cdots + a_k y_{i-k} + b_0 x_i + b_1 x_{i-1} + \cdots + b_l x_{i-l} \quad (i = n, \ldots, n + m) \tag{16}$$

where $n := \max\{k, l\}$. The states $z = \{x_j, y_j\}$ to be estimated are

$$z := (y_{n+m}, y_{n+m-1}, \ldots, y_{n-k}, x_{n+m}, x_{n+m-1}, \ldots, x_{n-l})^T,$$

and the vector $\tilde{z}$ of observations has the same form as z with $\{x_j, y_j\}$ replaced by $\{\tilde{x}_j, \tilde{y}_j\}$. The matrix which expresses the relation (16) in the form (15) is then given by

$$D(p) := \begin{bmatrix} -1 & a_1 & \cdots & a_k & & & & b_0 & b_1 & \cdots & b_l & & & & \\ & -1 & a_1 & \cdots & a_k & & & & b_0 & b_1 & \cdots & b_l & & & \\ & & \ddots & \ddots & \ddots & \ddots & & & & \ddots & \ddots & \ddots & \ddots & \\ & & & -1 & a_1 & \cdots & a_k & & & & b_0 & b_1 & \cdots & b_l \end{bmatrix}. \tag{17}$$

This matrix depends linearly on the parameters $p = (a_1, a_2, \ldots, a_k, b_0, b_1, \ldots, b_l)^T$ and has a special band Toeplitz structure. The model described corresponds formally to the case $i = 1$ in (8) so that the optimization problem becomes

$$\min \left\{ \psi(z) := \tfrac{1}{2}\|\tilde{z} - z\|^2 : (z, p) \text{ s.t. } f(z, p) := D(p)^T z = 0 \right\} \tag{18}$$

We refer to [104] where such bilinear models (16) with errors in the variables are investigated. A somewhat different approach which leads to a total least squares problem is described in [106, 106], see also [66] where further material can be found. For related problems where D depends nonlinearly on p see [75].

2.2 Explicit Models

In many cases the model equations (2) can be solved for $dimf$ of the state variables, say for the last $dimf$ components of $z =: (x, y)$ collected in y. Hence, they can be written — or are a priori given — in the *explicit* form

$$y = r(x, p) \quad \text{where} \quad r : R^{dimx} \times R^{dimp} \to R^{dimy}, \quad dimy = dimf. \tag{19}$$

This means that z splits up into the independent variables x and the dependent variables y also-called *responses*, and f has the form $f(z, p) = f(x, y, p) = y - r(x, p)$.

When both x and y are subject to errors then the model is denoted as *errors-in-variables model*. Here we have $\|\tilde{z}^i - z^i\|^2 = \|\tilde{x}^i - x^i\|^2 + \|\tilde{y}^i - y^i\|^2$. By solving the constraint $f(z^i, p) = y^i - r(x^i, p) = 0$ for y^i and replacing y^i by $r(x^i, p)$ problem (8) goes over into the *unconstrained nonlinear least squares problem*

$$\min \{ \psi(X, p) : (X, p) \}, \tag{20}$$

$$\psi(X, p) := \tfrac{1}{2}\sum_{i=1}^{m} \left\{ \|\tilde{x}^i - x^i\|^2 + \|\tilde{y}^i - r(x^i, p)\|^2 \right\} = \tfrac{1}{2}\left\{ \|\tilde{X} - X\|_F^2 + \|\tilde{Y} - R(X, p)\|_F^2 \right\}$$

with

$$X := \begin{bmatrix} - & x^{1^T} & - \\ & \vdots & \\ - & x^{m^T} & - \end{bmatrix}, \quad \tilde{Y} := \begin{bmatrix} - & \tilde{y}^{1^T} & - \\ & \vdots & \\ - & y^{m^T} & - \end{bmatrix}, \quad R(X, p) := \begin{bmatrix} r(x^1, p)^T \\ \vdots \\ r(x^m, p)^T \end{bmatrix}. \tag{21}$$

Problems of this type are also refered to as *total approximation problems*, see [110, 112] or, due to the orthogonality property mentioned above, as *orthogonal regression problems*, cf. [14].

When the independent variables x are free of errors, then the model is called *regression model*. Here we have $x^i = \tilde{x}^i = x^{*i}$, and (20) reduces to

$$\min \left\{ \psi(p) := \tfrac{1}{2}\sum_{i=1}^{m} \|\tilde{y}^i - r(\tilde{x}^i, p)\|^2 = \tfrac{1}{2}\|\tilde{Y} - R(\tilde{X}, p)\|_F^2 : p \right\} \tag{22}$$

as special case of (9) for $z_{fe} = x$, $z_e = y$ and the explicit model equation (19). The regression model is the classical standard model in nonlinear parameter estimation. Unlike the preceding models it requires to estimate only the parameters p but no states.

169

2.2.1 Explicit Models Linear in Some Parameters

In many practical explicit models some of the parameters occur linearly. A typical example is fitting spectral data by scalar functions of the type

$$y = r(x, a, b) = \sum_{j=1}^{l} a_j \cdot e^{-\left(\frac{x-b_j}{b_{2j}}\right)^2} + a_{l+1} + a_{l+2}x + a_{l+3}x^2$$

where y depends linearly on the $l + 3$ parameters $a = (a_j)$ but nonlinearly on the $2l$ parameters $b = (b_n)$.

The one-dimensional general case is characterized by $dimy = 1$, $p = (a, b)$ with $a \in R^l$, $b \in R^n$ and a model equation

$$y = r(x, a, b) = \sum_{j=1}^{l} \varphi_j(x, b) \cdot a_j, \tag{23}$$

i.e., r is a linear combination of ansatz functions $\varphi_j(.,.) : R^{dimx} \times R^n \to R$. Here we have $\tilde{Y} = \tilde{y}$ and $R(X, p) = R(X, a, b) = \Phi(X, b)a$ where $\tilde{y}$ and $\Phi(X, b)$ are defined as

$$\tilde{y} := \begin{bmatrix} \tilde{y}_i \\ \vdots \\ \tilde{y}_m \end{bmatrix}, \quad \Phi(X, b) := \left(\varphi_j(x^i, b)\right) = \begin{bmatrix} \varphi_1(x^1, b) & \cdots & \varphi_l(x^1, b) \\ \vdots & & \vdots \\ \varphi_1(x^m, b) & \cdots & \varphi_l(x^m, b) \end{bmatrix}. \tag{24}$$

The regression model belonging to (23) leads to the least squares problem

$$\min\left\{\psi(a, b) := \frac{1}{2}\sum_{i=1}^{m}\left[\tilde{y}_i - \sum_{j=1}^{l}\varphi_j(\tilde{x}^i, b)a_j\right]^2 = \frac{1}{2}\|\tilde{y} - \Phi(\tilde{X}, b)a\|^2 : (a, b)\right\}. \tag{25}$$

Problems of such a structure are denoted as *separable least squares problems*, see [41, 84].

In the multiresponse case $dimy > 1$ one allows for each component of $y = (y_k)$ an extra set of coefficients $a := a^k$ which are collected in the matrix $A := \left[a^1, \ldots, a^{dimy}\right]$, but one takes the same b for all k, i.e. the model equations become

$$y_k = r_k(x, a^k, b) = \sum_{j=1}^{l} \varphi_j(x, b) \cdot a_j^k \quad (k = 1, \ldots, dimy),$$

cf. [40, 53]. The resulting least squares problem for the regression model is then

$$\min\left\{\psi(A, b) := \frac{1}{2}\sum_{i=1}^{m}\sum_{k=1}^{dimy}\left[\tilde{y}_i^k - \sum_{j=1}^{l}\varphi_j(\tilde{x}^i, b)a_j^k\right]^2 = \frac{1}{2}\|\tilde{Y} - \Phi(\tilde{X}, b)A\|^2 : (A, b)\right\}. \tag{26}$$

When the ansatz functions φ_j are functions of x only, i.e., when b disappears, then the classical *linear model* is obtained, or more precisely the *model linear in the parameters*. Examples are polynomials or splines with fixed knots. In the regression case the resulting optimization problem is the linear least squares problem

$$\min\left\{\psi(a) := \frac{1}{2}\sum_{i=1}^{m}\left[\tilde{y}_i - \sum_{j=1}^{l}\varphi_j(\tilde{x}^i)a_j\right]^2 = \frac{1}{2}\|\tilde{y} - \Phi(\tilde{X})a\|^2 : a\right\}$$

with the so-called *observation matrix* $\Phi(\tilde{X}) = (\varphi_j(\tilde{x}^i))$ as matrix of coefficients. In the errors-in-variables case, however, the problem becomes a nonlinear one, namely

$$\min\{\psi(X, a) : (X, a)\},$$

$$\psi(X, a) := \frac{1}{2} \sum_{i=1}^{m} \left\{ \|\tilde{x}^i - x^i\|^2 + \left[\tilde{y}_i - \sum_{j=1}^{l} \varphi_j(x^i) a_j \right]^2 \right\} = \frac{1}{2} \left\{ \|\tilde{X} - X\|_F^2 + \|\tilde{y} - \Phi(X) a\|^2 \right\}.$$

The simplest explicit model is given by the relation

$$y = r(x, a) = \sum_{j=1}^{l} x_j \cdot a_j \tag{27}$$

which is bilinear in x and a; statisticians denote only this model as *linear*. Here we have $\varphi_j(x) = x_j$ and $\Phi(X) = X$ so that in the regression case the linear least squares problem

$$\min\left\{ \psi(a) := \frac{1}{2} \sum_{i=1}^{m} \left[\tilde{y}_i - \sum_{j=1}^{l} \tilde{x}_j^i a_j \right]^2 = \frac{1}{2} \|\tilde{y} - \tilde{X}a\|^2 : a \right\} \tag{28}$$

is obtained whereas the errors-in-variables case leads to the so-called *total least squares problem*

$$\min\{\psi(X, a) : (X, a)\}, \tag{29}$$

$$\psi(X, a) := \frac{1}{2} \sum_{i=1}^{m} \left\{ \|\tilde{x}^i - x^i\|^2 + \left[\tilde{y}_i - \sum_{j=1}^{l} x_j^i a_j \right]^2 \right\} = \frac{1}{2} \left\{ \|\tilde{X} - X\|_F^2 + \|\tilde{y} - Xa\|^2 \right\}$$

where both a and X have to be estimated, see [43] for basic facts and [107] for a thorough discussion of such problems including applications in different areas.

Let us end this section with the remark that all the models considered here are homogeneous with respect to the $\{a_j\}$. It is not difficult to add an inhomogenity $\varphi_0(x, b)$ on the right hand side of (23) which can be thought as generated by additional ansatz functions φ_j the coefficients of which has been frozen at fixed values.

3 Methods for Least Squares Estimation Problems

The optimization problems which belong to the estimation problems discussed in the preceding section are either unconstrained least squares problems

$$\min\left\{ \psi(u) = \frac{1}{2} \|F(u)\|^2 : u \in R^N \right\} \tag{30}$$

with $F : R^N \to R^M, M \geq N$, or else equality constrained problems

$$\min\left\{ \psi(u) = \frac{1}{2} \|F(u)\|^2 : u \in R^N \text{ s. t. } G(u) = 0 \right\} \tag{31}$$

with $F : R^N \to R^M$, $G : R^N \to R^L$, $M \geq N - L$ where $u = (\ldots, p)$ stands for the variables to be estimated and F and G define the objective function and the constraints.

The algorithms for solving such problems are iterative where at the k-th iteration step the next iterate $u^+ := u^{k+1}$ is obtained as

$$u^+ := u + s. \tag{32}$$

Here $u := u^k$ denotes the current iterate and $s := s^k$ is the so-called *step*.

3.1 Unconstrained Problems

3.1.1 Quadratic Models

In order to define the step s for the unconstrained problem (30) the function ψ is approximated in the neighbourhood of u by a quadratic *model function* μ according to

$$\psi(u+s) \approx \mu(u+s) := \psi + \nabla\psi^T s + \tfrac{1}{2}s^T H s, \quad H \approx \nabla^2 \psi \qquad (33)$$

where $\psi := \psi(u)$, $\nabla\psi := \nabla\psi(u)$ is the gradient, and s is determined as solution of

$$\min\left\{\mu(u+s) = \psi + \nabla\psi^T s + \tfrac{1}{2}s^T H s : s \in R^N\right\}; \qquad (34)$$

for simplicity the arguments $u = u^k$ are always omitted. Looking at the Taylor expansion

$$\psi(u+s) = \psi + \nabla\psi^T s + \tfrac{1}{2}s^T\nabla^2\psi s + R = \tfrac{1}{2}\|F\|^2 + F^T Js + \tfrac{1}{2}s^T[J^T J + F \circ F'']s + R \quad (35)$$

of ψ where $J := J(u) := F'(u) = (\partial F_i(u)/\partial u_j)$ denotes the $M \times N$ *Jacobian* of F it is seen that the simplest way for defining H would be to drop the term

$$S := F \circ F'' := \sum_{i=1}^{M} F_i \cdot \nabla^2 F_i \qquad (36)$$

of the Hessian $\nabla^2\psi$ containing products of the F_i and its second order derivatives. This leads to the so-called *Gauss-Newton* (=: GN) *model*

$$H := H_{GN} := J^T J, \quad \mu_{GN}(u+s) := \psi + \nabla\psi^T s + \tfrac{1}{2}s^T J^T J s = \tfrac{1}{2}\|F + Js\|^2. \qquad (37)$$

The minimizer of $\mu_{GN}(u+s)$, i.e., the solution of the linear least squares problem

$$\min\left\{\mu_{GN}(u+s) = \tfrac{1}{2}\|F + Js\|^2 : s \in R^N\right\} \qquad (38)$$

of minimal norm is then given by $s := s_{GN} = J^+ F$ where J^+ denotes the Moore-Penrose inverse of J, and the iteration (32) then becomes the *undamped Gauss-Newton method*

$$s^+ = s + J^+ F$$

which is the basic method for solving least squares problems. The GN approximation H_{GN} is always positive semidefinite, and it is positive definite if and only if $\text{rank}\, J = N$, i.e., if J has full column rank, and in this case s_{GN} is the unique solution of (38).

Let u^* be a stationary point of ψ such that $\nabla\psi(u^*) = 0$ and $\text{rank}\, J(u^*) = N$, let F' be Lipschitz continuous with a constant L in a neighbourhood $\mathcal{U}(u^*)$ of u^*, and suppose that

$$\kappa^* := L\|F(u^*)\|\|[J(u^*)^T J(u^*)]^{-1}\| = L\|F(u^*)\|\|J(u^*)^+\|^2 < 1 \qquad (39)$$

holds . Then for sufficiently good starting values u^0, the undamped GN method is well defined and converges Q-linearly to u^* with a convergence factor $\tau_{GN} \leq \kappa^*$. The *small residual condition* (39) requires the *residual* $\|F(u^*)\|$ to be sufficiently small compared to the nonlinearity of F measured by L. When $F(u^*) = 0$, i.e., when the overdetermined system $F(u) = 0$ is *consistent*, then the convergence is even Q-quadratic, see [92, 28, 36, 9] for a general discussion of the GN method. A sharper analysis shows that κ^* can be replaced by the smaller value $\kappa^{**} := \max |\lambda_j^*|$ with λ_j^* being the eigenvalues of the generalized eigenproblem $[F(u^*) \circ F''(u^*)]u = \lambda[J(u^*)^T J(u^*)]u$, cf. [113, 84] and also [109] for a discussion of the terms that build the Hessian. Note further that $\kappa^{**} \leq \kappa^* < 1$ implies $\nabla^2 \psi(u^*)$ to be positive definite, i.e., u^* is then even a strong local minimizer of ψ.

In order to obtain fast convergence also in case of *large residual problems* where (39) is not fullfilled, *Newton-type models* defined by

$$H_N := J^T J + B, \quad \mu_N(u+s) := \psi + \nabla \psi^T s + \tfrac{1}{2} s^T H_N s, \quad B \approx S := F \circ F'' \qquad (40)$$

which employ second order information have to be used. Note that when exact second derivatives are used H_N may be indefinite so that (34) with $H = H_N$ that defines the *Newton step* $s = s_N$ may not have a solution. Moreover, unlike (38) this problem is not a linear least squares problem and, hence, the numerically stable orthogonalization techniques do not apply for solving it, see [54, 43, 9] for a discussion of the linear algebra aspects and [20, 56] for techniques which avoid the explicit formation of $J^T J$ when solving the linear system $(J^T J + B)s + J^T F = 0$ which defines s_N.

There are many proposals for approximating S by applying quasi-Newton techniques using exact gradients $\nabla \psi = J^T F$, we refer to [71, 27, 1, 103]. The two search directions s_{GN}, s_N can also be used to construct *hybrid algorithms* where the decision about which direction should be taken is made adaptively during the iteration process, see [37] and also [27] where a widely used algorithm is described which makes use of both the Gauss-Newton and Newton model in a trust region approach.

However, in the opinion of the author bad performance of GN methods in most cases signals bad modelling and/or insufficient observation accuracy so, instead of using methods which can handle large residuals, it would be better to improve the model and the measurements. Many of the numerical examples that demonstrate poor behaviour of the GN method, see [47, 38], are, by the way, rather artificial and far from being realistic.

It is commonly accepted that good first order information, i.e., exact Jacobians $J(u) = F'(u)$ or consistent difference approximations $\delta F(u, h) \approx F'(u)$ with a reasonable chosen stepsize h should be used in solving least squares problems. There are, however, attempts to apply quasi-Newton techniques as Broyden's update to approximate J cheaply using only function values of F, see [46].

3.1.2 Globalization

In order to obtain convergence for bad starting values too, appropriate globalization techniques have to be applied.

The simplest way consists of using *line search methods* of the form

$$s = s(\alpha) = \alpha \cdot s_B, \quad i.e., \quad u^+ = u + \alpha \cdot s_B$$

with a basic step s_B that has to be a descent direction for ψ, say $s_B = s_{GN}$ in the full rank case, and a *stepsize* or *damping factor* $\alpha > 0$ which is chosen by a *line search algorithm* in such a way that ψ is sufficently decreased in the sense of

$$\underbrace{[\psi(u) - \psi(u + \alpha s_B)]}_{:=ared} \geq \delta\underbrace{[\mu(u) - \mu(u + \alpha s_B)]}_{:=pred}, \quad \delta > 0 \tag{41}$$

where $ared$ is the *actual reduction* obtained in ψ with the step $s(\alpha)$ and $pred$ is the *predicted reduction* predicted by the model μ; in (41) instead of the quadratic model μ also the linear model $\mu_1(u + s) := \psi + \nabla\psi^T s$ may be used. For a general treatment of such methods see [92, 28, 36]; line searches tailored to the least squares case are proposed in [62, 2].

When H is not positive definite then the unconstained problem (34) has no or arbitrary large solutions. A possibility to overcome this shortcoming of line search methods give the *trust region methods* where the step $s = s_\Delta$ is defined by the constrained problem

$$\min\left\{\mu(u + s) = \psi + \nabla\psi^T s + \tfrac{1}{2}s^T Hs : \|Ds\| \leq \Delta\right\} \tag{42}$$

that is always solvable. Here $D = D_k$ is a *scaling matrix* which has full rank N and generates the elliptic norm $\|s\|_D = \|Ds\|$ in which the step s is measured; the standard choice is a diagonal matrix $D = diag(d_j)$ with appropriately chosen weights $d_j > 0$. The number $\Delta = \Delta_k > 0$ is the so-called *trust region radius* that characterizes the region in which the model μ is considered to be a good approximation to ψ and which is updated depending on how $pred$ approximates $ared$.

In most cases the solution $s = s_\Delta$ of (42) can be found as solution $s = s(\lambda)$ of the regularized system

$$(H + \lambda D^T D)s + \nabla\psi = 0. \tag{43}$$

where λ has to be chosen such that $H + \lambda D^T D$ is positive definite and $\|s(\lambda)\| = \Delta$ holds which, in general, requires the repeated solution of (43) for different values of λ, see [39, 70] for details.

In the GN case (43) goes over into $(J^T J + \lambda D^T D)s + J^T F = 0$. These equations are the normal equations to the regularized linear least squares problem

$$\min\left\{\left\|\begin{bmatrix} J \\ \sqrt{\lambda}D \end{bmatrix}s + \begin{bmatrix} F \\ 0 \end{bmatrix}\right\|^2 = \|F + Js\|^2 + \lambda\|Ds\|^2 : s \in R^N\right\}. \tag{44}$$

which can be solved by orthogonalization methods, see [68, 27]. In the GN case we have always $s_\Delta = s(\lambda)$ with $\lambda \geq 0$, eventually as limit $\lambda \to 0$. When the step is directly controlled by the Lagrange variable λ and not by Δ then the method is called *Marquardt's method*, see [65] and [91] for a convergence analysis.

Convergence results for trust region methods are given in [69, 28, 36], see also [100] for a very general approach. Note that these methods work also when H is singular or indefinite.

Since the linear algebra for computing the trust region step is comparatively costly attempts have been made to reduce this work by, e.g., restricting s to a polygonial arc as is done in Powell's *dog leg method*, see [79], or to a low dimensional subspace, see [69, 55], or by using nonstandard scaling matrices that allow cheap factorizations, see [102, 97, 94].

A completely different possibility for obtaining global convergence consists in applying *continuation* or *homotopy methods*. In these methods the single problem (30) is replaced by a family of problems

$$\min\{\phi(u,\lambda) : x\}, \quad \lambda \in [0,1] \tag{45}$$

parametrized by the real *homotopy parameter* λ where ϕ is chosen in such a way that for $\lambda = 0$ problem (45) has the the known solution $u(0) = u^0$ whereas for $\lambda = 1$ the function ϕ coincides with ψ, i.e., that $\phi(u,1) = \psi(u) \; \forall u$ holds which implies that each solution $u(1) = u^*$ of (45) for $\lambda = 1$ solves the original problem (30). Examples are

$$\phi(u,\lambda) := \tfrac{1}{2}\|F(u) - (1-\lambda)F(u^0)\|^2$$

used in [26] and

$$\phi(u,\lambda) := \tfrac{1}{2}\left\{(1-\lambda)\alpha\|D(u-u^0)\|^2 + \|F(u) - (1-\lambda)F(u^0)\|^2\right\}$$

proposed in [86] where D is again a scaling matrix and $\alpha > 0$ is a weighting factor for the regularizing term $\|D(u-u^0)\|^2$. By applying a *path following algorithm* of predictor-corrector type for tracing the path $\{u = u(\lambda) : \lambda \in [0,1]\}$ (hopefully) implicitly defined by the solutions of (45) one wishes to arrive at $u(1) =: x^*$ which will be a solution of (30). For a general treatment of continuation methods for nonlinear equations see [92, 82, 3].

3.1.3 Nonquadratic Models

Since the objective functions ψ arising in parameter estimation are often strongly nonlinear, model functions which are more complex than quadratic functions could be used. The *conic models* are of type

$$\mu_{con}(u+s) := \psi + \frac{\nabla\psi^T s}{1 - a^T s} + \frac{1}{2}\frac{s^T H s}{(1 - a^T s)^2}$$

where $a \in R^N$ is an additional parameter of the model called the *horizon vector*, see [25, 4, 44, 63, 101, 88]. The term "conic" comes from the fact that the level sets of μ are conic sections. Since μ_{con} can be generated from the quadratic model μ by the transformation $s \to s/(1 - a^T s)$ these models are also called *collinear scalings*. It seems, however, that the additional costs raised by the more complicated model do not pay off in most cases.

Alternatives are the *tensor* [88] or *quadratically appended models* [45] of type

$$\mu_{tens}(u+s) := \tfrac{1}{2}\|F + Js + \tfrac{1}{2}T[s,s]\|^2$$

where $T(u)[.,.]$ is a tensor, i.e., a bilinear map from $R^N \times R^N$ into R^M which approximates F''. Usually the class of tensors T will be restricted to linear combinations

$$T[.,.] := \sum_{k=1}^{l} a^k [d^k \, d^{k^T}]$$

of so-called *rank one tensors* $T = add^T$ defined by $T[s^1, s^2] := a(s^{1T}d)(d^T s^2)$ where the vectors a^k, $d^k \in R^N$ are chosen such that additional interpolation conditions are met, see

[89, 88, 90] for a detailed discussion. Tensor models can be used in a trust region context, too, and methods based on such tensor models are able to converge superlinearly also in case of rank deficient Jacobians.

For a different approach to problems with rank deficient $J(u^*)$ see also [67, 48].

3.1.4 Separable Least Squares Problems

A least squares problems is called *(linearly) separable* if it is of type

$$\min \left\{ \psi(a, b) := \tfrac{1}{2}\|\tilde{y} - \Phi(b)a\|^2 : (a, b) \in R^l \times R^n \right\} \tag{46}$$

where $\tilde{y} \in R^M$, and $\Phi(.) : R^n \to R^{M \times l}$ is an arbitrary matrix function, cf. problem (25) of section 2.2.1 for an example. In the following $\Phi(b)$ is supposed to have full rank l.

For fixed b problem (46) is a linear least squares problem with respect to a which has the unique solution

$$a = a_{opt}(b) = \Phi(b)^+\tilde{y}, \qquad \Phi(b)^+ = [\Phi(b)^T\Phi(b)]^{-1}\Phi(b)^T. \tag{47}$$

By inserting this expression into the objective function the reduced problem

$$\min \left\{ \phi(b) := \tfrac{1}{2}\| \underbrace{[I - \Phi(b)\Phi(b)^+]\tilde{y}}_{:=G(b)} \|^2 = \tfrac{1}{2}\|G(b)\|^2 : b \in R^n \right\} \tag{48}$$

is obtained which contains only the nonlinearly occuring parameters b. Since $I - \Phi(b)\Phi(b)^+$ as a function of b is the projector onto $im\,\Phi(b)^\perp$, methods based on this reduction technique are called *variable projection methods*, see [41].

The Jacobian G' of G is given by

$$G'(b)s = \left\{ \mathcal{A}(b)[s] + \left(\mathcal{A}(b)[s]\right)^T \right\}\tilde{y}, \quad \mathcal{A}(b)[s] := -[I - \Phi(b)\Phi(b)^+](\Phi'(b)[s])\Phi(b)^+ \tag{49}$$

so that the GN model

$$\mu_{GN}(b + s) := \tfrac{1}{2}\|G + G's\|^2 \tag{50}$$

can be built and used for computing the step s as is done in [41]. However, a closer look at this model shows that

$$\|G + G's\|^2 = \|G + \mathcal{A}[s]\tilde{y}\|^2 + s^T T_G s, \quad u^T T_G v := G^T(\Phi'[u])\Phi^+\Phi^{+T}(\Phi'[v])^T G$$

holds. This means that the second term $(\mathcal{A}[.])^T\tilde{y}$ of G' in (49) does not contribute to the gradient $\nabla\phi = G'^T G$ and changes the Hessian $\nabla^2\phi$ only by the matrix T_G which is of order $\mathcal{O}(\|G\|^2)$. If $\|G\|$ is small, then this perturbation is smaller than the perturbation caused by the second order term $S_G = G \circ G''$ of $\nabla^2\phi$ which is of order $\mathcal{O}(\|G\|)$, cf. (35). Following the GN philosophy this term can be dropped too, which leads to the simplified model

$$\mu_K(b + s) := \tfrac{1}{2}\|G + As\|^2, \quad As := \mathcal{A}[s]\tilde{y} = -[I - \Phi\Phi^+](\Phi'[s])\Phi^+\tilde{y} \tag{51}$$

named after Kaufman [51], see also also [84] where a more direct derivation and generalizations are given. An efficient implementation is described in [59], extensions to equality and inequality constrained problems can be found in [52, 22] and [76], resp., and problems of type (26) with several a's are investigated in [40, 53].

Let us note that the implicit linear errors-in-variables problem (18) of section 2.1.2 can be reduced to the form (48) too, as has been done in a different way in [104]. By introducing the new variable $u := \tilde{z} - z$ problem (18) goes over into

$$\min \left\{ \tfrac{1}{2} \|u\|^2 : (u, p) \text{ s.t. } D(p)^T u = D(p)^T \tilde{z} \right\}.$$

For fixed p the solution u is the minimum-norm solution of the underdetermined system $D^T u = D^T \tilde{z}$ which is given by $u = u_{opt}(p) = D(p)^{T+} D(p)^T \tilde{z} = D(p) D(p)^+ \tilde{z}$. By inserting this expression into the objective function the reduced unconstrained problem

$$\min \left\{ \phi(p) := \tfrac{1}{2} \| \underbrace{D(p) D(p)^+ \tilde{z}}_{:=H(p)} |^2 = \tfrac{1}{2} \| H(p) \|^2 : p \right\}. \tag{52}$$

is obtained. This problem looks "complementary" to (48). Indeed, here the term corresponding to $(\mathcal{A}[.])^T \tilde{z}$ of G', cf. (49), can be dropped, which leads to the simplified model

$$\mu_K(p+s) := \tfrac{1}{2} \| G + Bs \|^2, \quad Bs := (\mathcal{A}[s])^T \tilde{z} = (D^+)^T (D'[s])^T [I - DD^+] \tilde{z}. \tag{53}$$

In [104] an efficient method for computing the QR-factorization of D is proposed for the case that D is the Toeplitz matrix (17). This factorization is exploited for computing ϕ and $\nabla \phi$ for use in a quasi-Newton minimization algorithm. However, at the same costs also a Kaufman-like GN method based on the model (53) can be used. This model is equivalent to the model which would be obtained by the method described in [52], but does not require a specific factorization for its formulation.

3.1.5 Application to Explicit Estimation Problems

In the *regression case* the least squares problem (22) to be solved is of the form (30) with $u = p$ and

$$F = F(p) = \begin{bmatrix} \tilde{y}^1 - r(\tilde{x}^1, p) \\ \vdots \\ \tilde{y}^m - r(\tilde{x}^m, p) \end{bmatrix}, \quad J = J(p) = - \begin{bmatrix} r_p(\tilde{x}^1, p) \\ \vdots \\ r_p(\tilde{x}^m, p) \end{bmatrix}$$

where r_p is the Jacobian of r with respect to the parameter p. The blocks of F and J are built by r and r_p, resp., evaluated at the m observations $\{\tilde{x}^i\}$. This row-wise block structure suggests to apply row-oriented orthogonalization methods for computing the GN or trust region step, see [23, 24].

In the *errors-in-variables case* the related least squares problem (20) yields a problem (30) whith $u := (X, p)$ and $F = F(X, p)$, $J = J(X, p)$ given by

$$\underbrace{F(X, p)}_{=F} = \begin{bmatrix} \tilde{x}^1 - x^1 \\ \vdots \\ \tilde{x}^m - x^m \\ \hline \tilde{y}^1 - r(x^1, p) \\ \vdots \\ \tilde{y}^m - r(x^m, p) \end{bmatrix}, \quad \underbrace{J(X, p)}_{=J} = - \left[\begin{array}{ccc|c} I & & & 0 \\ & \ddots & & \vdots \\ & & I & 0 \\ \hline r_x(x^1, p) & & & r_p(x^1, p) \\ & \ddots & & \vdots \\ & & r_x(x^m, p) & r_p(x^m, p) \end{array} \right]$$

where r_x denotes the Jacobian of r with respect to x.

In this case the $dimu = m \times dimx + p$ scalar variables contained in $u := (X, p)$ have to be estimated, so that not only the number $dimF = m \times dimy$ of scalar squares in the objective function, but also the number of unknowns is of order m which will in general be large. However, the Jacobian J has a very special block-angular structure which can be exploited when working with the GN model $\mu(u + \Delta u) = \frac{1}{2}\|F + J\Delta u\|^2$ where $\Delta u := (\Delta X, \Delta p)$ denotes the step s. This block-angular structure of J is typical for hierarchic optimization problems where "global" variables — the parameters p — and local variables — the states x^i — occur, and the coupling comes only from the global variables. It allows an efficient QR factorization, see [96] where essentially the method of [42] is used for computing the GN step Δu, compare also [24] for a discussion of methods for solving such block-angular problems.

The computation of the trust region step where regularized least squares problems (44) have to be solved is more difficult. In the case of diagonal scaling matrices D and $dimy = 1$ an efficient method is proposed in [14] and implemented in [15]. A different way for reducing the cost described in [96, 97, 94] consists of using nonstandard scaling matrices of the form

$$D = \left[\begin{array}{c|c} J_1 & D_2 \\ \hline 0 & D_3 \end{array}\right], \quad D_2 \in \{0, J_2, J_1 J_1^+ J_2\}$$

where J is naturally partitioned according to

$$J = [\, J_1 |\, J_2 \,] = \left[\begin{array}{c|c} J_{11} & J_{12} \\ \hline J_{21} & J_{22} \end{array}\right] \quad \text{where} \quad J_{11} = -I, \ J_{12} = 0,$$

and D_3 is a standard scaling matrix that regularizes the possibly rank deficient p-part J_2 of J. Note that J_1 has always full rank due to its upper block $J_{11} = -I$.

3.2 Equality Constrained Problems

3.2.1 Basic Techniques

Let us consider the equality constrained problem (31) and suppose that the constraints are regular in the sense of $rank\, G'(u) = L$. Then the conceptually simplest way for handling the constraints consists in reducing the problem to an unconstrained one by numerically solving the constraints for L of the variables as follows:

Suppose that $u^k = (x^k, y^k)$ with $x^k \in R^K$, $y^k \in R^L$, $K + L = N$ is feasible, i.e. that $G(u^k) = G(x^k, y^k) = 0$ holds, and let for simplicity of notation the Jacobian G_y of G with respect to y be nonsingular at u^k, otherwise a permutation of the variables has to be done. Then, due to the Implicit Function Theorem there exists a function

$$r = r_k : \mathcal{U}(x^k) \to \mathcal{U}(y^k) \text{ such that } G(x, r(x)) = 0 \ \forall x \in \mathcal{U}(x^k),$$

and these solutions $u = (x, r(x))$ of $G(u) = 0$ are locally unique. By replacing y in $F(x, y)$ by $r(x)$ the constrained problem (31) is reduced to the unconstrained problem

$$\min \left\{ \psi(x) := \frac{1}{2}\| \underbrace{F(x, r(x))}_{:=H(x)} \|^2 = \frac{1}{2}\|H(x)\|^2 : \ x \in R^K \right\}$$

locally in a neighbourhood $\mathcal{U}(u^k)$ of u^k. Then, in order to obtain the next iterate u^{k+1} all the techniques described in section 3.1 can be applied. This approach has the desirable property to yield only feasible iterates $\{u^k\}$, i.e., they satisfy $G(u^k) = G(x^k, y^k) = 0$. However, the price to be paid is that for computing $H(x)$ and its derivative $H'(x)$ at a point x the corresponding $y = r(x)$ has to be computed numerically by solving the defining equation $G(x, y) = 0$ with respect to y by, e.g., Newton's method. By implicit differention the Jacobian H' is obtained as $H'(x) = -G_y(x, r(x))^{-1} G_x(x, r(x))$.

The approach just described has been applied to the implicit estimation problems of section 2.1 in the context of GN methods in [16, 85, 95, 61].

An alternative to the feasible point methods just sketched are methods which work with nonfeasible iterates. The basic method of this type is the undamped *generalized GN method* where the step s is defined by the linearized problem

$$\min \left\{ \tfrac{1}{2} \|F + F's\|^2 : u \text{ s.t. } G + G's = 0 \right\} \tag{54}$$

which is uniquely solvable if $rank\binom{F'}{G'} = N$. When the residual $\|F(u^*)\|$ is sufficiently small then this method converges locally though no second order information is used in the objective function, see, e.g. [13, 115, 57] for a detailed description.

When line search methods are used for globalization then a *penalty fuction* is needed which measures as well the reduction in the objective function as the inconsistency in the constraints and allows to decide wether the step αs can be accepted. The functions taken are of the form

$$\Phi(u, \omega) := \tfrac{1}{2} \left\{ \|F(u)\|^\alpha + \omega \|G(u)\|^\beta \right\} \tag{55}$$

with fixed exponents $\alpha, \beta > 0$ and an appropriately chosen *penalty parameter* $\omega = \omega_k > 0$. In [57] the case $\alpha = \beta = 1$ is investigated whereas in [114, 115] the case $\alpha = \beta = 2$ is considered.

On the other hand the use of the linearized model (54) in the trust region approach causes some difficulties since adding the trust region constraint $\|Du\| \leq \Delta$ can lead to an empty feasible region. Thus some proposals for relaxing the linearized constraints $G + G's = 0$ has been made, see [18, 80, 35] and the literature cited there for the general, non Gauss-Newton case.

Instead of (55) also so-called *augmented Lagrangians*

$$\Phi(u, \omega, \lambda) := \tfrac{1}{2} \left\{ \|F(u)\|^2 + \omega \|G(u)\|^2 \right\} + \lambda^T G(u) \tag{56}$$

can be used for evaluating a candidate s for the step. As in (55) the problem consists in the right choice of the parameters $\omega = \omega_k > 0$ and $\lambda = \lambda^k \in R^L$.

Finally, line search or trust region techniques can directly be applied to penalty or augmented Lagrangian functions to give the next iterate. In [29] the damped GN method is applied to the quadratic function (55) when there $\alpha = \beta = 2$ is chosen. General optimization methods based on (56) are described in [21]. For the linear algebra necessary for computing the GN or trust region step see again [20, 56].

3.2.2 Application to Implicit Estimation Problems

The basic estimation problem (8) of section 2.1 is of type (31) with $u = (Z, p)$ and

$$F(Z,p) := F(Z) := \begin{bmatrix} \tilde{z}^1 - z^1 \\ \vdots \\ \tilde{z}^m - z^m \end{bmatrix}, \quad G(Z,p) := \begin{bmatrix} f(z^1, p) \\ \vdots \\ f(z^m, p) \end{bmatrix}.$$

The Jacobians of F and G have the simple structure

$$F' = -\begin{bmatrix} I \\ & \ddots \\ & & I \end{bmatrix}, \quad G' = \begin{bmatrix} f_z(z^1, p) & & \bigg| & f_p(z^1, p) \\ & \ddots & \bigg| & \vdots \\ & & f_z(z^m, p) & \bigg| & f_p(z^m, p) \end{bmatrix}. \tag{57}$$

which can be exploited when computing the GN step $\Delta u = (\Delta Z, \Delta p)$, see [42, 24, 95, 13].

When additional observations of critical states are available as in the estimation problem (14) of section 2.1.1 then certain rows of G' are built by the gradients $\nabla_y \varphi$ and $\nabla_p \varphi$ of the function φ which defines the turning point property, cf. (12) and (13). These gradients are given by

$$(\nabla_y \varphi)^T \Delta y + (\nabla_p \varphi)^T \Delta p = \tag{58}$$
$$-[B^{-T} e^{n+1}]^T \begin{bmatrix} \varphi_{yy}[v, \Delta y] + \varphi_{yp}[v, \Delta yp] \\ 0 \end{bmatrix} \quad \forall \, \Delta y \in R^{n+1}, \, \Delta p \in R^{dimp},$$

i.e. they require to compute all second order derivatives of f. In [87] a special modified block elemination technique for the constraints is used which goes back to [77, 78]. This technique is designed in such a way that the second order terms $\varphi_{yq}[v, .]$ with $q \in \{y, p\}$ occuring in (58) are replaced by second order directional derivatives $\varphi_{yq}[v, g]$ with fixed vectors g. Unlike the terms $\varphi_{yq}[v, .]$ the latter can easily be approximated by divided differences using a few additional function values of f.

4 Robust and Multiresponse Parameter Estimation

4.1 Robust Parameter Estimation

An essential shortcoming of least squares estimates is their high sensitivity with respect to outliers. If such outliers occur, i.e., if some individual errors $t_l := \tilde{z}^i_j - z^i_j$ are essentially greater than the other one's then, due to the big amount they contribute to the objective function when squared the information contained in the remaining "good" observations is lost. Hence it would make sense to use so-called *robust* criterion functions ϱ which do not grow so fast as $\varrho_2(t) := \frac{1}{2} t^2$ for large $|t|$.

4.1.1 Least Absolute Deviations Estimates

One possibility consists of using the ℓ_1 criterion

$$\varrho_1(t) := |t|$$

which defines the so-called ℓ_1 or *least absolute deviation estimates*. Since ϱ_1 is not differentiable at $t = 0$ methods of nonsmooth optimization have to be used for solving the corresponding optimization problems . We will not discuss these methods here, see [10] for the statistical background and [110, 111, 112] for numerical methods.

4.1.2 M Estimates

In order to avoid the difficulties caused by the jump of the derivative ϱ_1' at $t = 0$ and for statistical reasons so-called *M estimates* defined by smooth C^1 criterion functions which behave like $\frac{1}{2}t^2$ for small $|t|$ have been proposed. The most famous example of such a criterion function is the *Huber function* which is given by

$$\varrho(t) := \varrho_H(t) := \begin{cases} \frac{1}{2}t^2 & if \quad |t| \le k_H \\ k_H \cdot |t| - \frac{1}{2}k_H^2 & if \quad |t| > k_H \end{cases} \tag{59}$$

where $k_H > 0$ defines the breakpoints $t = \pm k_H$ at which ϱ changes with continuous derivative from the quadratic ℓ_2 function ϱ_2 to a linear one. There are, however, also other criterion functions used in the literature as the convex one proposed by Fair (1974) or the nonconvex one's introduced by Hampel (1973) and Beaton/Tukey (1974) the latter being constant for sufficiently large $|t|$, see [49, 83] where the statistical background is dicussed or, e.g., [118] for further references.

In the following we restrict ourself to the one-dimensional regression case $y = r(x, p)$ with $dimy = 1$, $dimp =: n$ and $\delta x^i = 0$ so that the robust regression problem to be solved becomes the minimization problem

$$\min\left\{\psi(p) := \sum_{i=1}^{m} \varrho(u_i(p)) : \ p\right\} \tag{60}$$

where $u = u(p) \in R^m$ denotes the *residual vector* defined as

$$u := u(p) := (\tilde{y}_i - r(\tilde{x}^i, p))^T = \tilde{y} - R(\tilde{X}, p),$$

c.f. (21) and (24). In the special case of the linear model (27), cf. also (28), the residual u is simply given as

$$u := u(p) := (\tilde{y}_i - \sum_{j=1}^{n} \tilde{x}_j^i \cdot p_j)^T = \tilde{y} - \tilde{X}p. \tag{61}$$

4.1.3 Scaling

Due to the constants occuring in the robust criterion functions as k_H in the Huber case, the M-estimates defined as solution of (60) are, in general, not scale invariant. Hence, in robust regression a so-called *scaling* is necessary where problem (60) is replaced by

$$\min\left\{\psi(p, s) := \sum_{i=1}^{m} \varrho\left(\frac{u_i(p)}{s}\right) : \ (p, s) \in R^n \times [0, \infty]\right\} \tag{62}$$

181

where $s > 0$ is the *scale factor*. When an appropriate value for s is not known a priori then it has to be determined simultaneously with p. A first possibility often used in practice is to set

$$s := \text{median}_{i=1,\ldots,m}\left\{|u_i(p) - \text{median}_{j=1,\ldots,m}\{r_j(p)\}|\right\}/0.6745. \tag{63}$$

A second possibility gives "Huber's proposal 2" where s is determined such that

$$\sum_{i=1}^{m}\left[\varrho'\left(\frac{u_i(p)}{s}\right)\right]^2 = a \tag{64}$$

holds . Here a is defined by

$$a := (m - n)\mathbf{E}(\chi(T)), \quad \chi(t) := t \cdot \varrho'(t) - \varrho(t)$$

where $\mathbf{E}(\chi(T))$ denotes the expectation with respect to the normal distribution. In case of the Huber function solving (62) under the constraint (64) is equivalent to solving the modified problem

$$\min\left\{\psi(p,s) := \sum_{i=1}^{m}\varrho\left(\frac{u_i(p)}{s}\right) \cdot s + a \cdot s : (p,s) \in R^n \times [0,\infty]\right\} \tag{65}$$

with a as above.

4.1.4   Linear Models

In the following the modified problem (65) restricted to the linear case (61) will be considered where ϱ is supposed to be a convex criterion function which satisfies the conditions statet in [118].

If some exceptional cases are excluded, see again [118], then solving (65) is equivalent to solving the system built by the necessary conditions

$$\nabla_p\psi(p,s) = \tilde{X}^T\varrho'\left(\frac{u(p)}{s}\right) = 0 \tag{66}$$

$$\varphi(p,s) := \frac{\partial}{\partial s}\psi(p,s) = a - \sum_{i=1}^{m}\chi\left(\frac{u_i(p)}{s}\right) = 0 \tag{67}$$

whith

$$\varrho'\left(\frac{u(p)}{s}\right) := \left(\varrho'\left(\frac{u_i(p)}{s}\right),\ldots,\varrho'\left(\frac{u_m(p)}{s}\right)\right)^T.$$

The methods for solving the above problem are again iterative of structure

$$p^+ := p + \alpha\Delta p, \qquad s+ := s + \alpha\Delta s.$$

In the so-called *single step methods* at first an p-step is performed for fixed s which reduces $\psi(p + \Delta p, s)$ as a function of Δp. For computing Δp the function $\psi(p + \Delta p, s)$ is approximated by the function

$$\mu(p + \Delta p, s) := \sum_{i=1}^{m}\sigma\left(\left(\frac{u_i(p)}{s}\right);\left(\frac{u_i(p + \Delta p)}{s}\right)\right)s + as =$$

$$\psi(p,s) - \Delta p^T\tilde{X}^T\varrho'\left(\frac{u(p)}{s}\right) + \frac{1}{2s}\Delta p^T\tilde{X}^T D\tilde{X}\Delta p \tag{68}$$

which is quadratic in Δp where

$$\sigma(t; t + \Delta t) := \varrho(t) + \varrho'(t)\Delta t + \tfrac{1}{2}d(t)\Delta t^2 \approx \varrho(t + \Delta t)$$

with an approximation $d(t) \geq 0$ to $\varrho''(t)$ and

$$D := diag\left(d\left(\frac{u_i(p)}{s}\right), \ldots, d\left(\frac{u_m(p)}{s}\right)\right).$$

In the simplest case which corresponds to the GN step in the least squares case, the step Δp is computed from $\min\{\mu(p + \Delta p, s) : \Delta p\}$ or the eqivalent necessary conditions, the "normal"-like equations

$$\frac{1}{s}\tilde{X}^T D \tilde{X}\Delta p - \tilde{X}^T \varrho'\left(\frac{u(p)}{s}\right) = 0. \tag{69}$$

Since these equations are the true normal equations to the least squares problem

$$\min\left\{\tfrac{1}{2}\|D^{1/2}\tilde{X}\Delta p - sD^{-1/2}\varrho'\left(\frac{u(p)}{s}\right)\| : \Delta p\right\}$$

with the weighting matrix D which depends on (p, s) such methods are also called *iteratively reweighted least squares*. Possible choices of $d(.)$ are

(i) Hubers H-approximation [50]:

$$d^H(t) := const. = \sup\{\varrho''(\tau) : \tau \in R\}$$

(ii) W-approximation [50]:

$$d^W(t) := \begin{cases} \dfrac{\varrho'(t)}{t} & if \quad t \neq 0 \\ \varrho''(t) & if \quad t = 0 \end{cases}$$

(iii) Newton-approximation:

$$d^N(t) := \varrho''(t)$$

(iv) NLQ-approximation [118]:

$$d^{NLQ}(t) := \begin{cases} \dfrac{\varrho'(t)^2}{2\varrho(t)} & if \quad t \neq 0 \\ \varrho''(t) & if \quad t = 0 \end{cases}$$

(v) Byrd's approximation [117]:

$$d^B(t) := \min\left\{d^W(t), \max\{\varrho''(\tau) : |t - \tau| \leq \delta\}\right\}.$$

Since the H- and W-approximations are so-called global majorants no line search is needed in the p-step, i.e., $p^+ = p + \Delta p$. Otherwise a line search has to be performed for obtaining global convergence, but trust-region-like algorithms are also possible, see [118, 17, 117, 119]. A different strategy which works in the space of residuals is investigated in [74] in case of fixed $s = 1$ and the Newton-approximation.

When ϱ is the Huber function then the piecewise quadratic structure of ψ can be exploited to construct finite algorithms, see [19, 64] for details. Moreover, in this case the linear algebra can be made more efficient by using update techniques when changing from D_k to D_{k+1} in (69), see [32, 74, 117].

Having computed an improved p^+ a s-step is performed in order to reduce $\psi(p^+, s^+)$ as function of the real variable $s^+ > 0$. In [50] the simple iteration

$$[s^+]^\omega = \frac{1}{a} \sum_{i=1}^m \chi\left(\frac{u_i(p^+)}{s}\right) [s]^\omega$$

has been proposed where $\omega > 1$ is a fixed relaxation parameter, cf. (67). This iteration is also used in recent papers as [99, 34]. Numerical tests show, however, that it is advantageous to perform an "exact" minimization with respect to s in an inner iteration by solving the scalar equation

$$\varphi(p^+, s^+) = \frac{\partial}{\partial s}\psi(p^+, s^+) = a - \sum_{i=1}^m \chi\left(\frac{u_i(p^+)}{s^+}\right) = 0$$

with respect to s^+, see [118, 117]. For the Huber function there exist efficient finite algorithms for doing this, see [118, 99].

Instead of the single step methods just described also *simultaneous methods* can be used where the pair (p, s) is simultaneously improved by, e.g., applying a Newton-like step to the coupled system (66), (67), cf. [32] and also [119] where approximations to ψ which are rational functions with respect to s are investigated. A method based on implicitly solving (67) for $s = s(p)$ is described in [99].

Let us remark here that only the first steps has been made for computing M-estimates in case of nonlinear models. We refer to [31, 34, 60, 116, 33] and the literature cited there.

4.2 Multiresponse Parameter Estimation

Let us consider the explicit regression model $y = r(x, p)$ of section 2.2 in the multiresponse case $dimy > 1$, cf. (19). In *multiresponse parameter estimation* the parameter p is determined as solution of

$$\min\left\{\psi(p) := \tfrac{1}{2}\det[U(p)^T U(p)] : p\right\}, \quad U(p) := \tilde{Y} - R(\tilde{X}, p) \tag{70}$$

see [7, 8]. If $dimy = 1$ then this problem reduces to the least squares estimation problem, but for $dimy > 1$ the objective function has not the simple structure of a sum of squares.

When applying minimization techniques to (70) the determinant $\det[U^T U]$ has to be differentiated, see [6]. Using the derivatives the second order Taylor approximation can be formed and, following the GN philosophy, all terms containing second order derivatives of r can be omitted. This leads to a quadratic approximation

$$\mu(p + s) = \psi + \nabla\psi^T s + \tfrac{1}{2}s^T Hs$$

where

$$\nabla\psi = A^T a - B^T b, \quad H = A^T A - B^T B$$

with certain vectors a, b and matrices A, B which differs from the GN approximation by the "-" sign which may lead to an indefinite matrix H. This model can principally be used as well in a line search as in a trust region approach, see [7, 8, 7, 8, 58] for a general discussion and possibilities for realizing the required linear algebra.

Let us end this overview article with the remark that nonlinear parameter estimation is both mathematically interesting and practically useful, and a lot of further results can be expected in future.

References

[1] M. Al-Baali and R. Fletcher. Variational methods for nonlinear least squares problems. *J. Oper. Res. Soc.*, 36:405–421, 1985.

[2] M. Al-Baali and R. Fletcher. An efficient line search for nonlinear least squares. *J. Optim. Theory Appl.*, 48:359–377, 1986.

[3] E. L. Allgower and K. Georg. *Numerical Continuation Methods: An Introduction.* Springer, Berlin, 1990.

[4] K. A. Ariyawansa. Deriving collinear scaling algorithms as extensions of quasi-Newton methods and the local convergence of DFP- and BFGS-related collinear scaling algorithms. *Math. Programming*, 49:23–48, 1990.

[5] Y. Bard. *Nonlinear Parameter Estimation.* Academic Press, New York, 1974.

[6] D. M. Bates. The derivative of $|X^T X|$ and its uses. *Technometrics*, 25:373–376, 1983.

[7] D. M. Bates and D. G. Watts. A generalized Gauss-Newton procedure for multi-response parameter estimation. *SIAM J. Sci. Statist. Comput.*, 8:49–55, 1987.

[8] D. M. Bates and D. G. Watts. *Nonlinear Regression Analysis and its Applications.* Wiley, New York, 1988.

[9] Å. Björck. Solution of equations in R^n: Least squares methods. In P. G. Ciarlet and J. L. Lions, editors, *Handbook of Numerical Analysis*, volume I, pages 589–652. North Holland, Amsterdam, 1990.

[10] P. Bloomfield. *Least Absolute Deviations: Theory, Applications, and Algorithms.* Birkhäuser, Boston, 1983.

[11] H. G. Bock. Numerical treatment of inverse problems in chemical reaction kinetics. In K. H. Ebert, P. Deuflhard, and W. Jäger, editors, *Modelling of Chemical Reaction Systems*, pages 102–125. Springer, Berlin, 1981.

[12] H. G. Bock. Recent advances in parameter identification techniques for O.D.E. In P. Deuflhard and E. Hairer, editors, *Numerical Treatment of Inverse Problems in Differential and Integral Equations*, pages 95–121. Birkhäuser, Boston, 1983.

[13] H. G. Bock. Randwertproblemmethoden zur Parameterschätzung in Systemen nichtlinearer Differentialgleichungen (Habilitationsschrift). Bonner Math. Schriften 183, Universität Bonn, 1987.

[14] P. T. Boggs, R. H. Byrd, and R. B. Schnabel. A stable and efficient algorithm for nonlinear orthogonal regression. *SIAM J. Sci. Statist. Comput.*, 6:1052–1078, 1987.

[15] P. T. Boggs, J. R. Donaldson, R. H. Byrd, and R. B. Schnabel. Algorithm 676. ODRPACK: Software for weighted orthogonal distance regression. *ACM Trans. Math. Software*, 15:348–364, 1989.

[16] H. I. Britt and R.H. Luecke. The estimation of parameters in nonlinear implicit models. *Technometrics*, 15:233–247, 1973.

[17] R. H. Byrd. Algorithms for robust regression. In M. J. D. Powell, editor, *Nonlinear Optimization*. Academic Press, London, 1982.

[18] R. H. Byrd, R. S. Schnabel, and G. A. Shultz. A trust region algorithm for nonlinearly constrained optimization. *SIAM J. Numer. Anal.*, 24:1152–1170, 1987.

[19] D.I. Clark and M. R. Osborne. Finite algorithms for Huber's M-estimator. *SIAM J. Sci. Statist. Comput.*, 7:72–85, 1986.

[20] T. F. Coleman and C. Hempel. Computing a trust region step for a penalty function. *SIAM J. Sci. Statist. Comput.*, 11:180–201, 1990.

[21] A. R. Conn, N. I. M. Gould, and P .L. Toint. A globally convergent augmented Lagrangian algorithm for optmization with general constraints and simple bounds. *SIAM J. Numer. Anal.*, 28:545–572, 1991.

[22] C. Corradi. A note on the solution of separable nonlinear least-squares problems with separable nonlinear equality constraints. *SIAM J. Numer. Anal.*, 18:1134–1138, 1981.

[23] M. G. Cox. The least squares solution of overdetermined linear equations having band or augmented band structure. *IMA J. Numer. Anal.*, 1:3–22, 1981.

[24] M. G. Cox. The least-squares solution of linear equations with block-angular observation matrix. Report DITC 139/89, National Physical Laboratory, Teddington, 1989.

[25] W. C. Davidon. Conic approximations and collinear scalings for optimizers. *SIAM J. Numer. Anal.*, 17:268–281, 1980.

[26] N. de Villiers and D. Glasser. A continuation method for nonlinear regression. *SIAM J. Numer. Anal.*, 18:1139–1154, 1981.

[27] J. Dennis, D. Gay, and R. Welsch. An adaptive nonlinear least squares algorithm. *ACM Trans. Math. Software*, 7:348–368, 1981.

[28] J. E. Dennis and R. B. Schnabel. *Numerical Methods for Unconstrained Optimization and Nonlinear Equations*. Prentice Hall, Englewood Cliffs, NJ, 1983.

[29] P. Deuflhard and V. Apostolescu. An underrelaxed Gauss-Newton method for equality constrained nonlinear least squares. In A. V. Balakrishnan and M. Thoma, editors, *Optimization Techniques*, pages 22–32. Springer, Berlin, 1978.

[30] N. R. Draper and H. Smith. *Applied Regression Analysis*. Wiley, New York, second edition, 1981.

[31] R. Dutter and P. J. Huber. Algorithms for non-linear Huber estimation. *J. Statist. Comput. Simulation*, 9:79–113, 1981.

[32] H. Ekblom. A new algorithm for the Huber estimator in linear models. *BIT*, 28:123–132, 1988.

[33] H. Ekblom, G. Li, and K. Madsen. Algorithms for non-linear Huber regression with variable scale. Report NI-90-10, Institute for Numerical Analysis, Technical University of Denmark, Lyngby, 1990.

[34] H. Ekblom and K. Madsen. Algorithms for nonlinear Huber estimation. *BIT*, 29:60–76, 1989.

[35] M. El-Alem. A global convergence theory for the Celis-Dennis-Tapia trust-region algorithm for constrained optimization. *SIAM J. Numer. Anal.*, 28:266–290, 1991.

[36] R. Fletcher. *Practical Methods of Optimization*. J. Wiley, New York, second edition, 1987.

[37] R. Fletcher and C. Xu. Hybrid methods for nonlinear least squares. *IMA J. Numer. Anal.*, 7:371–389, 1987.

[38] C. Fraley. Computational behaviour of Gauss-Newton methods. *SIAM J. Sci. Statist. Comput.*, 10:515–532, 1989.

[39] D. M. Gay. Computing optimal locally constrained steps. *SIAM J. Sci. Statist. Comput.*, 2:186–197, 1981.

[40] G. H. Golub and R. LeVeque. Extensions and uses of the variable projection algorithm for solving nonlinear least squares problems. Report SU 326, Dept. of Computer Science, Stanford University, Stanford, 1978.

[41] G. H. Golub and V. Pereyra. The differentiation of pseudo-inverses and nonlinear least squares problems whose variables separate. *SIAM J. Numer. Anal.*, 10:413–432, 1973.

[42] G. H. Golub and R. J. Plemmons. Large-scale geodetic least-squares adjustments by dissection and orthogonal decomposition. *Linear Algebra Appl.*, 34:3–27, 1980.

[43] G. H. Golub and C. F. van Loan. *Matrix Computations.* J. Hopkins University Press, BaltiMoré-London, second edition, 1989.

[44] H. Gourgon and J. Nocedal. A conic algorithm for optimization. *SIAM J. Sci. Statist. Comput.*, 6:253–267, 1985.

[45] A. Griewank. On solving nonlinear equations with simple singularities or nearly singular solutions. *SIAM Rev.*, 27:537–563, 1985.

[46] A. Griewank and L. Sheng. On the Gauss-Broyden method for nonlinear least squares. In J. C. Diaz, editor, *Mathematics for Large Scale Computing*, pages 1–33. Marcel Dekker, New York, 1989.

[47] K. L. Hiebert. An evaluation of mathematical software that solves nonlinear least squares problems. *ACM Trans. Math. Software*, 7:1–16, 1981.

[48] A. Hoy. Numerische Lösung singulärer nichtlinearer Gleichungssysteme. Dissertation B, Fachbereich Mathematik und Informatik, Martin-Luther-Universität, Halle, 1991.

[49] P. J. Huber. *Robust Statistics.* J. Wiley, New York, 1981.

[50] P. J. Huber and R. Dutter. Numerical solution of robust regression problems. In G. Brushmann, editor, *COMPSTAT 1974.* Physica-Verlag, Wien, 1974.

[51] L. Kaufman. Variable projection method for solving separable nonlinear least squares problems. *BIT*, 15:49–57, 1975.

[52] L. Kaufman and V. Pereyra. A method for separable nonlinear least squares problems with separably nonlinear equality constraints. *SIAM J. Numer. Anal.*, 15:12–20, 1978.

[53] L. Kaufman and G. Sylvester. Separable nonlinear least squares with multiple right hand sides. Manuscript, presented at ICIAM 91 Washington, 1991.

[54] A. Kiełbasiński and H. Schwetlick. *Numerische lineare Algebra. Eine computerorientierte Einführung.* Deutscher Verlag der Wissenschaften, Berlin, 1988. (Also: Harri Deutsch Verlag, Thun-Frankfurt, 1988).

[55] O. Knoth. Marquardt-ähnliche Verfahren zur Minimierung nichtlinearer Funktionen. Dissertation A, Sektion Mathematik, Martin-Luther-Universität, Halle, 1983.

[56] O. Knoth. The computation of the trust-region step for Gauss-Newton-like models. Manuscript, submitted for publicaion, 1989.

[57] O. Knoth. A globalization scheme for the Gauss-Newton method. *Numer. Math.*, 56:591–607, 1989.

[58] O. Knoth. A Gauss-Newton trust-region algorithm for solving multi-response parameter estimation problems. Manuscript, submitted for publicaion, 1990.

[59] F. T. Krogh. Efficient implementation of a variable projection algorithm for nonlinear least squares problems. *Comm. ACM*, 17:167–169, 1974.

[60] G. Li and K. Madsen. Robust parameter estimation. In D. F. Griffiths and G. A. Watson, editors, *Numerical Analysis 1987*, pages 176–191. Longman, 1988.

[61] P. Lindström. Algorithms for nonlinear least squares—particularly problems with constraints (PhD Thesis). Report UMINF-106.83, Inst. of Information Processing, University of Umeå, Umeå, Sweden, 1983.

[62] P. Lindström and P.-Å. Wedin. A new linesearch algorithm for nonlinear least squares problems. *Math. Programming*, 29:268–296, 1984.

[63] D. G. Liu and J. Nocedal. Algorithms with conic termination for nonlinear optimization. *SIAM J. Sci. Statist. Comput.*, 10:1–17, 1989.

[64] K. Madsen and H. B. Nielsen. Finite algorithms for robust linear regression. *BIT*, 30:682–699, 1990.

[65] D. W. Marquardt. An algorithm for least squares estimation of nonlinear parameters. *SIAM J. Appl. Math.*, 11:431–441, 1963.

[66] J. G. McWhirter, editor. *Mathematics in Signal Processing II*, Oxford, 1990. Oxford University Press.

[67] R. Menzel. On solving nonlinear least-squares problems in case of rankdeficient Jacobians. *Computing*, 34:63–72, 1985.

[68] J. Moré. The Levenberg-Marquardt algorithm: Implementation and theory. In G. A. Watson, editor, *Numerical Analysis. Proceedings of the 1977 Dundee Conference*, pages 105–116. Springer, Berlin, 1978.

[69] J. Moré. Recent developments in algorithms and software for trust region methods. In A. Bachem, M. Grötschel, and B. Korte, editors, *Mathematical Programming— The State of the Art*, pages 258–287. Springer, Berlin, 1983.

[70] J. J. Moré and D. C. Sorensen. Computing a trust region step. *SIAM J. Sci. Statist. Comput.*, 4:553–572, 1983.

[71] L. Nazareth. Some recent approaches to solving large residual nonlinear least squares problems. *SIAM Rev.*, 22:1–11, 1980.

[72] U. Nowak and P. Deuflhard. Towards parameter identification for large chemical reaction systems. In P. Deuflhard and E. Hairer, editors, *Numerical Treatment of Inverse Problems in Differential and Integral Equations*, pages 13–26. Birkhäuser, Boston, 1983.

[73] U. Nowak and P. Deuflhard. Numerical identification of selected rate constants in large chemical reaction systems. *Appl. Numer. Math.*, 1:59–75, 1985.

[74] D. P. O'Leary. Robust regression computation using iteratively reweighted least squares. *SIAM J. Matrix Anal. Appl.*, 11:466–480, 1990.

[75] M. R. Osborne and G. K. Smyth. A modified Prony algorithm for fitting functions defined by difference equations. *SIAM J. Sci. Statist. Comput.*, 12:362–382, 1991.

[76] T. A. Parks. *Reducible Nonlinear Programming Problems*. PhD thesis, Houston Univ., Dept. of Mathematics, Houston, 1985.

[77] G. Pönisch and H. Schwetlick. Computing turning points of curves implicitly defined by nonlinear equations depending on a parameter. *Computing*, 26:107–121, 1981.

[78] G. Pönisch and H. Schwetlick. Some types of inverse problems for nonlinear equations having coupled turning points. Preprint 07-01-90, Sektion Mathematik, Technische Universität Dresden, Dresden, 1990.

[79] M. J. D. Powell. A hybrid method for nonlinear equations. In P. Rabinowitz, editor, *Numerical Methods for Nonlinear Algebraic Equations*, pages 87–114. Gordon and Breach, New York, 1970.

[80] M. J. D. Powell and Y. Yuan. A trust region algorithm for equality constrained optimization. *Math. Programming*, 49:189–211, 1986.

[81] D. A. Ratkowsky. *Nonlinear Regression Analysis: A Unified Practical Approach*. M. Dekker, New York, 1983.

[82] W. C. Rheinboldt. *Numerical Analysis of Parametrized Nonlinear Equations*. Wiley, New York, 1986.

[83] P. J. Rousseeuw and A. M. Leroy. *Robust Regression and Outlier Detection*. Wiley, New York, 1987.

[84] A. Ruhe and P.-Å Wedin. Algorithms for separable nonlinear least squares problems. *SIAM Rev.*, 22:318–337, 1980.

[85] W. H. Sachs. Implicit multifunctional nonlinear regression analysis. *Technometrics*, 18:161–173, 1976.

[86] D. E. Salane. A continuation approach for solving large residual nonlinear least squares problems. *SIAM J. Sci. Statist. Comput.*, 8:655–671, 1987.

[87] S. Schleiff. Ein Parameterschätzproblem für nichtlineare Gleichungen mit Rückkehr-
punkten. Diplomarbeit, Fachbereich Mathematik und Informatik, Martin-Luther-
Universität, Halle, 1991.

[88] R. B. Schnabel. Conic methods for unconstrained minimization and tensor methods for nonlinear equations. In A. Bachem, M. Grötschel, and B. Korte, editors,
Mathematical Programming—The State of the Art, pages 417–438. Springer, Berlin,
1983.

[89] R. B. Schnabel and P. Frank. Tensor methods for nonlinear equations. *SIAM J.
Numer. Anal.*, 21:815–843, 1984.

[90] R. B. Schnabel and P. Frank. Solving systems of nonlinear equations by tensor methods. In A. Iserles and M. J. D. Powell, editors, *The State of the Art in Numerical
Analysis*, pages 245–271. Clarendon Press, Qxford, 1987.

[91] H. Schwetlick. Über die Konvergenz regularisierter Gauss-Newton-Verfahren. *Zh.
Vychisl. Mat. i Mat. Fiz.*, 13:1371–1382, 1973.

[92] H. Schwetlick. *Numerische Lösung nichtlinearer Gleichungen*. Deutscher Verlag der
Wissenschaften, Berlin, 1979. (Also: Oldenbourg Verlag, München, 1979).

[93] H. Schwetlick. Effective methods for computing turning points of curves implicitly
defined by nonlinear equations. In A. Wakulicz, editor, *Computational Mathematics*,
volume 13 of *Banach Center Publications*, pages 623–645. PWN (Polish Scientific
Publ.), Warsaw, 1984.

[94] H. Schwetlick. Nonstandard scaling matrices in trust region methods. In E. L.
Allgower and K. Georg, editors, *Computational Solution of Nonlinear Systems of
Equations*, volume 26 of *Lectures in Applied Mathematics*, pages 587–604. American
Mathematical Society, Providence, RI, 1990.

[95] H. Schwetlick, W. Schellong, and V. Tiller. Gauss-Newton-like methods for nonlinear least squares with equality constraints — local convergence and applications to
parameter estimation in implicit models. *statistics*, 16:167–178, 1985.

[96] H. Schwetlick and V. Tiller. Numerical methods for estimating parameters in nonlinear models with errors in the variables. *Technometrics,* 27:17–24, 1985.

[97] H. Schwetlick and V. Tiller. Nonstandard scaling matrices for trust region Gauss-
Newton methods. *SIAM J. Sci. Statist. Comput.*, 10:654–670, 1989.

[98] G. A. F. Seber and C. J. Wild. *Nonlinear Regression*. Wiley, New York, 1989.

[99] D.F. Shanno and D. M. Rocke. Numerical methods for robust regression: Linear
models. *SIAM J. Sci. Statist. Comput.*, 7:86–97, 1986.

[100] G. A. Shultz, R. B. Schnabel, and R. H. Byrd. A family of trust-region-based algorithms for unconstrained minimization witf strong global convergence properties.
SIAM J. Numer. Anal., 22:47–67, 1985.

[101] D. C. Sorensen. The Q-superlinear convergence of a collinear scaling algorithm for unconstrained optimization. *SIAM J. Numer. Anal.*, 17:88–114, 1980.

[102] D. C. Sorensen. Newton's method with a model trust region modification. *SIAM J. Numer. Anal.*, 19:409–426, 1982.

[103] E. Spedicato and M. T. Vespucci. Numerical experiments with variations of the Gauss-Newton algorithm for nonlinear least squares. *J. Optim. Theory Appl.*, 57:323–339, 1988.

[104] J. M. ten Vregelaar. An algorithm for computing estimates for parameters of an ARMA-model from noisy mesurements of inputs and outputs. Memorandum COSOR 87-13, Dept. of Mathematics and Computing Science, Eindhoven University of Technology, Eindhoven, 1987.

[105] B. van Domselaar and P. W. Hemker. Nonlinear parameter estimation in initial value problems. Report NW18/75, Mathematisch Zentrum, Amsterdam, 1975.

[106] S. van Huffel and J. Vandewalle. Comparison of total least squares and instrumental variable methods for parameter estimation of transfer function models. *Int. J. Control*, 50:1039–1056, 1989.

[107] S. van Huffel and J. Vandewalle. *The Total Least Squares Problem. Computational Aspects and Analysis.* SIAM, Philadelphia, 1991.

[108] J. M. Varah. A spline least squares method for numerical parameter estimation in differential equations. *SIAM J. Sci. Statist. Comput.*, 3:28–46, 1982.

[109] J. M. Varah. Relative sizes of the Hessian terms in nonlinear parameter estimation. *SIAM J. Sci. Statist. Comput.*, 11:174–179, 1990.

[110] G. A. Watson. The numerical solution of total ℓ_p approximation problems. In D. F. Griffiths, editor, *Numerical Analysis. Proceedings of the 1983 Dundee Conference*, pages 221–238. Springer, Berlin, 1984.

[111] G. A. Watson. On a class of algorithms for total approximation. *J. Optim. Theory Appl.*, 45:219–231, 1985.

[112] G. A. Watson. Methods for best approximation and regression problems. In A. Iserles and M. J. D. Powell, editors, *The State of the Art in Numerical Analysis*, pages 139–164. Clarendon Press, Oxford, 1987.

[113] P.-Å. Wedin. The non-linear least squares problem from a numerical point of view. I. Geometrical properties. Report, Dept. of Computer Science, Lund University, Lund, Sweden, 1972.

[114] P.-Å. Wedin. On the use of a quadratic merit function for constrained nonlinear least squares. Report UMINF-135.87, Inst. of Information Processing, University of Umeå, Umeå, Sweden, 1987.

[115] P.-Å. Wedin and P. Lindström. Methods and software for nonlinear least squares problems (revised version July 1988). Report UMINF-133.87, Inst. of Information Processing, University of Umeå, Umeå, Sweden, 1988.

[116] R. Wolke. Iterative Verfahren zur numerischen Berechnung von M-Schätzungen. Dissertation A, Sektion Mathematik, Martin-Luther-Universität, Halle, 1985.

[117] R. Wolke. Iteratively reweighted least squares: A comparison of several single step algorithms for linear models. Manuscript, submitted for publication, 1991.

[118] R. Wolke and H. Schwetlick. Iteratively reweighted least squares: Algorithms, convergence, and numerical comparisons. *SIAM J. Sci. Statist. Comput.*, 9:907–921, 1988.

[119] R. Wolke and H. Schwetlick. Rational approximations in algorithms for iteratively reweighted least squares. Manuscript, submitted for publication, 1991.

Hubert Schwetlick, Fachbereich Mathematik und Informatik
Martin-Luther-Universität Halle-Wittenberg
D-(O)-4010 Halle, Germany

I H SLOAN

Unconventional methods for boundary integral equations in the plane

Abstract. The traditional methods for the solution of boundary integral equations are the Galerkin and collocation methods. This paper introduces some less familiar methods, characterized by unconventional quadrature rules. One is the qualocation method, a generalisation of the collocation method in which the order of convergence is increased. Another is a method which is similar in principle but fully discrete, and hence easily implemented. The main theoretical results are sketched, and the ideas behind the proofs indicated.

1. Introduction

The aim of this paper is to introduce in a simple way some recently proposed methods for boundary integral equations in the plane. The methods, which include the qualocation method and variants which are fully discrete, are characterised by unconventional quadrature rules – for example, we shall meet the composite 3/7, 4/7 rule. In the qualocation method, introduced in [11] and further developed in [15, 12, 5, 16], the approximate problem is formulated as a semi–discrete Petrov–Galerkin method, with a trial space of smoothest splines, and a special quadrature rule replacing the outer integral. The quadrature rule is designed to yield a higher order of convergence than the collocation method, which is the simplest special case. In the fully discrete variant introduced in [13] and further developed in [9,7,14], the aim is to obtain a stable fully–discrete method with any desired order of convergence.

At the present stage of development these methods rely on Fourier techniques for the design of the quadrature rules and the analysis of the methods. For this reason rigorous proofs are available only for smooth curves and uniform meshes, or, in the case of [7], for an interval. That may make the range of application seem rather narrow, given that most of the boundary contours of practical interest have corners. However, numerical experiments for regions with corners (in which the methods are used in conjunction with empirically determined mesh gradings) are, as we shall see, rather encouraging, even if currently lacking the support of theory.

In the next section boundary integral equations are introduced. Then in section 3 the classical Galerkin and collocation methods are defined, and the principal theoretical results of these methods are stated for smooth curves.

194

In section 4 the qualocation method is introduced, and typical theoretical results for smooth curves stated. The method of proof, via Fourier methods and perturbation theory, is sketched in section 5. The fully discrete version is outlined in section 6.

The last section gives some brief numerical results. For smooth curves the numerical results provide reassurance that the predicted orders of convergence can be seen in practice. For curves with corners the results may point to theoretical results which are as yet out of reach.

2. The setting

Let Γ be the piecewise–smooth boundary, without cusps, of a simply connected open region in the plane. Much of the time we shall, for simplicity, concentrate on the logarithmic–kernel boundary integral equation of the first kind,

$$-\frac{1}{\pi}\int_\Gamma \log|t - s|z(s)d\ell_s + \int_\Gamma k(t,s)z(s)d\ell_s = g(t), \quad t \in \Gamma, \tag{2.1}$$

where $|t - s|$ is the Euclidean distance between t and s, $k(t,s)$ is a smooth kernel, and $d\ell$ is the element of arc length. We shall assume that the solution of (2.1), if it exists, is unique.

An equation of the form (2.1) arises, for example, from the Laplace equation

$$\Delta\phi = 0 \quad \text{in} \quad \Omega \tag{2.2}$$

with Dirichlet boundary conditions

$$\phi = g \quad \text{on} \quad \Gamma = \partial\Omega, \tag{2.3}$$

if one seeks to represent ϕ in the 'single–layer' form

$$\phi(t) = -\frac{1}{\pi}\int_\Gamma \log|t - s|z(s)d\ell_s, \tag{2.4}$$

where z, the 'single–layer density', is an unknown function on Γ. Letting t approach Γ, and using the well known continuity in t of the integral in (2.4) under appropriate assumptions on z, one obtains (2.1) with $k = 0$.

A similar boundary integral equation also arises from the 'direct' boundary integral approach to the same problem: using Green's formula for the harmonic function ϕ, namely

$$\frac{1}{\pi}\int_\Gamma \frac{\partial}{\partial n_s}(\log|t - s|)\phi(s)d\ell_s - \frac{1}{\pi}\int_\Gamma \log|t - s|\frac{\partial\phi}{\partial n_s}d\ell_s = \phi(t), \quad t \in \Gamma, \tag{2.5}$$

which holds except at the corner points of Γ (see for example [17]), one obtains, by use of the boundary condition (2.3), an integral equation for $\partial\phi/\partial n$,

$$-\frac{1}{\pi}\int_\Gamma \log|t-s|\frac{\partial\phi}{\partial n_s}d\ell_s = g(t) - \frac{1}{\pi}\int_\Gamma \frac{\partial}{\partial n_s}(\log|t-s|)g(s)d\ell_s, \quad t \in \Gamma, \qquad (2.6)$$

which is of the form (2.1) with $z = \partial\phi/\partial n$ and $k \equiv 0$, and with g replaced by the more complicated expression on the right of (2.6). Integral equations of the form (2.1) also arise in potential flow, scattering of acoustic waves, and many other problems.

A first step towards a numerical solution of (2.1) is to define a convenient parametrisation of the curve. Since Γ is closed, we assume it is parametrised by $t = \nu(x)$, with

$$\nu : \mathbf{R} \to \Gamma, \quad \text{and} \quad \nu \; 1-\text{periodic}.$$

If Γ is a C^∞ curve then we shall also assume $|\nu'| \neq 0$ and $\nu \in C^\infty$. Then (2.1) can be written as

$$-2\int_0^1 \log|\nu(x)-\nu(y)|u(y)dy + 2\pi \int_0^1 k(\nu(x),\nu(y))u(y)dy = f(x), \quad x \in \mathbf{R}, \qquad (2.7)$$

where

$$f(x) = g(\nu(x)), \qquad (2.8)$$

and

$$u(x) = \frac{1}{2\pi}z(\nu(x))|\nu'(x)|, \qquad (2.9)$$

so that the new unknown function u incorporates the Jacobian of the transformation, as well as a convenient normalisation factor. We write the parametrised equation (2.7) as

$$Lu = f, \qquad (2.10)$$

where L is the integral operator on the left of (2.7).

We shall consider only approximations of the form $u \approx u_h \in S_h$, where S_h is a space of 1-periodic smoothest splines on a partition Π_h,

$$\Pi_h : 0 = x_0 < x_1 < \cdots < x_{N-1} < x_N = 1, \qquad (2.11)$$

with

$$h_k = x_{k+1} - x_k, \qquad k = 0, \cdots, N-1,$$

and

$$h = \max h_k.$$

The splines will be assumed to be of order $r \geq 1$; that is $v \in S_h$ satisfies $v \in C^{r-2}(\mathbf{R})$ and $v|_{(x_k, x_{k+1})} \in \mathbf{P}_{r-1}$. For example $r = 1$ corresponds to piecewise–constant functions, $r = 2$ to continuous piecewise linears, and $r = 4$ to cubic splines. (If $r = 1$ the value at a point of discontinuity is taken to be the mean of the left–hand and right–hand limits.) Because S_h is periodic, its dimension for all values of r is exactly N.

3. The Galerkin and collocation methods

Before defining the newer methods, it is convenient to look first at two well established methods : the Galerkin and collocation methods. In the Galerkin method for (2.10) one determines an approximation $u_h \in S_h$ by requiring

$$(Lu_h, \chi) = (f, \chi) \qquad \forall \chi \in S_h, \tag{3.1}$$

where

$$(v, z) = \int_0^1 v(x)\overline{z}(x)dx, \quad v, z \in L_2(0, 1), \tag{3.2}$$

and $\overline{z}$ denotes the complex conjugate of z. In the collocation method one chooses a point t_k in each sub–interval $[x_k, x_{k+1})$, $k = 0, \cdots, N - 1$, and determines $u_h \in S_h$ by

$$Lu_h(t_k) = f(t_k), \qquad k = 0, 1, \cdots, N - 1. \tag{3.3}$$

Of course in the collocation method one must choose the points $\{t_k\}$ appropriately. This topic is discussed systematically, and for a wide range of boundary integral equations, by Wendland [17]. For the equation (2.1), it is known that if r is even (for example, the piecewise–linear case) one should 'collocate at the break–points', i.e. choose

$$t_k = x_k, \qquad k = 0, \cdots, N - 1.$$

For the case r odd, on the other hand, collocation at the breakpoints gives an unstable method, and instead 'collocation at the midpoints', that is

$$t_k = \frac{x_k + x_{k+1}}{2}, \qquad k = 0, \cdots, N - 1,$$

is recommended. On the other hand if the logarithmic term in (2.1) were replaced by the Cauchy singular integral

$$\frac{1}{\pi} \int_\Gamma \frac{1}{t - s} z(s) ds, \tag{3.4}$$

then the advice for the even r and odd r cases would be interchanged.

The collocation method is easier to implement than the Galerkin method, and for that reason is very often preferred in practice. Let $\{v_0, \cdots, v_{N-1}\}$ be a basis for S_h; for example, for $N \geq r$ the basis functions may conveniently be taken to be B–splines. With the approximate solution u_h written as

$$u_h = \sum_{m=0}^{N-1} a_m v_m,$$

the Galerkin method gives the equations

$$\sum_{m=0}^{N-1} (Lv_m, v_k)a_m = (f, v_k), \quad k = 0, \cdots, N - 1,$$

whereas the collocation method gives

$$\sum_{m=0}^{N-1} Lv_m(t_k)a_m = f(t_k), \quad k = 0, \cdots, N - 1.$$

The matrix elements (Lv_m, v_k) for the Galerkin method are double integrals, whereas those for the collocation method are merely single integrals.

The Galerkin and collocation methods are both optimal, in the sense that under reasonable conditions both can give an L_2 error of order $\|u_h - u\|_{L_2} = O(h^r)$. In the light of the remarks in the preceding paragraph, one might think that there is little to be said in favour of the Galerkin method. Yet there is a real sense in which the Galerkin method is the more accurate of the two methods. Suppose that, instead of looking just at the approximation u_h on Γ, one computes, for fixed $t \in \Omega$,

$$\phi_h(t) = -\frac{1}{\pi} \int_\Gamma \log |t - s| z_h(s) d\ell s$$

$$= -2 \int_0^1 \log |t - \nu(y)| u_h(y) dy, \tag{3.5}$$

with $u_h(y) = (2\pi)^{-1} z_h(\nu(y))|\nu'(y)|$, as an approximation to $\phi(t)$ (given by (2.4)). Also, until further notice we shall assume that Γ is a C^∞ curve. Then for r even it follows from the results in [1] that the collocation method on an arbitrary mesh yields

$$\phi_h(t) - \phi(t) = O(h^{r+1}), \tag{3.6}$$

if u is sufficiently smooth, which is one order better than the L_2 order of convergence. On the other hand the Galerkin method under the same conditions yields

$$\phi_h(t) - \phi(t) = O(h^{2r+1}), \tag{3.7}$$

198

which is much better again.

The higher orders of convergence seen in (3.6) and (3.7) derive from 'negative–norm' convergence results which are known for the collocation and Galerkin methods. For each $s \in \mathbb{R}$ we may define a (Sobolev) norm by

$$\|v\|_s = (|\hat{v}(0)|^2 + \sum_{k \neq 0} |k|^{2s} |\hat{v}(k)|^2)^{1/2}, \tag{3.8}$$

where

$$\hat{v}(k) = \int_0^1 e^{-2\pi i k x} v(x) dx, \qquad k \in \mathbb{Z} \tag{3.9}$$

is the kth Fourier coefficient of v. The Sobolev space H^s may be defined as the closure of the space of 1–periodic C^∞ functions on $\mathbb{R}$ with respect to the norm (3.8). Note that if $s = 0$ the norm is just the L_2 norm, and if s is a positive integer the sum over k in (3.8) is essentially the square of the L_2 norm of the sth derivative of v. In a similar way, the norm for negative values of s may be thought of as the L_2 norm associated with the sth anti–derivative of v.

The 'best' (in the sense of highest order) result obtained by Arnold and Wendland [1] in their analysis of the collocation method for even r, when specialised to the logarithmic–kernel equation (2.1), is

$$\|u_h - u\|_{-1} \leq Ch^{r+1} \|u\|_r, \tag{3.10}$$

which shows the higher–order convergence foreshadowed in (3.6), but not in a norm which can be immediately observed.

However, one can capture this higher–order negative norm convergence in applications whenever one is interested not directly in u, but rather in an inner product (u, w), where w is a function with an appropriate degree of smoothness. This is because of the inequality

$$|(v, w)| \leq \|v\|_{-\alpha} \|w\|_\alpha, \qquad \alpha \in \mathbb{R}, \tag{3.11}$$

which follows easily from the definition of the Sobolev norm and the Cauchy–Schwarz inequality. From this follows

$$|(u_h, w) - (u, w)| = |(u_h - u, w)| \leq \|u_h - u\|_{-\alpha} \|w\|_\alpha, \qquad \alpha \in \mathbb{R}.$$

In particular, setting $\alpha = 1$ we have, using (3.10),

$$|(u_h, w) - (u, w)| \leq Ch^{r+1} \|u\|_r \|w\|_1. \tag{3.12}$$

As a special case we have, for $t \in \Omega$,

$$
\begin{aligned}
|\phi_h(t) - \phi(t)| &= \left| -2 \int_0^1 \log |t - \nu(y)| (u_h(y) - u(y)) dy \right| \\
&= |(u_h - u, w_t)| \\
&\leq C_t h^{r+1} \|u\|_r,
\end{aligned}
\tag{3.13}
$$

where

$$
w_t(y) = -2 \log |t - \nu(y)|,
$$

and where $C_t = C\|w_t\|_1$, which is finite because w_t is a C^∞ function for $t \in \Omega$. Thus (3.6) has been shown to follow from the negative norm result (3.10).

Under similar conditions the Galerkin method (with r either even or odd) has the negative–norm estimate

$$
\|u_h - u\|_{-r-1} \leq Ch^{2r+1} \|u\|_r,
\tag{3.14}
$$

which yields the $O(h^{2r+1})$ order of convergence for $\phi_h(t)$ predicted in (3.7).

Finally, if r is odd then theoretical results for the midpoint collocation method are (with one exception) known only under the assumption that the mesh defined by the partition Π_h is uniform, that is

$$
h_k = h = 1/N.
\tag{3.15}
$$

Under this assumption Saranen and Wendland [10] and Arnold and Wendland [2] have shown that (3.10) still holds, while Saranen [8] has established the higher–order result

$$
\|u_h - u\|_{-2} \leq Ch^{r+2} \|u\|_{r+1}.
\tag{3.16}
$$

From this it follows that

$$
\phi_h(t) - \phi(t) = O(h^{r+2})
\tag{3.17}
$$

if u is sufficiently smooth, one order better than in the case of even r.

The solitary exception mentioned above is a recent treatment by Chandler [3,4] of piecewise–constant (i.e. $r = 1$) midpoint collocation, for equation (2.1) in a space of functions whose integral is zero, and an almost arbitrary mesh. The techniques used in that interesting work are quite different to the Fourier methods used in [6,10,2,8] for the analysis of spline collocation. It is these Fourier methods which have played the seminal role in the development of the qualocation method, to which we now turn.

4. The qualocation method

The qualocation method is motivated by the desire to find an approximation which has a better order of convergence than the collocation method, while being little more difficult to implement. The name qualocation is intended to suggest a quadrature–based modification of the collocation method; for, as we shall see, the collocation method is a special case.

The qualocation method for the equation

$$Lu = f \tag{4.1}$$

is an approximation of the form $u_h \in S_h$ which satisfies

$$(Lu_h, \chi)_h = (f, \chi)_h \quad \forall \chi \in T_h, \tag{4.2}$$

where T_h is a linear space of the same dimension as S_h, and where $(v, z)_h$ is analogous to the exact inner product (v, z), but with the integral replaced by a special J–point quadrature rule on each sub–interval $[x_k, x_{k+1}]$:

$$(v, z)_h = Q_h(v\bar{z}), \tag{4.3}$$

where

$$Q_h(g) = \sum_{k=0}^{N-1} h_k \sum_{j=1}^{J} w_j g(x_k + h_k \xi_j), \tag{4.4}$$

with

$$0 \le \xi_1 < \xi_2 < \cdots < \xi_J < 1, \tag{4.5}$$

and

$$\sum_{j=1}^{J} w_j = 1, \qquad w_1, \cdots, w_J > 0. \tag{4.6}$$

Note that Q_h is the composition of a J–point rule defined on $[0, 1]$,

$$Q(g) = \sum_{j=1}^{J} w_j g(\xi_j), \tag{4.7}$$

on each sub–interval. The points $\{\xi_j\}$ and weights $\{w_j\}$ in this rule have yet to be specified.

Letting $\{v_0, \cdots, v_{N-1}\}$ be a basis for S_h and $\{z_0, \cdots, z_{N-1}\}$ a basis for T_h, the equations to be used in practice are

$$u_h = \sum_{m=0}^{N-1} a_m v_m, \tag{4.8}$$

where the coefficients $\{a_m\}$ satisfy

$$\sum_{m=0}^{N-1}(Lv_m, z_k)_h a_m = (f, z_k)_h, \qquad k = 0, \ldots, N-1. \tag{4.9}$$

Compared with the Galerkin method, the work involved in setting up the matrix is much reduced. To compare with the collocation method, observe that in a well organised calculation one calculates Lv_m once for each basis function v_m at each quadrature point, giving J times the number of operations required to set up the collocation matrix.

If $(\cdot, \cdot)_h$ were replaced in the definition (4.2) by the exact inner product $(\cdot, \cdot)$ then the method would reduce to the Petrov–Galerkin method (i.e. the Galerkin method with possibly different test and trial spaces T_h and S_h). Thus the qualocation method may be thought of as a semi–discrete Petrov–Galerkin method, in which a quadrature rule is used to approximate the outer integral.

There is one choice of the quadrature rule Q that leads to a very familiar method : if we take Q to be just the one–point rule

$$Qg = g(\xi) \tag{4.10}$$

for some ξ satisfying $0 \le \xi < 1$, then it is easy to see that the method is equivalent mathematically to the collocation method. For if $\{z_0, \cdots, z_{N-1}\}$ is a basis of T_h then (4.2) yields in this case

$$\sum_{k=0}^{N-1} h_k(Lu_h - f)(x_k + h_k\xi)z_\ell(x_k + h_k\xi) = 0, \quad \ell = 0, \cdots, N-1,$$

which, provided the $N \times N$ matrix $\{z_\ell(x_k + h_k\xi)\}$ is non–singular, is equivalent mathematically to the collocation equations

$$(Lu_h - f)(x_k + h_k\xi) = 0, \quad k = 0, \cdots, N-1. \tag{4.11}$$

Only in the case $J = 1$ is the choice of the test space T_h unimportant. In every other case the test space plays a significant role. In the earliest work of Sloan [11], as also in Sloan and Wendland [15], T_h was taken to be a space of trigonometric polynomials, to keep the mathematical development as simple as possible. In the present work, however, we follow Chandler and Sloan [5] in taking $T_h = S'_h$, the space of smoothest splines of order r'. In a formal sense the trigonometric polynomial results are then recovered by letting $r' \to \infty$. The spline choice for the test space is likely to be thought preferable in practice, because of the possibility in the spline case of using a local (B–spline) basis, and because their local character makes the splines more amenable to techniques such as mesh grading (see section 6).

We now restrict attention to the case of a uniform mesh, that is (3.15) is assumed, because it is only for this case that theoretical results are available.

How, then, should the quadrature rule Q be chosen? As indicated previously, the central idea in the qualocation method is to choose the quadrature rule to increase the maximum order of negative–norm convergence over that achieved by the collocation method ((3.10) for r even, (3.16) for r odd).

Since the choice $J = 1$ in the quadrature rule (4.7) merely recovers the collocation method, the obvious choice to look at next is $J = 2$. How should we choose the three free parameters ξ_1, ξ_2 and w_1? (Recall that $w_2 = 1 - w_1$ when $J = 2$.) It seems natural to require the rule to be symmetric. Perhaps the simplest choice, and the one considered first (see [11]), is

$$\xi_1 = 0, \qquad \xi_2 = \frac{1}{2}, \tag{4.12}$$

so that the quadrature points are the breakpoints and midpoints of each interval. That leaves just one parameter, namely the weight $w = w_1$ associated with the breakpoints, still to be determined.

The quadrature rule specified by (4.12) is similar to Simpson's rule, and indeed would be exactly equivalent to the composite Simpson rule if w were given the value $1/3$. However, the surprising fact, already established in [11], is that $w_1 = 1/3$ is generally not the best value. For example for the case $r = 2$ (i.e. the piecewise–linear trial space) it is shown in [11] that higher order convergence is achieved for the logarithmic–kernel integral equation if and only if w has the (curious!) value $3/7$. That result was obtained for the case of the trigonometric polynomial test space T_h, but the same property is now known to hold for the spline test spaces considered here.

To be precise, we state as theorems some of the results obtained in [5]. In that paper the class of integral equations is much more general than we consider here : it includes, for example, equations in which the principal part of the integral operator is the Cauchy singular operator (3.4) or the identity operator. The results are stated there for a range of norms, not just for the highest one that achieves the maximum possible order, and they also cover situations in which the exact solution has reduced smoothness. However, the price of generality too often is lesser comprehension, so we settle here for some important special cases.

Assumption A.   The equation to be solved is (2.7), with $k \in C^{\infty}(\Gamma \times \Gamma)$; Γ is a C^{∞} curve and $\nu \in C^{\infty}$; the partition Π_h is uniform; and the test space is S'_h, the space of smoothest splines on Π_h of order r'.

Assumption B. r' has the same parity as r (i.e. r' is even if r is even, odd if r is odd).

Theorem 1. *Assume that A and B hold, and that the quadrature rule Q has the form*

$$Qg = wg(0) + (1 - w)g(\tfrac{1}{2}), \qquad (4.13)$$

with $0 \le w < 1$.

(i) *The qualocation equation (4.2) has a unique solution $u_h \in S_h$ for all h sufficiently small.*

(ii) *If r is even then u_h satisfies (3.10); it satisfies also*

$$\|u_h - u\|_{-3} \le Ch^{r+3}\|u\|_{r+2} \qquad (4.14)$$

if and only if

$$w = \frac{2^r - 1}{2^{r+1} - 1}. \qquad (4.15)$$

(iii) *If r is odd then u_h satisfies (3.16); it satisfies also*

$$\|u_h - u\|_{-4} \le Ch^{r+4}\|u\|_{r+3} \qquad (4.16)$$

if and only if

$$w = \frac{2^{r+1} - 1}{2^{r+2} - 1}. \qquad (4.17)$$

The theorem says that, for r either even or odd, the Simpson–like quadrature rule (4.13) can achieve an order of convergence, in an appropriate negative norm and for u sufficiently smooth, that is two powers of h higher than the collocation method, provided w has the value specified in (4.15) or (4.17). For example, if $r = r' = 2$ (i.e. the piecewise–linear case) we have

$$\|u_h - u\|_{-3} \le Ch^5\|u\|_4, \qquad (4.18)$$

provided w has that curious value 3/7. Note too that there is no restriction on r' other than its parity.

In the next section we sketch the theoretical arguments that lead to these results. But first we indicate a result for a different kind of 2–point rule Q, which is similar to 2–point

Gauss quadrature, except that the two zeros of the 2nd degree Legendre polynomial are replaced by the two symmetrically located zeros on $[0,1]$ of the function

$$G_\alpha(x) = \sum_{\ell=1}^{\infty} \frac{1}{\ell^\alpha} \cos 2\pi \ell x, \qquad (4.19)$$

for appropriate values of $\alpha > 1$.

Theorem 2. *Assume that A and B hold, and that the quadrature rule has the form*

$$Qg = \frac{1}{2}g(\xi) + \frac{1}{2}g(1-\xi), \qquad (4.20)$$

with $0 < \xi < \frac{1}{2}$.

(i) *The qualocation equation (4.2) has a unique solution $u_h \in S_h$ for all h sufficiently small.*

(ii) *If r is even then u_h satisfies (3.10); it satisfies also (4.14) if and only if ξ is the unique zero of G_{r+1} in $(0, \frac{1}{2})$.*

(iii) *If r is odd then u_h satisfies (3.16); if $r' \geq 3$ it satisfies also (4.16) if and only if ξ is the unique zero of G_{r+2} in $(0, \frac{1}{2})$.*

For example, for the piecewise-linear case $r = r' = 2$ the fifth–order convergence result (4.18) holds if and only if $\xi = 0.2308296503$, the unique zero of G_3 in $(0, \frac{1}{2})$. The corresponding value for the 2–point Gauss rule is $\xi = 0.2113248654$; according to the theorem, this value gives only third–order convergence.

5. A sketch of the proof of theorems 1 and 2

Our aim here is to give a brief introduction to the ideas behind the proof in [5] of results which include theorems 1 and 2.

The main work is to prove the result for the case in which k in (2.1) is identically zero and Γ is a circle. The result for the general case can then be inferred from that for the circle by perturbation arguments which are now considered standard (see [1,2,10]).

For the case of a circle we are able to use Fourier series methods that have been used many times [6,10,2,8] in the analysis of collocation on a uniform mesh. If Γ is a circle of radius a and k is identically zero, then if the circle is parametrised in the obvious way the operator L_0 defined by the left side of (2.7) is

$$L_0 u(x) = -2 \int_0^1 \log |2a \sin \pi(x-y)| u(y) dy, \qquad x \in \mathbb{R}. \qquad (5.1)$$

This takes an extraordinarily simple form if u is expressed in Fourier series form: it follows easily from the well known Fourier cosine series

$$-\log(2|\sin \pi x|) \sim \sum_{k=1}^{\infty} \frac{1}{k} \cos 2\pi kx$$

that

$$L_0 u(x) \sim -2(\log a)\hat{u}(0) + \sum_{k\neq 0} \frac{1}{|k|}\hat{u}(k)e^{2\pi ikx};$$

this holds even in the pointwise sense if u is sufficiently smooth (e.g. $u \in L_2 = H^0$ is sufficient). The analysis can be carried out with any value of a other than 1. (The choice $a = 1$ would obviously make L_0 not invertible.) A particularly convenient choice, which we shall adopt from now on, is $a = \exp(-\frac{1}{2})$, since $L_0 u$ then has the very simple form, if $u \in L_2$,

$$L_0 u(x) = \hat{u}(0) + \sum_{k\neq 0} \frac{1}{|k|}\hat{u}(k)e^{2\pi ikx}, \quad x \in \mathbb{R}. \tag{5.2}$$

The method we now have to analyse is: find $u_h \in S_h$ such that

$$(L_0 u_h, \chi)_h = (L_0 u, \chi)_h \quad \forall \ \chi \in S'_h, \tag{5.3}$$

where on the right we have used $f = L_0 u$.

How should we choose bases for S_h and S'_h? B–splines are usually recommended for computation, but for the analysis there is a better choice, one that gives a diagonal matrix in (5.3). Assuming $N \geq r$, let b be the B–spline of order r with knots $\{kh\}$ and support $[0, rh]$, extended 1–periodically to the real line. Then a suitable basis of S_h for computations would be $\{b_0, \cdots, b_{N-1}\}$, where

$$b_k(x) = b(x - k/h), \quad k = 0, \cdots, N - 1, \quad x \in \mathbb{R},$$

while the basis recommended for the analysis is $\{\psi_\mu : \mu \in \Lambda_N\}$, where

$$\psi_\mu = a_\mu \sum_{k=0}^{N-1} e^{2\pi i\mu kh} b_k, \quad \mu \in \Lambda_N, \tag{5.4}$$

with

$$\Lambda_N = \left\{\mu \in \mathbb{Z} : \ -\frac{N}{2} < \mu \leq \frac{N}{2}\right\}, \tag{5.5}$$

and a_μ is a normalisation constant to be specified below.

An important property of ψ_μ, immediately apparent from the definition, is

$$\psi_\mu(x + h) = e^{2\pi i\mu h}\psi_\mu(x), \quad \mu \in \Lambda_N, \quad x \in \mathbb{R}. \tag{5.6}$$

206

It is this property that guarantees a diagonal matrix. Mathematically, the basis functions ψ_μ belong to the different (1–dimensional) irreducible representations of the translation symmetry group of the exact operator L_0 and the approximating spaces, consisting of the translations by multiples of h. Or, to put it more simply, ψ_μ behaves under translation like the trigonometric polynomial ϕ_μ defined by

$$\phi_\mu(x) = e^{2\pi i \mu x}, \qquad \mu \in \Lambda_N. \tag{5.7}$$

In fact the relation between ψ_μ and ϕ_μ is very close: making now a convenient choice of a_μ, it can be shown (see [5]) that

$$\psi_\mu(x) = \begin{cases} 1, & \mu = 0, \\ \displaystyle\sum_{k \equiv \mu} \left(\frac{\mu}{k}\right)^r e^{2\pi i k x}, & \mu \in \Lambda_N^*, \end{cases} \tag{5.8}$$

where $\Lambda_N^* = \Lambda_N \backslash \{0\}$, and where $k \equiv \mu$ means that $k - \mu$ is a multiple of N. (If $r = 1$, and elsewhere if the Fourier series is not absolutely convergent, the series is to be understood as the limit of the symmetric partial sums.) Separating out the first term in the sum, we can write (5.8) as

$$\psi_\mu(x) = \phi_\mu(x)[1 + \Delta(Nx, \tfrac{\mu}{N})], \quad \mu \in \Lambda_N^*, \tag{5.9}$$

where

$$\Delta(\xi, y) = y^r \sum_{\ell \neq 0} \frac{1}{(\ell + y)^r} e^{2\pi i \ell \xi}. \tag{5.10}$$

The function Δ is 1–periodic in ξ, i.e.

$$\Delta(\xi + 1, y) = \Delta(\xi, y), \tag{5.11}$$

from which we again recover the translation property (5.6), since

$$\begin{aligned} \psi_\mu(x + h) &= \phi_\mu(x + h)[1 + \Delta(N(x + h), \tfrac{\mu}{N})] \\ &= e^{2\pi i \mu h} \phi_\mu(x)[1 + \Delta(Nx, \tfrac{\mu}{N})] \\ &= e^{2\pi i \mu h} \psi_\mu(x). \end{aligned}$$

The expression (5.9) makes it clear that ψ_μ is indeed very much like ϕ_μ, especially when r is large. The term Δ in (5.9) is of course just the small correction that is needed to make the right hand side a C^{r-2} spline rather than a C^∞ trigonometric polynomial.

In exactly the same way we define a basis $\{\psi'_\mu : \mu \in \Lambda_N\}$ of S'_h by

$$\psi'_\mu(x) = \begin{cases} 1, & \mu = 0, \\ \phi_\mu(x)[1 + \Delta'(Nx, \tfrac{\mu}{N})], & \mu \in \Lambda_N^*, \end{cases} \tag{5.12}$$

where Δ' is given by (5.10) with r replaced by r'. Thus ψ'_μ too has the translation property,

$$\psi'_\mu(x + h) = e^{2\pi i \mu h} \psi'_\mu(x), \quad \mu \in \Lambda_N, \quad x \in \mathbf{R}. \tag{5.13}$$

The effect of L_0 on ψ_μ is easily studied, given that ψ_μ has the simple Fourier expansion (5.8). Recalling (5.2), we have

$$L_0\psi_\mu(x) = \begin{cases} 1, & \mu = 0, \\ \displaystyle\sum_{k \equiv \mu} \left(\frac{\mu}{k}\right)^r \frac{1}{|k|} e^{2\pi i k x}, & \mu \in \Lambda_N^* \end{cases} \tag{5.14}$$

From now on the parity of r makes a difference to the argument. We therefore confine our attention to the case r even (which includes the piecewise–linear case), leaving the case of odd r to the reader (or see [5]). For r even and $\mu \in \Lambda_N^*$ we can write

$$L_0\psi_\mu(x) = \frac{1}{|\mu|} \sum_{k \equiv \mu} \left|\frac{\mu}{k}\right|^{r+1} e^{2\pi i k x}$$

$$= \frac{1}{|\mu|} \phi_\mu(x) \left[1 + \Omega(Nx, \frac{\mu}{N})\right], \tag{5.15}$$

where

$$\Omega(\xi, y) = |y|^{r+1} \sum_{\ell \neq 0} \frac{1}{|\ell + y|^{r+1}} e^{2\pi i \ell \xi}. \tag{5.16}$$

From the property

$$\Omega(\xi + 1, y) = \Omega(\xi, y), \tag{5.17}$$

one sees that $L_0\psi_\mu$ has the same translation property as ψ_μ,

$$L_0\psi_\mu(x + h) = e^{2\pi i \mu h} L_0\psi_\mu(x). \tag{5.18}$$

It is now a very easy exercise, from the definition of $(\cdot, \cdot)_h$ and the translation properties (5.13) and (5.18), to show that in the uniform mesh case $(L_0\psi_\nu, \psi'_\mu)_h$ vanishes for $\nu \neq \mu$: all that is needed is the elementary sum

$$\sum_{k=0}^{N-1} e^{2\pi i (\mu - \nu) k h} = \begin{cases} N & \text{if } \mu \equiv \nu, \\ 0 & \text{otherwise.} \end{cases} \tag{5.19}$$

Using the explicit expressions (5.12) and (5.15) for the non–vanishing case, one then obtains

$$(L_0\psi_\nu, \psi'_\mu)_h = \begin{cases} 1 & \text{if } \mu = \nu = 0, \\ \dfrac{1}{|\mu|} D(\dfrac{\mu}{N}) & \text{if } \mu = \nu \neq 0, \\ 0 & \text{otherwise,} \end{cases} \tag{5.20}$$

where

$$D(y) = \sum_{j=1}^{J} w_j [1 + \Omega(\xi_j, y)][1 + \overline{\Delta'(\xi_j, y)}]. \tag{5.21}$$

The function D determines all of the properties of the qualocation rule. Note that it contains the point and weights of the quadrature rule Q in a very direct way.

Because we have an equation to solve, it will be necessary, clearly, to divide by $D(\mu/N)$. The key to numerical stability, therefore, is that $D(y)$ be bounded away from zero. This is not a trivial question, since it is well known that the collocation method (which is of course included as a special case) can be unstable: for our present case of even r, midpoint collocation is known to be unstable. Nevertheless, it is shown in [5] that in every other situation the qualocation method is stable: it is stable if $J \geq 2$, and if $J = 1$ and r is even then it is stable if $\xi = \xi_1$ is not the midpoint. The argument begins

$$|D(y)| \geq \mathrm{Re}\,D(y) = \sum_{j=1}^{J} w_j \mathrm{Re}[1 + \Omega(\xi_j, y)][1 + \overline{\Delta'(\xi_j, y)}], \tag{5.22}$$

and then proceeds by showing that each term in the sum is non–negative, and strictly positive if $\xi_j \neq 1/2$. We shall from now on take it that the stability property

$$\inf \left\{ |D(y)| : y \in [-\tfrac{1}{2}, \tfrac{1}{2}] \right\} > 0 \tag{5.23}$$

has been established.

We turn at last to the error analysis of the qualocation equation (5.3). If u_h is written as the linear combination

$$u_h = \sum_{\mu \in \Lambda_N} c_\mu \psi_\mu,$$

then it follows from (5.8) that

$$c_\mu = \hat{u}_h(\mu), \qquad \mu \in \Lambda_N,$$

so that

$$u_h = \sum_{\mu \in \Lambda_N} \hat{u}_h(\mu) \psi_\mu. \tag{5.24}$$

Thus (5.3) is equivalent to

$$\sum_{\nu \in \Lambda_N} (L_0 \psi_\nu, \psi'_\mu)_h \hat{u}_h(\nu) = (L_0 u, \psi'_\mu)_h, \qquad \mu \in \Lambda_N,$$

or, using (5.20),

$$\hat{u}_h(0) = (L_0 u, \psi_0')_h, \tag{5.25}$$

and

$$\frac{1}{|\mu|} D(\frac{\mu}{N}) \hat{u}_h(\mu) = (L_0 u, \psi_\mu')_h, \quad \mu \in \Lambda_N^*, \tag{5.26}$$

so that the equations for different μ are uncoupled. We now write u (assuming $u \in H^s$ with $s > \frac{1}{2}$, so that u is continuous) as

$$u(x) = \sum_{\nu \in \Lambda_N} \hat{u}(\nu) e^{2\pi i \nu x} + \tilde{u}_N(x), \tag{5.27}$$

where $\tilde{u}_N$ is the 'tail' of the Fourier series,

$$\tilde{u}_N(x) = \sum_{\nu \notin \Lambda_N} \hat{u}(\nu) e^{2\pi i \nu x}. \tag{5.28}$$

Then

$$L_0 u(x) = \hat{u}(0) + \sum_{\nu \in \Lambda_N^*} \frac{\hat{u}(\nu)}{|\nu|} e^{2\pi i \nu x} + L_0 \tilde{u}_N(x),$$

and hence, on using (5.12),

$$(L_0 u, \psi_\mu')_h = \begin{cases} \hat{u}(0) + \cdots & \text{if } \mu = 0, \\ \dfrac{\hat{u}(\mu)}{|\mu|} \displaystyle\sum_j w_j [1 + \overline{\Delta'(\xi_j, \frac{\mu}{N})}] + \cdots & \text{if } \mu \in \Lambda_N^*, \end{cases}$$

where, here and elsewhere, the trailing dots indicate terms depending only on $\tilde{u}_N$. It now follows from (5.25) and (5.26) that

$$\hat{u}_h(0) = \hat{u}(0) + \cdots \tag{5.29}$$

and

$$\hat{u}_h(\mu) = D\left(\frac{\mu}{N}\right)^{-1} \hat{u}(\mu) \sum_j w_j [1 + \overline{\Delta'(\xi_j, \frac{\mu}{N})}] + \cdots \quad \text{if } \mu \in \Lambda_N^*, \tag{5.30}$$

leading to the error expressions

$$\hat{u}(0) - \hat{u}_h(0) = \cdots \tag{5.31}$$

and

$$\hat{u}(\mu) - \hat{u}_h(\mu) = \frac{E(\frac{\mu}{N})}{D(\frac{\mu}{N})} \hat{u}(\mu) + \cdots \quad \text{if } \mu \in \Lambda_N^*, \tag{5.32}$$

where

$$E(y) = D(y) - \sum_j w_j[1 + \overline{\Delta'(\xi_j, y)}]$$

$$= \sum_{j=1}^{J} w_j \Omega(\xi_j, y)[1 + \overline{\Delta'(\xi_j, y)}]. \qquad (5.33)$$

Now at last we see how the error might be controlled. For $\mu \neq 0$ the error consists of two parts, of which the second, shown in (5.32) by trailing dots, comes from the Fourier components $\hat{u}(k)$ with $|k| \geq N/2$. That part can be controlled by requiring the Fourier components of u to decay sufficiently rapidly, or, more precisely, by requiring u to belong to H^s with a large enough value of s. A proper treatment of that contribution to the error is of course essential in proving the error estimates indicated in the theorem, but we shall not trouble with it any further here.

The first term of (5.32), however, is of a different character, since it places an absolute limit to the order of convergence that can be achieved in any norm. For if we consider a fixed $\mu \neq 0$ and then let $N \rightarrow \infty$ it follows from the first term of (5.32) that the error in the μth Fourier component, unless $\hat{u}(\mu)$ happens to be zero, is governed by the behaviour of the function $E(y)$ for y near zero.

The novel idea in the qualocation method is that the behaviour of $E(y)$ in the vicinity of zero can be controlled by adjusting the quadrature rule. Putting the argument as simply as possible, it follows from (5.33) and the explicit expressions for Ω and Δ' that

$$E(y) = |y|^{r+1} \sum_j w_j \sum_{\ell \neq 0} \frac{e^{2\pi i \ell \xi_j}}{|\ell + y|^{r+1}} + O(|y|^{r+r'+1}). \qquad (5.34)$$

Thus no matter how the quadrature rule is chosen (provided it gives a stable rule!) we will have $E(y) = O(|y|^{r+1})$, and hence

$$\hat{u}(\mu) - \hat{u}_h(\mu) = O(h^{r+1}), \qquad (5.35)$$

giving the order of convergence we have seen in (3.10) for the collocation method.

Can we do better than this? Obviously the answer is yes : all we need do is to force the coefficient of $|y|^{r+1}$ to be zero. Even better, in our theorems 1 and 2 we are concerned only with symmetric quadrature rules, i.e. rules with the property that if ξ is a quadrature point then so is $1 - \xi$, with the same associated weight. For such rules we have

$$\sum_j w_j e^{2\pi i \ell \xi_j} = \sum_j w_j \cos 2\pi \ell \xi_j,$$

from which it follows easily that

$$E(y) = |y|^{r+1}2\sum_j w_j \sum_{\ell=1}^{\infty} \frac{\cos 2\pi\ell\xi_j}{\ell^{r+1}} + O(|y|^{r+3}),$$

so that we achieve $E(y) = O(|y|^{r+3})$, and hence

$$\hat{u}(\mu) - \hat{u}_h(\mu) = O(h^{r+3}), \tag{5.36}$$

i.e. two orders of improvement, whenever the leading term vanishes, or equivalently, whenever the quadrature rule satisfies

$$\sum_{j=1}^{J} w_j \sum_{\ell=1}^{\infty} \frac{\cos 2\pi\ell\xi_j}{\ell^{r+1}} = 0. \tag{5.37}$$

The quadrature rules specified in the second part of theorems 1 and 2 are designed precisely to satisfy this condition. In the case of the rule (4.20), used in theorem 2, the condition (5.37) reduces immediately to

$$\sum_{\ell=1}^{\infty} \frac{\cos 2\pi\ell\xi}{\ell^{r+1}} = 0,$$

which is exactly the condition that ξ be a zero of G_{r+1}. In the case of the rule (4.13) the condition (5.37) becomes

$$w\sum_{\ell=1}^{\infty} \frac{1}{\ell^{r+1}} + (1-w)\sum_{\ell=1}^{\infty} \frac{(-1)^\ell}{\ell^{r+1}} = 0,$$

or, by the use of a standard zeta function trick,

$$\left[w - (1-w)\left(1 - \frac{1}{2^r}\right)\right]\sum_{\ell=1}^{\infty} \frac{1}{\ell^{r+1}} = 0.$$

This is satisfied if and only if the weight w is given by (4.15).

For the remainder of the proof, and in particular for the arguments leading to the Sobolev norm estimates in the theorems, the reader is referred to [5].

Before leaving the matter of proof, we may observe that the argument is little altered if the definition of L_0 is generalised to

$$L_0 u(x) = \hat{u}(0) + \sum_{k\neq 0} |k|^\beta \hat{u}(k)e^{2\pi i kx}, \tag{5.38}$$

or to

$$L_0 u(x) = \hat{u}(0) + \sum_{k \neq 0} (\text{sign } k) |k|^\beta \hat{u}(k) e^{2\pi i k x}, \tag{5.39}$$

for β an arbitrary real number. This, together with smooth perturbations, is the class considered in [5]. Of course β now appears as a parameter in the quadrature rules, as well as in the final error estimates. For details see [5].

6. Fully discrete variants

The matrix in the Galerkin method, we may recall, requires two levels of exact integration, whereas the collocation and qualocation matrices require only one. The methods now to be described require no exact integrals at all, and so are particularly convenient for implementation.

Again we assume that Γ is a C^∞ curve, and take Π_h to be uniform. In the method of Sloan and Burn [13] the first step is to replace the exact integral $Lu(x)$, see (2.7), by its rectangle rule approximation

$$\begin{aligned}
L_h u(x) &= \left[-2h \sum_{m=0}^{N-1} \log |\nu(x) - \nu(mh)| \right. \\
&\quad \left. + 2\pi h \sum_{m=0}^{N-1} k(\nu(x), \nu(mh)) \right] u(mh) \\
&=: \sum_{m=0}^{L-1} A_m(x) u(mh). \tag{6.1}
\end{aligned}$$

One then proceeds as in the qualocation method, but with L replaced by L_h: one seeks $u_h \in S_h$ such that

$$(L_h u_h, \chi)_h = (f, \chi)_h \quad \forall \quad \chi \in T_h, \tag{6.2}$$

where $(\cdot, \cdot)_h$ is defined by (4.3)-(4.6).

But now the points and weights in the quadrature rule Q are chosen differently – for in replacing L by L_h we have committed a numerical crime, and the quadrature rule now has the added burden of correcting that crime.

The equations to be solved in practice are, if $\{z_0, \cdots, z_{N-1}\}$ is a basis for T_h,

$$\sum_{m=0}^{N-1} (A_m, z_k)_h u_h(mh) = (f, z_k)_h, \quad k = 0, \cdots, N - 1. \tag{6.3}$$

Since one solves for the values of u_h only at the quadrature points of the rectangle rule, the choice of a trial space is of importance only when one wants to interpolate between the points. For the theoretical analysis, however, the choice is significant. In [13] a trigonometric trial space

$$S_h = \text{span } \{\phi_\mu : \mu \in \Lambda_N\} \tag{6.4}$$

is assumed, to simplify the analysis.

How should the quadrature rule Q be chosen? Zero is excluded as a quadrature point, since it would lead to $\log 0$ appearing as a term in each element of the matrix. In [13] general quadrature rules satisfying this condition are analysed, and then specific results are stated for quadrature rules of the symmetric two–point form

$$Qg = \tfrac{1}{2}g(\xi) + \tfrac{1}{2}g(1-\xi), \quad 0 < \xi < \tfrac{1}{2}. \tag{6.5}$$

The following theorem, proved by Fourier techniques analogous to those in section 5 for the case of a circle, and then extended by perturbation arguments to general smooth curves, shows that the choice $\xi = \tfrac{1}{6}$ has very special properties. (This theorem is proved in [13] for the case of a circle, but the extension to general curves in that paper is achieved only under more stringent conditions. That the same theorem as for the circle holds for general curves with no additional restrictions has been shown in [9].)

Theorem 3. *Assume that A holds, that the trial space is given by (6.4), and that the quadrature rule Q is given by (6.5). Let r' be even.*

(i) *Equation (6.2) has a solution $u_h \in S_h$ for all h sufficiently small.*

(ii) *For $s \geq -1$, u_h satisfies*

$$\|u_h - u\|_s \leq Ch^1 \|u\|_{s+1}. \tag{6.6}$$

It satisfies also

$$\|u_h - u\|_s \leq Ch^3 \|u\|_{s+3} \tag{6.7}$$

if and only if $\xi = \tfrac{1}{6}$.

The theorem tells us, for instance, that negative norm convergence of order $O(h^3)$ is achievable if $u \in H^2$, and that convergence of this order can be achieved also in the L_2 or higher Sobolev norms if u has the necessary additional smoothness.

If, on the other hand, u is of lesser smoothness then it may be better to use an alternative version proposed by Saranen and Sloan [9], in which the right side of (6.2) is replaced by the exact integral, so that the method becomes : find $u_h \in S_h$ such that

$$(L_h u_h, \chi)_h = (f, \chi) \quad \forall \; \chi \in T_h. \tag{6.8}$$

214

It is shown in [9] that theorem 3 still holds, but with the condition $s \geq -1$ replaced by $s \geq -r' - 1$, and with (6.7) replaced by

$$\|u_h - u\|_s \leq Ch^{\min(r',3)}\|u\|_{s+\min(r',3)}. \tag{6.9}$$

This allows negative–norm estimates of the same $O(h^3)$ order to be obtained even when the exact solution is of very low regularity, provided we use at least a cubic–spline test space. The piecewise–linear test space, on the other hand, gives only $O(h^2)$ convergence.

The original paper [13] also provides a framework within which fully discrete methods of still higher order can be sought, but does not guarantee the existence of the corresponding quadrature rules, because non–linear algebraic equations have to be solved for the points and weights. It is now known that in fact some such rules do exist [14]: for example a quadrature rule of more elaborate form than (6.4) exists which gives a fully discrete $O(h^5)$ method if used in (6.2) with $r' \geq 4$; and another rule of this form exists, albeit with some negative weights, which gives an $O(h^5)$ method even if $r' = 2$. However, further information on this will have to wait until another occasion.

7. Numerical examples

We here consider just two examples, to illustrate two different aspects: the first illustrates the fact that for smooth curves the orders of convergence really are exactly as predicted; and the second demonstrates that the methods might have a role even where the theory is currently lacking.

Example 1.[5]   We take Γ to be a circle of radius $\frac{1}{2}$ centred at $(0,0)$, and take $k \equiv 0$ and $g(t) = g(t_1, t_2) = t_1$ in (2.1), and seek to approximate the single–layer potential (2.4) at the point $\tau = (0.1, 0.2)$. (Because g is in this case the restriction to Γ of a harmonic function $\tilde{g}$ in the interior, the exact solution in this case is $\phi(\tau_1, \tau_2) = \tilde{g}(\tau_1, \tau_2) = \tau_1$.) The approximations considered are versions of the qualocation method for piecewise–linear trial and test spaces, i.e. $r = r' = 2$.

Table 1. Errors and computed convergence rates for example 1.

N	(i) collocation	(ii) qualocation $\frac{3}{7}, \frac{4}{7}$ rule	(iii) qualocation symm.2-pt.	(iv) discrete Galerkin
4	6.42(-3)	2.20(-3)	1.90(-3)	2.42(-3)
8	5.06(-4) 3.66	1.31(-5) 7.39	4.54(-6) 8.71	5.55(-5) 5.44
16	5.98(-5) 3.08	3.71(-7) 5.14	1.12(-7) 5.33	6.05(-6) 3.19
32	7.37(-6) 3.02	1.12(-8) 5.04	3.27(-9) 5.10	7.33(-7) 3.04
64	9.18(-7) 3.00	3.50(-10) 5.00	1.00(-10) 5.02	9.09(-8) 3.01
88	3.53(-7) 3.00	7.11(-11) 5.00	2.03(-11) 5.00	3.49(-8) 3.00

In Table 1 we show the errors for the following methods: (i) collocation, (ii) qualocation with the $\frac{3}{7}, \frac{4}{7}$ rule of theorem 1, (iii) qualocation with the higher–order symmetric 2–point rule of theorem 2 and (iv) the discrete Galerkin method using 2–point Gauss quadrature for the outer integral (which is of exactly the form considered in theorem 2, but with the 'wrong' value of ξ). We also show the apparent orders of convergence α obtained from the ratios of successive errors, by assuming the error to be of the form Ch^α. It is very clear that the two versions of the qualocation method are indeed of order $O(h^5)$, as predicted by (4.18), whereas the discrete Galerkin method, like the collocation method, is of order $O(h^3)$.   In a similar way the numerical experiments of [13] with the fully discrete method of theorem 3 show clearly the predicted $O(h^3)$ order of convergence when $\xi = 1/6$; and moreover for values of ξ other than $1/6$ show just as clearly an asymptotic $O(h)$ order of convergence.

Example 2.[13] In this case Γ is taken to be the curve whose polar coordinate specification is

$$r = 2\theta \left(\frac{3\pi}{2} - \theta \right), \quad 0 \le \theta \le \frac{3\pi}{2},$$

which has a re–entrant right–angled corner at $(0,0)$. Further, $g(t) = g(t_1, t_2) = (t_2 - 1)^2$, and the single layer potential (2.4) is evaluated at $(17/70, 281/70)$. To weaken the effect of the singularity at the corner, the curve was parametrised by taking

$$\theta = \theta(x) = \frac{3\pi}{2} \left(3x^2 - 2x^3 \right), \quad 0 \le x \le 1,$$

so that $\theta(x) \approx Cx^2$ for small x, and

$$\theta(1 - x) = \frac{3\pi}{2} - \theta(x).$$

Thus in effect we are 'grading the mesh' near the corner (with grading exponent 2). The method of [13] was then applied, to give the results shown in Table 2, in which the apparent order of convergence is better than $O(h^{5/2})$. On the other hand values of ξ other than $1/6$ were shown by experiment in [13] to give only $O(h)$ convergence. The apparent rate of convergence in this case is perhaps good enough to encourage further investigation of qualocation and its variants allied with mesh grading in the presence of corners.

Table 2. Errors and computed convergence rates for example 2.

N	error	
8	-2.48(-2)	
16	1.19(-2)	
32	1.98(-3)	2.59
64	3.07(-4)	2.69
128	4.74(-5)	2.70
256	7.71(-6)	2.62

Acknowledgements. The writer acknowledges the support of the Australian Research Council, and of the Technical University of Vienna, where this paper was written.

References:

[1] D.N. Arnold and W.L. Wendland, 'On the asymptotic convergence of collocation methods', *Math. Comp.*, **41** (1983), 349–381.

[2] D.N. Arnold and W.L. Wendland, 'The convergence of spline collocation for strongly elliptic equations on curves', *Numer. Math.* **47** (1985), 317–341.

[3] G.A. Chandler, 'Midpoint collocation for Cauchy singular integral equations', submitted.

[4] G.A. Chandler, 'Discrete norms for the convergence of boundary element methods', in *Proceedings of the Workshop on Theoretical and Numerical Aspects of Geometric Variational Inequalities*, Canberra 1990, to appear.

[5] G.A. Chandler and I.H. Sloan, 'Spline qualocation methods for boundary integral equations', *Numer. Math.* **58** (1990), 537–567.

[6] F.R. de Hoog, 'Product integration techniques for the numerical solution of integral equations', Ph.D. thesis, Australian National University (Canberra), 1974.

[7] S. Prössdorf, J. Saranen and I.H. Sloan, 'A discrete method for the logarithmic–kernel integral equation on an arc', submitted.

[8] J. Saranen, 'The convergence of even degree spline collocation solution for potential problems in smooth domains of the plane', *Numer. Math.* **53** (1988), 499–512.

[9] J. Saranen and I.H. Sloan, 'Quadrature methods for logarithmic kernel integral equations on closed curves', submitted.

[10] J. Saranen and W.L. Wendland, 'On the asymptotic convergence of collocation methods with spline functions of even degree', *Math. Comp.* **45** (1985), 91–108.

[11] I.H. Sloan, 'A quadrature–based approach to improving the collocation method', *Numer. Math.* **54** (1988), 41–56.

[12] I.H. Sloan, 'Superconvergence in the collocation and qualocation methods', in *ICNM88: Numerical Mathematics, Singapore 1988*, ed. R. Agarwal, ISNM86, Birkhäuser, Basel (1988), 429–441.

[13] I.H. Sloan and B.J. Burn, 'An unconventional quadrature method for logarithmic–kernel integral equations on closed curves', *J. Integral Eqns. and Applics.*, to appear.

[14] I.H. Sloan and B.J. Burn, 'An unconventional quadrature method for logarithmic–kernel integral equations on closed curves II', in preparation.

[15] I.H. Sloan and W.L. Wendland, 'A quadrature based approach to improving the collocation method for splines of even degree', *Zeit. Anal. Anwendungen* **8** (1989), 361–376.

[16] W.L. Wendland, 'Qualocation, the new variety of boundary element methods', Wiss. Z. d TU Karl–Marx–Stadt **31** (1989), 276–284.

[17] W.L. Wendland, 'Boundary element methods for elliptic problems', in *Mathematical Theory of Finite and Boundary Element Methods*, by A.H. Schatz, V. Thomée and W.L. Wendland, Birkhäuser, Basel (1990), 219–276.

V THOMÉE
Numerical methods for hyperbolic integro-differential equations

Abstract. Stability and error estimates are derived for fully discrete numerical methods for time-dependent partial integro-differential equations of hyperbolic type. Special attention is paid to the storage requirements in the time-stepping.

1. Introduction.

The object of this paper is to present work expounded in more detail in Lin, Thomée, and Wahlbin [3] and Pani, Thomée, and Wahlbin [4] concerning numerical methods for the solution of the hyperbolic integro-differential equation

$$u_{tt} + A(t)u = \int_0^t B(t,s)u(s)ds + f(t), \quad \text{in } \Omega \times J, \tag{1.1a}$$

together with initial and boundary conditions,

$$
\begin{aligned}
u &= 0, \quad \text{on } \partial\Omega \times J, \\
u(x,0) &= u_0(x), \quad u_t(x,0) = u_1(x), \quad \text{in } \Omega.
\end{aligned}
\tag{1.1b}
$$

Here Ω is a bounded domain in R^d with smooth boundary $\partial\Omega$, J denotes the interval $[0,T]$ with a fixed upper limit T, $A(t)$ is a selfadjoint, uniformly positive definite, uniformly elliptic second order differential operator, and $B(t,s)$ is a second order partial differential operator, both with smooth coefficients. Problems of this nature, and non-linear versions thereof, occur, e.g., in visco-elasticity, cf. Renardy, Hrusa, and Nohel [5] and references therein.

The numerical solutions will be obtained by first discretizing in space by a Galerkin finite element method, and then applying a finite difference and quadrature scheme for the time stepping. They will be sought, for discrete time levels $t_n = nk$, with k the time step, in a finite dimensional space $S_h \subset H_0^1 = H_0^1(\Omega)$ belonging to a family such that, for a fixed given integer $r \geq 2$, we have

$$\min_{\chi \in S_h} \{\|v - \chi\| + h\|v - \chi\|_1\} \leq Ch^i\|v\|_i, \quad \text{for } v \in H_0^1 \cap H^i, \ 1 \leq i \leq r, \tag{1.2}$$

where $\| \cdot \|$ and $\| \cdot \|_i$ denote the norms in $L_2 = L_2(\Omega)$ and $H^i = H^i(\Omega)$, respectively. Special such methods have been applied and analyzed earlier for equations of the above type, with a first order operator B in the memory term, in Yanik and Fairweather [10] (cf. also Cannon, Lin, and Xie [2]).

One of the difficulties involved in a time-stepping scheme of the type indicated is that, normally, all the values of the discrete solution have to be retained at all previous time levels, causing great demands for data storage. This is in contrast to the situation for a purely hyperbolic equation where only a fixed low number of time levels is involved at each time step, and the data can be discarded as the computation goes along. As a way around this difficulty, in the case of discretization in time only of a parabolic integro-differential equation, it was proposed in Sloan and Thomée [6] that the quadrature be based on fewer points, thus reducing the number of time levels at which the data need to be saved. Adopting this point of view, we shall consider below, for the present hyperbolic problem, several quadrature methods of this kind, for which the accuracy of the scheme equals that of the corresponding scheme for the pure differential equation.

The program just described has been carried out for discretization in both space and time of integro-differential equations of parabolic type in Thomée and Zhang [9] and Zhang [11], [12] (cf. also [8] for a survey). In these papers the elliptic operator was assumed time independent, which permitted the use of a spectral argument at a crucial point in the proofs. In contrast, in the present work we rely on energy arguments, which are not correspondingly restrictive. For this reason, this approach may be used to improve the results also in the parabolic case, see [4].

The starting point of our investigation is the spatially discrete version of (1.1), which will be treated in Section 2 below. Here, we shall use in our analysis a variant of the Ritz projection, invented in Cannon and Lin [1], cf. also [3], which is specially designed to take into account the memory term, and which we refer to as the Ritz-Volterra projection. In Section 3 we discuss a time stepping scheme based on a first order accurate, three level, backward difference approximation of u_{tt}, and show stability and error estimates in the natural discrete energy norm associated with the hyperbolic equation. The error estimate will contain an as yet undetermined term which depends on the choice of the quadrature rule used. In Section 4 we complete the error analysis by bounding the quadrature error for three different choices of quadrature formulas which are consistent with the order of accuracy of the difference approximation, thus showing a total error bound of optimal order $O(h^r + k)$. They are based on the rectangle rule with time step k, the trapezoidal rule with a time step of order $O(k^{1/2})$, and Simpson's rule with a time step of order $O(k^{1/4})$; they require storage of the solution at a number of time levels of orders $O(k^{-1})$, $O(k^{-1/2})$, and $O(k^{-1/4})$, respectively. In Section 5 we derive similar results to those of Sections 3 and 4, now of second order of accuracy in time, for a symmetric approach to the differential equation part of (1.1), combined with quadrature rules based on midpoint and Simpson's type rules, with storage requirements of orders $O(k^{-1})$ and $O(k^{-1/2})$, respectively. In this case the stability result will be expressed in terms of averages $(U^n + U^{n+1})/2$, which will slightly complicate the construction of the quadrature rules.

2. Discretization with respect to space.

As a starting point for the discretization of (1.1) we formulate its semidiscrete analogue, based on a weak form of the initial boundary value problem. Letting $(\cdot, \cdot), A(t; \cdot, \cdot)$ and $B(t, s; \cdot, \cdot)$ denote the inner product in L_2 and the bilinear forms on $H_0^1 \times H_0^1$ defined by the differential operators $A(t)$ and $B(t, s)$, we define the semidiscrete solution of (1.1) as the function $u_h : J \to S_h$ such that

$$(u_{h,tt}, \chi) + A(t; u_h, \chi) = \int_0^t B(t, s; u_h(s), \chi)ds + (f(t), \chi), \text{ for } \chi \in S_h, t \in J, \tag{2.1}$$

$$u_h(0) = u_{0h}, \quad u_{h,t}(0) = u_{1h},$$

where u_{0h} and u_{1h} are appropriate approximations of u_0 and u_1 in S_h.

A principal tool used in [2], [3] for the analysis of the semidiscrete problem was a generalization of the elliptic, or Ritz, projection called the Ritz-Volterra projection $W : C(J; H_0^1) \to C(J; S_h)$, defined by

$$A(t; (W - u)(t), \chi) = \int_0^t B(t, s; (W - u)(s), \chi) ds, \quad \text{for } \chi \in S_h, \ t \in J. \quad (2.2)$$

The following estimate may be proved for the error in this projection:

Proposition 2.1. *Under the assumption (1.2) we have, for any $j \geq 0$,*

$$\|D_t^j(W - u)(t)\| + h\|D_t^j(W - u)(t)\|_1 \leq Ch^r \sum_{l=0}^{j} \{\|D_t^l u(t)\|_r + \int_0^t \|D_t^l u\|_r ds\}, \quad t \in J.$$

PROOF: We show the result for $j = 0$; for positive j, see [3]. Setting $\rho = W - u$ we first bound $\|\rho\|_1$. With $R_h = R_h(t) : H^1 \to H^1$ denoting the standard Ritz projection, corresponding to $B \equiv 0$ in (2.2), and for which the error bound is well-known (without the integral term on the right), it suffices to estimate $\eta = W - R_h u \in S_h$. Using (2.2) we have, with $c > 0$,

$$c\|\eta(t)\|_1^2 \leq A(t; \eta(t), \eta(t)) = \int_0^t B(t, s; \rho(s), \eta(t)) ds \leq C\|\eta(t)\|_1 \int_0^t \|\rho\|_1 ds,$$

so that

$$\|\eta(t)\|_1 \leq C \int_0^t \|\rho\|_1 ds.$$

Hence

$$\|\rho(t)\|_1 \leq \|(R_h u - u)(t)\|_1 + C \int_0^t \|\rho\|_1 ds \leq Ch^{r-1}\|u(t)\|_r + C \int_0^t \|\rho\|_1 ds,$$

from which the desired estimate follows by Gronwall's lemma.

The estimate for ρ in L_2 is derived by a modification of the standard duality argument. Thus, for any $\phi \in L_2$, we have, with $\psi = \psi(t) = A(t)^{-1}\phi$ and using (2.2) once more,

$$(\rho(t), \phi) = A(t; \rho(t), \psi) = A(t; \rho(t), \psi - R_h(t)\psi) - \int_0^t B(t, s; \rho(s), R_h(t)\psi) ds$$

$$= A(t; (R_h u - u)(t), \psi) - \int_0^t B(t, s; \rho(s), R_h(t)\psi - \psi) ds - \int_0^t B(t, s; \rho(s), \psi) ds$$

whence

$$|(\rho(t), \phi)| \leq C\{\|R_h(t)u - u\| + \int_0^t (h\|\rho\|_1 + \|\rho\|) ds\}\|\psi\|_2.$$

Since $\|\psi\|_2 \leq C\|\phi\|$ the desired estimate easily follows by Gronwall's lemma and the known estimates for $\|R_h(t)u - u\|$ and $\|\rho\|_1$.

The following error estimate for the semidiscrete problem was shown in Cannon and Lin [1], see Lin, Thomée, and Wahlbin [3] (cf. also [10] in the case that B is of first order).

Theorem 2.1. *We have for the solutions of (1.1) and (2.1), with $\theta = u_h - W$,*

$$\|u_h(t) - u(t)\| + h\|u_h(t) - u(t)\|_1 \leq C(\|\theta(0)\|_1 + \|\theta_t(0)\|) + C(u)h^r, \quad t \in J.$$

PROOF: With our above notation we write $e = u_h - u = \theta + \rho$. Proposition 2.1 shows

$$\|\rho(t)\| + h\|\rho(t)\|_1 \leq C(u)h^r, \tag{2.3}$$

and it remains to estimate θ. We have by our definitions

$$(\theta_{tt}, \chi) + A(t; \theta, \chi) = \int_0^t B(t, s; \theta(s), \chi)ds - (\rho_{tt}, \chi), \quad \forall \chi \in S_h, \ t \in J.$$

Setting $\chi = \theta_t \in S_h$ this yields

$$\frac{1}{2}\frac{d}{dt}\{\|\theta_t\|^2 + A(t; \theta, \theta)\} \leq \|\rho_{tt}\| \, \|\theta_t\| + \frac{1}{2}A_t(t; \theta, \theta)$$

$$+ \frac{d}{dt}\int_0^t B(t, s; \theta(s), \theta(t))ds - B(t, t; \theta(t), \theta(t)) - \int_0^t B_t(t, s; \theta(s), \theta(t))ds.$$

By integration we infer after trivial estimates that

$$\|\theta_t\|^2 + c\|\theta\|_1^2 \leq \|\theta_t(0)\|^2 + C\|\theta(0)\|_1^2 + \int_0^t \|\rho_{tt}\| \, \|\theta_t\| ds + \frac{1}{2}c\|\theta\|_1^2 + C\int_0^t \|\theta\|_1^2 ds,$$

from which Gronwall's lemma and simple arguments give

$$\|\theta_t\| + \|\theta\|_1 \leq C\{\|\theta_t(0)\| + \|\theta(0)\|_1 + \int_0^t \|\rho_{tt}\|ds\} \leq C(\|\theta_t(0)\| + \|\theta(0)\|_1) + C(u)h^r.$$

Together with (2.3) this completes the proof.

It is not difficult to see that if we choose, e.g., $u_{0h} = R_h(0)u_0, u_{1h} = R_h(0)u_1$, then

$$\|\theta_t(0)\| + \|\theta(0)\|_1 \leq C(u)h^r,$$

so that the error bound in Theorem 2.1 reduces to $C(u)h^r$.

3. Schemes based on backward differencing.

The completely discrete methods we shall consider in this section will be defined by replacing the time derivative in (2.1) by a three-level backward difference quotient, and using a quadrature rule in the memory term of the form

$$\sigma^n(g) = \sum_{j=0}^{n-1} \omega_{nj}g(t_j) \approx \int_0^{t_n} g(s)ds, \quad t_j = jk, \tag{3.1}$$

where ω_{nj} are nonnegative quadrature coefficients and U^j is the approximation of $u_h(t_j)$. Thus, we define $U^n \in S_h$ by

$$(\bar{\partial}^2 U^n, \chi) + A_n(U^n, \chi) = \sigma^n(B_n(U, \chi)) + (f^n, \chi), \quad \text{for } \chi \in S_h, \quad n \geq 2, \qquad (3.2)$$

with U^0 and U^1 given. Here $\bar{\partial} U^n = (U^n - U^{n-1})/k$ denotes a backward difference quotient, further $A_n(\psi, \chi) = A(t_n; \psi, \chi)$, and $\sigma^n(B_n(U, \chi))$ is shorthand notation for $\sum_{j=0}^{n-1} \omega_{nj} B(t_n, t_j; U^j, \chi)$.

As mentioned above, one of the difficulties involved in such a time-stepping scheme is that if $\omega_{nj} \neq 0$ for $j \leq n$, then all the values of U^j have to be retained, causing great demands for data storage. We shall therefore consider below several quadrature methods for which many of the ω_{nj} vanish, but for which the accuracy of the scheme nevertheless equals that of the difference quotient used to approximate $u_{h,tt}$. We shall assume that the quadrature formulas have *persistent dominated weights*, i.e. that the ω_{nj} are such that for some sequence of nonnegative numbers $\omega_j, j = 0, 1, 2, \cdots$, independent of n,

$$\omega_{nj} \leq \omega_j, \quad \text{for } 0 \leq j < n, \quad \text{with } \sum_{j=0}^{n-1} \omega_j \leq C, \quad \text{for } t_n \in J, \qquad (3.3)$$

and

$$\sum_{i=j+2}^{n} |\omega_{ij} - \omega_{i-1,j}| \leq \omega_j, \quad \text{for } 0 \leq j < n - 1, t_n \in J. \qquad (3.4)$$

The latter condition means that the quadrature coefficients associated with a particular time level t_j do not change too much as the upper limit of the integral increases. For brevity we shall refer to schemes satisfying (3.3) and (3.4) as $\omega-stable$ below.

We shall first derive a stability estimate in an energy norm which is a discrete analogue of the standard such norm associated with the wave equation, namely

$$|||\phi^n||| = (\|\bar{\partial}\phi^n\|^2 + \|\phi^n\|_1^2)^{1/2}, \quad \text{for } n \geq 1.$$

For the purpose of later error estimates, we shall derive our stability estimate for a slight generalization of (3.2), namely

$$(\bar{\partial}^2 U^n, \chi) + A_n(U^n, \chi) = \sigma^n(B_n(U, \chi)) + (f^n, \chi) + F^n(\chi), \quad \text{for } \chi \in S_h, \quad n \geq 2, \ (3.5)$$

where F^n is a linear functional on H_0^1, with norm $\| \cdot \|_*$. For ease of exposition we shall assume below that F^0 and F^1 is also given, with $F^0 = 0$, even though they do not appear in (3.2) or (3.5). It is also convenient to assume σ^1 included in the above definition, and $\sigma^0 = 0$.

Theorem 3.1. *Let the quadrature rule (3.1) be* $\omega-stable$. *Then we have for the solution of (3.5), for* $n \geq 2$, $t_n \leq T$,

$$|||U^n||| \leq C\{\|U^0\|_1 + |||U^1||| + k \sum_{m=2}^{n} \|f^m\| + k \sum_{m=1}^{n} \|\bar{\partial}F^m\|_*\}.$$

PROOF: We choose $\chi = \bar{\partial}U^n$ in (3.5) to obtain

$$(\bar{\partial}^2 U^n, \bar{\partial}U^n) + A_n(U^n, \bar{\partial}U^n) = \sigma^n(B_n(U, \bar{\partial}U^n)) + (f^n, \bar{\partial}U^n) + F^n(\bar{\partial}U^n)$$

$$\tag{3.6}$$

$$= I_1^n + I_2^n + I_3^n.$$

Note that

$$(\bar{\partial}^2 U^n, \bar{\partial}U^n) = \frac{1}{2}\bar{\partial}\|\bar{\partial}U^n\|^2 + \frac{k}{2}\|\bar{\partial}^2 U^n\|^2$$

and

$$A_n(U^n, \bar{\partial}U^n) = \frac{1}{2}\bar{\partial}(A_n(U^n, U^n)) - \frac{1}{2}(\bar{\partial}A_n)(U^{n-1}, U^{n-1}) + \frac{k}{2}A_n(\bar{\partial}U^n, \bar{\partial}U^n),$$

where $\bar{\partial}A$ denotes the backward difference quotient of $A(t_n; \cdot, \cdot)$ with respect to its first argument. Multiplying both sides of (3.6) by $2k$ and summing from $n = 2$ to N we therefore obtain,

$$|||U^N|||^2 \le C\{|||U^1|||^2 + k\sum_{n=2}^{N}(I_1^n + I_2^n + I_3^n)| + k\sum_{n=2}^{N}|||U^{n-1}|||^2\}.$$

Now letting $|||U|||_N = \max_{1 \le n \le N}|||U^n|||$, we have

$$k|\sum_{n=2}^{N}I_2^n| \le k\sum_{n=2}^{N}\|f^n\| \, |||U|||_N.$$

Further, since

$$I_3^n = F^n(\bar{\partial}U^n) = \bar{\partial}(F^n(U^n)) - (\bar{\partial}F^n)(U^{n-1}),$$

we find after summation, using also the fact that $F^N = k\sum_{n=1}^{N}\bar{\partial}F^n$,

$$k|\sum_{n=2}^{N}I_3^n| \le |F^N(U^N) - F^1(U^1)| + k\sum_{n=2}^{N}|(\bar{\partial}F^n)(U^{n-1})| \le 2k\sum_{n=1}^{N}\|\bar{\partial}F^n\|_*|||U|||_N.$$

In order to estimate the sum in I_1^n we note that, with obvious notation,

$$I_1^n = \sigma^n(B_n(U, \bar{\partial}U^n))$$
$$= \bar{\partial}(\sigma^n(B_n(U, U^n))) - \sigma^n(\bar{\partial}B_n(U, U^{n-1})) - (\bar{\partial}\sigma^n)(B_{n-1}(U, U^{n-1})) = I_{10}^n + I_{11}^n + I_{12}^n.$$

Here

$$k\sum_{n=2}^{N}|I_{11}^n| \le Ck\sum_{n=2}^{N}\sum_{j=0}^{n-1}\omega_j\|U^j\|_1\|U^{n-1}\|_1 \le C\sum_{j=0}^{N-1}\omega_j\|U^j\|_1|||U|||_N.$$

Further,

$$k|I_{12}^n| \leq C|\omega_{n,n-1}|\|U^{n-1}\|_1^2 + C\sum_{j=0}^{n-2}|\omega_{nj} - \omega_{n-1,j}|\|U^j\|_1\|U^{n-1}\|_1,$$

and, after summation, interchanging the order of summation in the double sum, and using (3.3) and (3.4),

$$k\sum_{n=2}^{N}|I_{12}^n| \leq C\{\sum_{j=1}^{N-1}\omega_j\|U^j\|_1 + \sum_{n=2}^{N}\sum_{j=0}^{n-2}|\omega_{nj} - \omega_{n-1,j}|\|U^j\|_1\}\||U\||_N$$

$$\leq C\{\sum_{j=1}^{N-1}\omega_j\|U^j\|_1 + \sum_{j=0}^{N-2}\sum_{n=j+2}^{N}|\omega_{nj} - \omega_{n-1,j}|\|U^j\|_1\}\||U\||_N$$

$$\leq C\{\sum_{j=0}^{N-1}\omega_j\|U^j\|_1\||U\||_N.$$

Thus

$$k|\sum_{n=2}^{N}I_1^n| \leq |\sigma^N(B_N(U,U^N)) - \sigma^1(B_1(U,U^1))| + C\sum_{n=0}^{N-1}\omega_n\|U^n\|_1\||U\||_N.$$

$$\leq C\sum_{n=0}^{N-1}\omega_n\|U^n\|_1\||U\||_N.$$

Altogether our estimates show

$$\||U^N\||^2 \leq C\||U^1\||^2 + C\{\|U^0\|_1 + \|U^1\|_1$$

$$+ k\sum_{n=2}^{N}\|f^n\| + k\sum_{n=1}^{N}\|\bar{\partial}F^n\|_* + \sum_{n=2}^{N-1}\omega_n\||U^n\|| + k\sum_{n=1}^{N-1}\||U^n\||\}\||U\||_N.$$

Since this estimate also holds for $N = 1$, we easily conclude that, for $N \geq 1$,

$$\||U\||_N \leq C\{\|U^0\|_1 + \||U^1\|| + k\sum_{n=2}^{N}\|f^n\| + k\sum_{n=1}^{N}\|\bar{\partial}F^n\|_*\} + C\sum_{n=1}^{N-1}(\omega_n + k)\||U\||_n.$$

The proof is now completed by an application of the following easily proven nonstandard discrete version of Gronwall's lemma, together with (3.3).

Lemma 3.1. Let $\{\eta_n\}$ be a sequence of nonnegative numbers satisfying

$$\eta_n \leq \alpha_n + \sum_{j=0}^{n-1}\beta_j\eta_j, \quad \text{for } n \geq 0,$$

225

where $\{\alpha_j\}$ is a nondecreasing sequence and β_j are nonnegative. Then

$$\eta_n \le \alpha_n \exp\left(\sum_{j=0}^{n-1} \beta_j\right), \quad \text{for } n \ge 1.$$

We shall now turn to a preliminary error estimate, in which the choice of the initial values and the quadrature formula remain to be specified. For notational convenience we define a linear functional $q_B^n(W)$ on H_0^1 which is associated with the quadrature error

$$q^n(g) = \sigma^n(g) - \int_0^{t_n} g(s)\,ds$$

by $q_B^n(W)(\phi) = q^n(B_n(W, \phi))$, for $\phi \in H_0^1$.

Theorem 3.2. *Assume that the quadrature rule (3.1) is ω-stable. Then we have, with W the Ritz-Volterra projection defined by (2.2) and $\theta^n = U^n - W(t_n)$, for $n \ge 2, t_n \le T$,*

$$\|U^n - u(t_n)\| \le C\{\|\theta^0\|_1 + \|\|\theta^1\|\| + k \sum_{m=1}^n \|\bar\partial q_B^m(W)\|_*\} + C(u)(h^r + k).$$

PROOF: As in the proof of Theorem 2.1 we write the error as $U^n - u(t_n) = \theta^n + \rho^n$. Since $\rho^n = \rho(t_n)$ is estimated by (2.3) it only remains to estimate θ^n. From our definitions (3.2) and (2.2) we have

$$(\bar\partial^2 \theta^n, \chi) + A_n(\theta^n, \chi) = \sigma^n(B_n(\theta, \chi)) - (\bar\partial^2 \rho^n + \tau^n, \chi) + q_B^n(W)(\chi),$$

where $\tau^n = \bar\partial^2 u(t_n) - u_{tt}(t_n)$. We may now apply Theorem 3.1 to obtain

$$\|\|\theta^n\|\| \le C\{\|\theta^0\|_1 + \|\|\theta^1\|\| + k \sum_{m=2}^n (\|\bar\partial^2 \rho^m\| + \|\tau^m\|) + k \sum_{m=1}^n \|\bar\partial q_B^m(W)\|_*\},$$

where we have used that $q_B^0(W) = 0$. Here

$$k \sum_{m=2}^n (\|\bar\partial^2 \rho^m\| + \|\tau^m\|) \le C(u)(h^r + k),$$

and hence, for $n \ge 2$,

$$\|\theta^n\| \le C\|\|\theta^n\|\| \le C\{\|\theta^0\|_1 + \|\|\theta^1\|\| + k \sum_{m=1}^n \|\bar\partial q_B^m(W)\|_*\} + C(u)(h^r + k),$$

which completes the proof.

One may show that choosing $U^0 = R_h(0)u_0, U^1 = R_h(0)(u_0 + k u_1)$ gives

$$\|\theta^0\|_1 + \|\|\theta^1\|\| = k^{-1}\|\theta^1\| + \|\theta^1\|_1 \le C(u)(h^r + k),$$

so that the global error is $O(h^r + k)$, provided this is the case for the quadrature error.

We remark that in order to obtain optimal order error estimates in L_2 we need to require, somewhat excessively, that the initial data be approximations to W of superconvergent order $O(h^r)$ in H^1. It is possible to avoid this by carrying out the discrete stability analysis instead in the discrete energy norm $(\|\bar{\partial}\phi^n\|_*^2 + \|\phi^n\|)^{1/2}$.

4. The global quadrature error.

In this section we shall estimate the quadrature dependent term in $\bar{\partial}q_B^n(W)$ occurring in our preliminary error estimates above, for various particular quadrature rules.

The simplest quadrature rule of type (3.1) which is consistent with the order of accuracy of the backward Euler scheme is the rectangle rule

$$\sigma^n(g) = k \sum_{j=0}^{n-1} g(t_j), \tag{4.1}$$

which thus corresponds to choosing $\omega_{nj} = k$, for $0 \leq j \leq n - 1$.

Proposition 4.1. *The rectangle rule (4.1) is ω-stable and satisfies*

$$k \sum_{m=1}^{n} \|\bar{\partial}q_B^m(W)\|_* \leq C(u)k. \tag{4.2}$$

PROOF: In this case $\omega_{nj} \leq \omega_j = k$ for $0 \leq j \leq n - 1$ and $\sum_{j=0}^{n-1} \omega_j = nk = t_n \leq T$. Also, the sum in (3.4) vanishes, so that the rule is ω-stable.

We may represent the quadrature error as

$$q^m(g) = \sum_{j=0}^{m-1} \left\{ kg(t_j) - \int_{t_j}^{t_{j+1}} g(s)ds \right\} = \int_0^{t_m} \psi_0(s)D_s g(s)ds,$$

where $\psi_0(s) = s - t_{j+1}$ for $s \in [t_j, t_{j+1}]$. We therefore have, for $\phi \in H_0^1$,

$$q^m(B_m(W, \phi)) - q^{m-1}(B_{m-1}(W, \phi))$$
$$= k \int_0^{t_{m-1}} \psi_0(s)D_s(\bar{\partial}_1 B)(t_m, s; W(s), \phi)ds + \int_{t_{m-1}}^{t_m} \psi_0(s)D_s B(t_m, s; W(s), \phi)ds.$$

Since $\psi_0(s) = O(k)$ we obtain after using also Proposition 2.2 that

$$|\bar{\partial}q^m(B_m(W, \phi))| \leq Ck \max_{s \leq t_m}(\|W(s)\|_1 + \|W_t(s)\|_1)\|\phi\|_1 \leq C(u)k\|\phi\|_1,$$

from which (4.2) follows.

We shall now discuss a quadrature rule which is based on the use of the trapezoidal rule on intervals of length $O(k^{1/2})$, and somewhat modified at the end of the intervals of integration. Thus, let $\mu = [k^{-1/2}]$, where $[x]$ denotes the integral part of x. Set $k_1 = \mu k$ and $\bar{t}_j = jk_1$, and let j_n be the largest integer such that $\bar{t}_{j_n} < t_n$. In approximating the integral term over $[0, t_n]$ we shall now apply the trapezoidal rule with step size k_1 on $[0, \bar{t}_{j_n}]$ and the rectangle rule with step size k' on the remaining part $[\bar{t}_{j_n}, t_n]$. More precisely, we set

$$\sigma^n(g) = \sum_{j=0}^{n-1} \omega_{nj} g(t_j) = \frac{k_1}{2} \sum_{j=1}^{j_n} (g(\bar{t}_j) + g(\bar{t}_{j-1})) + k \sum_{j=\mu j_n}^{n-1} g(t_j) = \sigma_1^n(g) + \sigma_0^n(g). \quad (4.3)$$

Note that this rule requires storage at $O(k^{-1/2})$ time levels as opposed to $O(k^{-1})$ for the rectangle rule. For this choice we have the following.

Proposition 4.2. *The modified trapezoidal rule (4.3) is ω–stable and satisfies (4.2).*

PROOF: Setting

$$\omega_j = \begin{cases} k_1, & \text{for } j \equiv 0 \pmod{\mu}, \\ k, & \text{otherwise}, \end{cases}$$

we have that $\omega_{nj} \le \omega_j$ for $j \le n-1$, and $\sum_{j=0}^{n-1} \omega_j \le j_n k_1 + nk \le 2T$, so that the quadrature formula has dominated weights. We further note that for fixed j, ω_{nj} only changes its value once as n increases, and this happens when n for the first time has passed a multiple of μ. With $j \in [(q-1)\mu, q\mu)$ we therefore have

$$\sum_{n=j+2}^{N} |\omega_{nj} - \omega_{n-1,j}| \le |\omega_{q\mu+1,j} - \omega_{q\mu,j}| = \begin{cases} \frac{1}{2}k_1 - k, & \text{if } j \equiv 0 \pmod{\mu}, \\ k, & \text{otherwise}, \end{cases}$$

so that the dominating weights are persistent.

This time we have for the corresponding quadrature error

$$q^m(g) = \{\sigma_1^m(g) - \int_0^{\bar{t}_{jm}} g \, ds\} + \{\sigma_0^m(g) - \int_{\bar{t}_{jm}}^{t_m} g \, ds\} = \int_0^{\bar{t}_{jm}} \psi_1 D_s^2 g \, ds + \int_{\bar{t}_{jm}}^{t_m} \psi_0 D_s g \, ds,$$

where $\psi_0(s)$ is defined as above and

$$\psi_1(s) = \begin{cases} (s - \bar{t}_{j-1})(s - \bar{t}_{j-\frac{1}{2}}), & \text{for } s \in [\bar{t}_{j-1}, \bar{t}_{j-\frac{1}{2}}], \\ (s - \bar{t}_j)(s - \bar{t}_{j-\frac{1}{2}}), & \text{for } s \in [\bar{t}_{j-\frac{1}{2}}, \bar{t}_j]. \end{cases}$$

We therefore obtain

$$q^m(B_m(W, \phi)) - q^{m-1}(B_{m-1}(W, \phi))$$

$$= k \int_0^{\bar{t}_{jm-1}} \psi_1(s) D_s^2(\bar{\partial}_1 B)(t_m, s; W(s), \phi) ds + \int_{\bar{t}_{jm-1}}^{\bar{t}_{jm}} \psi_1(s) D_s^2 B(t_m, s; W(s), \phi) ds$$

$$+ k \int_{\bar{t}_{jm}}^{t_{m-1}} \psi_0(s) D_s(\bar{\partial}_1 B)(t_m, s; W(s), \phi) ds - \int_{\bar{t}_{jm-1}}^{\bar{t}_{jm}} \psi_0(s) D_s B(t_{m-1}, s; W(s), \phi) ds$$

$$+ \int_{t_{m-1}}^{t_m} \psi_0(s) D_s B(t_m, s; W(s), \phi) ds.$$

We remark that in the case $\bar{t}_{j_m-1} = \bar{t}_{j_m}$ the second and fourth terms will vanish. If $\bar{t}_{j_m-1} \neq \bar{t}_{j_m}$, which happens exactly when $\bar{t}_{j_m} = t_{m-1}$, then $\bar{t}_{j_m-1} = \bar{t}_{j_m-1}$, and the third term will be zero. Since $\psi_0(s) = O(k)$ and $\psi_1(s) = O(k_1^2)$ we conclude easily from (4.3) that

$$\|q_B^m(W) - q_B^{m-1}(W)\|_* \leq C(u)(kk_1^2 + k^2 + (k_1^2 + k)(\bar{t}_{j_m} - \bar{t}_{j_m-1})).$$

Because $k_1^2 \leq k$ and $\bar{t}_{j_m}$ is nondecreasing in m, summation now yields (4.2).

We shall end this section by presenting a quadrature rule introduced in [11], which is based on using Simpson's rule on subintervals of length $O(k^{1/4})$. For this rule the number of values of U^j which need to be stored in J is of order $O(k^{-1/4})$, thus further reducing the storage requirement as compared to the above modified trapezoidal rule.

In order to define our rule, let now $\mu = [k^{-1/4}]$ and set $k_i = \mu^i k$, for $0 \leq i \leq 3$. Note that thus $k_i = O(k^{(4-i)/4})$. The rule σ^n then uses Simpson's rule on as many intervals of length $2k_3$ which can be fitted into $[0, t_{n-1}]$. In the remaining interval, which is of length at most $O(k^{1/4})$, we use the trapezoidal rule on as many intervals of length k_2 as can be fitted in, leaving an interval of length at most $O(k^{1/2})$. Here we apply the trapezoidal rule based on intervals of length $k_1 = O(k^{3/4})$, and finally, the rectangle rule on the remaining basic intervals of length $k_0 = k$.

For given n, we thus introduce the quadrature points $\bar{t}_j^n$ as follows: Let j_{3n} be the largest even integer with $j_{3n}k_3 < t_n$, and set $\bar{t}_j^n = jk_3$ for $0 \leq j \leq j_{3n}$. Now let $\bar{t}_j^n = \bar{t}_{j_{3n}}^n + (j - j_{3n})k_2$, for $j_{3n} < j \leq j_{2n}$, where j_{2n} is the largest integer such that $\bar{t}_{j_{2n}}^n < t_n$, and set $\bar{t}_j^n = \bar{t}_{j_{2n}}^n + (j - j_{2n})k_1$, for $j_{2n} < j \leq j_{1n}$, where j_{1n} is the largest integer such that $\bar{t}_{j_{1n}}^n < t_n$, and finally $\bar{t}_j^n = \bar{t}_{j_{1n}}^n + (j - j_{1n})k$, for $j_{1n} < j \leq j_{0n}$, where $\bar{t}_{j_{0n}}^n = t_n$. We thus have

$$[0, t_n] = \bigcup_{j=1}^{j_{3n}} [\bar{t}_{j-1}^n, \bar{t}_j^n] \bigcup_{j=j_{3n}+1}^{j_{2n}} [\bar{t}_{j-1}^n, \bar{t}_j^n] \bigcup_{j=j_{2n}+1}^{j_{1n}} [\bar{t}_{j-1}^n, \bar{t}_j^n] \bigcup_{j=j_{1n}+1}^{j_{0n}} [\bar{t}_{j-1}^n, \bar{t}_j^n],$$

and our modified Simpson's rule is

$$\sigma^n(g) = \frac{k_3}{3} \sum_{j=1}^{j_{3n}/2} [g(\bar{t}_{2j}^n) + 4g(\bar{t}_{2j-1}^n) + g(\bar{t}_{2j-2}^n)] \qquad (4.4)$$

$$+ \frac{k_2}{2} \sum_{j=j_{3n}+1}^{j_{2n}} [g(\bar{t}_j^n) + g(\bar{t}_{j-1}^n)] + \frac{k_1}{2} \sum_{j=j_{2n}+1}^{j_{1n}} [g(\bar{t}_j^n) + g(\bar{t}_{j-1}^n)] + k \sum_{j=j_{1n}}^{j_{0n}-1} g(\bar{t}_j^n).$$

One may then show the following proposition.

Proposition 4.3. *The modified Simpson's rule (4.4) is ω-stable and satisfies (4.2).*

5. Schemes based on a symmetric difference approximation.

Because of the nonsymmetric choice of the discretization of u_{tt}, the backward differencing schemes considered above are only first order accurate in time. We shall now discuss schemes that achieve second order accuracy by symmetry around the point t_n. The method will be based on the standard symmetric discretization of the wave equation,

$$\frac{U^{n+1} - 2U^n + U^{n-1}}{k^2} + A(\frac{U^{n+1} + 2U^n + U^{n-1}}{4}) = 0.$$

Our basic stability result will now be expressed in terms of the averages $(U^n + U^{n-1})/2$. Therefore, in order to apply it in our error analysis we shall approximate the integral term in (1.1) by a quadrature formula using only such averages. We thus set

$$\sigma^n(g) = \sum_{j=0}^{n-1} \omega_{nj} g(t_{j+\frac{1}{2}}) \approx \int_0^{t_n} g(s)ds, \quad \text{with } t_{j+\frac{1}{2}} = (j + \frac{1}{2})k, \tag{5.1}$$

and say again that σ^n is ω−stable if the quadrature weights ω_{nj} satisfy our above assumptions (3.3), (3.4). In order to apply this to our discrete function U^n, we introduce its continuous piecewise interpolant $\tilde{U}$ in time, so that, in particular, $\tilde{U}(t_{j+\frac{1}{2}}) = U^{j+\frac{1}{2}} = (U^j + U^{j+1})/2$. With $\bar{\partial}U^n$ as before, $\partial U^n = (U^{n+1} - U^n)/k$ the forward difference quotient, and

$$\hat{U}^n = (U^{n+1} + 2U^n + U^{n-1})/4 = (U^{n+\frac{1}{2}} + U^{n-\frac{1}{2}})/2,$$

we now define our completely discrete scheme by

$$(\partial\bar{\partial}U^n, \chi) + A_n(\hat{U}^n, \chi) = \sigma^n(B_n(\tilde{U}, \chi)) + (f^n, \chi), \quad \text{for } \chi \in S_h, \ n \geq 1,$$

with U^0 and U^1 given, where thus (5.1) is applied with $g(t_{j+\frac{1}{2}}) = B(t_n, t_{j+\frac{1}{2}}; U^{j+\frac{1}{2}}, \chi)$. As earlier we shall need a stability result for the more general equation

$$(\partial\bar{\partial}U^n, \chi) + A_n(\hat{U}^n, \chi) = \sigma^n(B_n(\tilde{U}, \chi)) + (f^n, \chi) + F^n(\chi), \quad \text{for } \chi \in S_h, \ n \geq 1, \tag{5.2}$$

where $F^n, n = 1, 2, \cdots$, are linear functionals on H_0^1. For notational convenience we also set $F^0 = \sigma^0 = 0$. Our stability result will be expressed in terms of a discrete energy norm, now defined by

$$|||\phi^{n+\frac{1}{2}}||| = (||\partial\phi^n||^2 + ||\phi^{n+\frac{1}{2}}||_1^2)^{1/2}.$$

Theorem 5.1. *If the quadrature rule (5.1) is ω−stable we have for the solution of (5.2)*

$$|||U^{n+\frac{1}{2}}||| \leq C\{|||U^{\frac{1}{2}}||| + k\sum_{m=1}^n ||f^m|| + k\sum_{m=1}^n ||\bar{\partial}F^m||_*\}, \quad \text{for } n \geq 1, t_{n+1} \leq T.$$

PROOF: We choose $\chi = \bar{\partial}U^{n+\frac{1}{2}} = (\partial U^n + \bar{\partial}U^n)/2$ in (5.2) to obtain

$$(\partial\bar{\partial}U^n, \bar{\partial}U^{n+\frac{1}{2}}) + A_n(\hat{U}^n, \bar{\partial}U^{n+\frac{1}{2}}) \tag{5.3}$$
$$= \sigma^n(B_n(\tilde{U}, \bar{\partial}U^{n+\frac{1}{2}})) + (f^n, \bar{\partial}U^{n+\frac{1}{2}}) + F^n(\bar{\partial}U^{n+\frac{1}{2}}) = I_1^n + I_2^n + I_3^n.$$

Note that
$$(\partial\bar\partial U^n, \bar\partial U^{n+\frac{1}{2}}) = (\partial\bar\partial U^n, (\partial U^n + \bar\partial U^n)/2) = \frac{1}{2}\bar\partial\|\partial U^n\|^2$$

and
$$A_n(\hat U^n, \bar\partial U^{n+\frac{1}{2}}) = \frac{1}{2}\bar\partial(A_n(U^{n+\frac{1}{2}}, U^{n+\frac{1}{2}})) - \frac{1}{2}(\bar\partial A_n)(U^{n-\frac{1}{2}}, U^{n-\frac{1}{2}}).$$

We now multiply both sides of (5.3) by $2k$ and sum from $n = 1$ to N to obtain

$$\|\partial U^N\|^2 + c\|U^{N+\frac{1}{2}}\|_1^2 \le \|\partial U^0\|^2 + C\|U^{\frac{1}{2}}\|_1^2$$
$$+ 2k|\sum_{n=1}^{N}(I_1^n + I_2^n + I_3^n + \frac{1}{2}(\bar\partial A_n)(U^{n-\frac{1}{2}}, U^{n-\frac{1}{2}}))|.$$

Hence

$$\||U^{N+\frac{1}{2}}\||^2 \le C\{\||U^{\frac{1}{2}}\||^2 + k|\sum_{n=1}^{N}(I_1^n + I_2^n + I_3^n)| + k\sum_{n=1}^{N}\||U^{n-\frac{1}{2}}\||^2\}.$$

Defining $\||U\||_N = \max_{0 \le n \le N}\||U^{n+\frac{1}{2}}\||$, we obtain

$$k|\sum_{n=1}^{N}I_2^n| \le k\sum_{n=1}^{N}\|f^n\|\||U\||_N.$$

The terms I_1^n and I_3^n are treated as in the proof of Theorem 3.1, with U^n replaced by $U^{n+\frac{1}{2}}$ and t_j by $t_{j+\frac{1}{2}}$. Therefore, we have similarly to there, that

$$k|\sum_{n=1}^{N}(I_1^n + I_3^n)| \le C\{\sum_{n=1}^{N-1}\omega_n\||U^{n+\frac{1}{2}}\|| + \||U^{\frac{1}{2}}\|| + k\sum_{n=1}^{N}\|\bar\partial F^n\|_*\}\||U\||_N.$$

The proof is then completed as in Theorem 3.1.

The following is the corresponding preliminary error estimate.

Theorem 5.2. If (5.1) is ω-stable then, for $n \ge 1$, $t_{n+1} \le T$, with $\theta = U - W$,

$$\|U^{n+\frac{1}{2}} - u(t_{n+\frac{1}{2}})\| \le C\{\|\partial\theta^0)\| + \|\theta^{\frac{1}{2}}\|_1 + k\sum_{m=1}^{n}\|\bar\partial q_B^n(W)\|_*\} + C(u)(h^r + k^2).$$

In this case a possible choice of discrete initial data, matching the $O(h^r + k^2)$ error, is $U^0 = R_h(0)u_0, U^1 = R_h(0)(u_0 + ku_1 + k^2u_2/2)$, where $u_2 = u_{tt}(0) = -A(0)u_0 + f(0)$.

We close by presenting two ω-stable quadrature formulas for which the global quadrature error matches the other error terms. We begin with the midpoint rule

$$\sigma^n(g) = k\sum_{j=0}^{n-1}g(t_{j+\frac{1}{2}}). \tag{5.5}$$

For this rule, the storage requirement is $O(k^{-1})$.

We also present a quadrature formula which uses fewer time steps, thus reducing the storage requirements without sacrificing accuracy. This formula will be based on Simpson's formula, with larger time steps than k in the main part of the interval of integration.

Since we have assumed the quadrature formula to be expressed in terms of the averages $U^{n+\frac{1}{2}} = (U^n + U^{n+1})/2$, we shall need to shift by $k/2$ the intervals for which we use Simpson's rule. Thus let $\mu = [k^{-1/2}]$ and $k_1 = \mu k$. Choose j_n to be the largest even integer such that $j_n k_1 < t_n$. We shall then apply Simpson's rule with step k_1 on the interval $[k/2, j_n k_1 + k/2]$. On the remaining intervals $[0, k/2], [j_n k_1 + k/2, j_n k_1 + k]$, and $[j_n k_1 + k, t_n]$ we use a rectangle rule on the two former intervals of length $k/2$ and the midpoint rule with step k on the latter. If we thus denote the quadrature points by

$$\bar{t}_j^n = \begin{cases} jk_1 + \frac{1}{2}k, & \text{for } 0 \le j \le j_n, \\ j_n k_1 + (j - j_n + \frac{1}{2})k, & \text{for } j_n < j \le J_n = j_n + (n - 1 - j_n\mu), \end{cases}$$

we set

$$\sigma^n(g) = \frac{k}{2}g(\bar{t}_0^n) + \frac{k_1}{3}\sum_{j=1}^{j_n/2}\{g(\bar{t}_{2j}^n) + 4g(\bar{t}_{2j-1}^n) + g(\bar{t}_{2j-2}^n)\} + \frac{k}{2}g(\bar{t}_{j_n}^n) + k\sum_{j=j_n+1}^{n-1}g(\bar{t}_j^n). \quad (5.7)$$

The storage requirement for this rule is $O(k^{-1/2})$, and we have the following.

Proposition 5.1. *The midpoint rule (5.5) and the modified Simpson's rule (5.6) are both ω-stable and satisfy*

$$k\sum_{m=1}^{n}\|\bar{\partial}q_B^m(W)\|_* \le C(u)k^2.$$

We finally remark that the above results show error estimates of the form

$$\|U^{n+\frac{1}{2}} - u(t_{n+\frac{1}{2}})\| \le C(u)(h^r + k^2),$$

under various hypotheses. If one desires instead approximations to $u(t)$ at $t = t_n$, it is clear that $\hat{U}^n = (U^{n+\frac{1}{2}} + U^{n-\frac{1}{2}})/2$ achieves this without loss of accuracy.

REFERENCES

1. J.R. Cannon and Y. Lin, *Nonclassical H^1 projection and Galerkin methods for nonlinear parabolic integro-differential equation*, Calcolo **25** (1988), 187-201.
2. J.R. Cannon, Y. Lin, and C.-Y. Xie, *Galerkin methods and L^2-error estimates for hyperbolic integro-differential equations*, manuscript.
3. Y. Lin, V. Thomée, and L.B. Wahlbin, *Ritz-Volterra projections to finite element spaces and applications to integro-differential and related equations*, SIAM J. Numer. Anal. (to appear).
4. A.K. Pani, V. Thomée, and L.B. Wahlbin, *Numerical methods for hyperbolic and parabolic integro-differential equations*, Research Report CMA-R39-90, Australian National University.

5. M. Renardi, W.J. Hrusa, and J.A. Nohel, "Mathematical Problems in Viscoelasticity," Pitman Monographs and Surveys in Pure and Applied Mathematics No. 35, Wiley, 1987.

6. I.H. Sloan and V. Thomée, *Time discretization of an integro-differential equation of parabolic type*, SIAM J. Numer.Anal. **23** (1986), 1052-1061.

7. V. Thomée, "Galerkin Finite Element Methods for Parabolic Problems," Lecture Notes in Mathematics No. 1054, Springer Verlag, 1984.

8. V. Thomée, *On the numerical solution of integro-differential equations of parabolic type*, International Series of Numerical Mathematics **86** (1988), 477-493, Birkhäuser Verlag, Basel.

9. V. Thomée and N.-Y. Zhang, *Error estimates for semidiscrete finite element methods for parabolic integro-differential equations*, Math.Comp. **53** (1989), 121-139.

10. E.G. Yanik and G. Fairweather, *Finite element methods for parabolic and hyperbolic integro-differential equations*, Nonlinear Anal., Theory, Methods and Appl. **12** (1988), 785-809.

11. N.-Y. Zhang, "On the Discretization in Time and Space of Parabolic Integro-Differential Equations," Thesis, Department of Mathematics, Göteborg, 1990.

12. N.-Y. Zhang, *On fully discrete Galerkin approximations for partial integro-differential equations of parabolic type*, to appear.

L N TREFETHEN
Pseudospectra of matrices

1 Spectra

The notion of eigenvalues developed historically in connection with hermitian matrices and their infinite-dimensional counterparts, self-adjoint linear operators. Physically, eigenvalues may be the first thing one observes in such a system. The natural frequencies of oscillation of a string or a drum, for example, present themselves to the ear immediately. The great twentieth century example is quantum mechanics, with its remarkable discovery that atoms and molecules occupy energy states that can be interpreted as eigenfunctions of a self-adjoint Schrödinger operator. It was quantum mechanics that brought the ideas of matrices, operators, and eigenvalues to the central position in science and mathematics that they occupy today.

Speaking more generally, the matrices and operators that have been most studied over the years are *normal.* This means that their eigenvectors can be taken to be orthogonal, or equivalently, that they can be unitarily diagonalized. Along with the hermitian matrices, the normal matrices include all those that are skew-hermitian, unitary, or circulant, as well as others besides.

The history of spectral theory for non-normal matrices and operators is sparser. Most books on functional analysis make no assumption of normality in presenting the fundamental definitions. Yet when it comes to applications, it is surprising how little has been done with non-normal systems. The "small matrix" case, exemplified by such developments as the Jordan canonical form and its application to ordinary differential equations, is of course well understood and widely used. But applications of spectral ideas for strongly non-normal operators are rather rare, and in fact, among the largest consumers appears to be numerical analysis. Numerical analysts come at operators by way of "large matrices," that is, matrices whose dimensions are determined by a potentially unbounded discretization parameter rather than by the physics. Eigenvalues are typically applied to analyze stability ($|\lambda| \leq 1$?) and convergence rates ($|\lambda| \ll 1$?).

The point of these remarks is that the notion of eigenvalues, as applied to non-normal matrices and operators, is not so clearly sanctified by history as you may think. In the past few years I have become convinced that there is a reason for this. Eigenvalues and eigenvectors are an imperfect tool for analyzing non-normal matrices and operators, a tool that has often been abused. Physically, it is not always the eigenmodes that dominate what one observes in a highly non-normal system. Mathematically, eigenanalysis is not always an efficient means to the end that really matters: understanding *behavior.* The essence of the eigenvalue idea is a normal one, whose appropriateness in the non-normal case has been accepted largely by analogy. For non-normal systems eigenvalues may still be useful to a greater or lesser degree, like a nail file when a screwdriver can't be found, but they are rarely exactly right.

Here is an example. If u is a vector and A is a matrix of the same dimension, the differential equation $u_t = Au$ has the solution $u(t) = e^{tA}u(0)$. Consider in particular the 2×2 matrices

$$A' = \begin{pmatrix} -1 & 1 \\ 0 & -1 \end{pmatrix}, \qquad A'' = \begin{pmatrix} -1 & 5 \\ 0 & -2 \end{pmatrix}. \tag{1}$$

Figure 1 shows the 2-norms $\|e^{tA'}\|$ and $\|e^{tA''}\|$ as functions of t, which control the growth or decay that solutions $u(t)$ may exhibit. Quiz for the reader: which curve is which? Which of these matrices allows growth in norm during the transient phase?

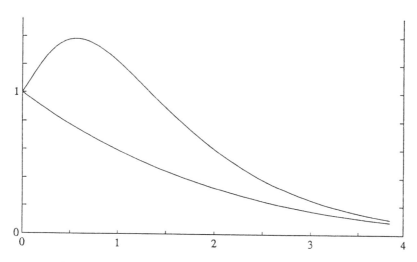

Figure 1. $\|e^{tA}\|$ vs. t for the matrices A' and A'' of (1). Which matrix corresponds the upper curve? For the answer, see the text.

The answer is: A''. The function $\|e^{tA'}\|$, the lower curve in the plot, decays monotonically, but $\|e^{tA''}\|$ grows for a time before eventually decaying. Now there is nothing deep about this result. Nevertheless, it may be surprising to some who are used to making predictions on the basis of eigenvalues. The eigenvalue -1 of A' is defective, according to one pattern of thinking, and therefore some growth must be expected in the transient phase before eventual geometric decay. On the other hand the eigenvalues of A'' are negative and distinct, so no growth should occur. Figure 1 proves that such reasoning is false. Defective eigenvalues are neither necessary nor sufficient for transient growth of $\|e^{tA}\|$. In fact eigenvalues of any kind, defective or not, do not necessarily say much about the behavior of e^{tA} or of A^n or of any other matrix process in the transient phase. Their significance is asymptotic as $t \to \infty$ or $n \to \infty$.

What makes this gap between the eigenvalues of A and the behavior of $\|e^{tA}\|$ possible is the fact that the eigenvectors are not orthogonal. When two eigenvectors are far from orthogonal, a linear combination of them may have large coefficients but small norm, thanks to cancellation. If the coefficients drift out of phase with increasing t so that the cancellation is lost, the norm of the linear combination may increase even though each individual eigenvector component is decaying monotonically. This is the explanation of the upper curve in Figure 1.

For the simplest quantitative approach to such phenomena, let V denote any matrix of eigenvectors of A and

$$\kappa(V) = \|V\|\,\|V^{-1}\|$$

its condition number (we define $\kappa(V) = \infty$ if A is not diagonalizable). Then it is easy to derive the bound

$$e^{t\alpha(A)} \leq \|e^{tA}\| \leq \kappa(V)e^{t\alpha(A)}, \tag{2}$$

where $\alpha(A)$ denotes the spectral abscissa of A, i.e., the maximum of the real parts of its eigenvalues. The analogous bound for matrix powers is

$$\rho(A)^n \leq \|A^n\| \leq \kappa(V)\rho(A)^n, \tag{3}$$

where $\rho(A)$ is the spectral radius of A, and the analogous bound for arbitrary functions $f(z)$ analytic in a neighborhood of the spectrum $\Lambda(A)$ is

$$\|f\|_{\Lambda(A)} \leq \|f(A)\| \leq \kappa(V)\|f\|_{\Lambda(A)}, \tag{4}$$

where $\|f\|_{\Lambda(A)} = \sup_{z \in \Lambda(A)} |f(z)|$. If A is normal, $\kappa(V)$ can be taken to be 1, and all three inequalities (2)–(4) coalesce into equalities. But if A is non-normal, there is always a gap. In general, any quantitative prediction about the behavior of a non-normal matrix, if based on eigenvalues, can be valid only up to a factor such as $\kappa(V)$ that quantifies the non-normality.

For the matrices A' and A'' we have $\kappa(V') = \infty$ and $\kappa(V'') = 10.1$; obviously (2) need not be sharp. For the matrices of Section 3 below, a typical value is $\kappa(V) \approx 10^{10}$. Here and throughout, eigenvector matrices V are taken to be normalized by columns $\|v\| = 1$, and decimal equalities such as "$\kappa = 10.1$" are to be interpreted as accurate to the precision given.

This paper is meant for those who are surprised by Figure 1. It is also meant for those who are not surprised. The fact is, many of us understand the pitfalls of eigenvalue analysis clearly enough when the spotlight is on, but tend to make mistakes as soon as the emphasis moves elsewhere. Nor are numerical analysts alone in such carelessness; one finds mistakes in the use of eigenvalues also, for example, in the literature of hydrodynamic stability in fluid mechanics. Our intuitions have been molded too greatly by the normal case.

236

2 Pseudospectra

If spectra may be misleading, what should one consider instead? Where in the complex plane does a non-normal matrix or operator "live"? I must state immediately that there is no fully satisfactory answer to this question. For a normal matrix, virtually any way of making the question precise leads to the spectrum. If A is not normal, however, no set in the complex plane has all the properties one would like. The spectrum is too small; the field of values (= numerical range) is too large; the disk about 0 of radius $\|A\|$ is even larger. See the discussion of "spectral sets" in [38].

The purpose of this paper is to present examples of *pseudospectra*, another imperfect answer that has the advantage of being a natural extension of the idea of spectra. The following ideas can be generalized to arbitrary closed operators in a Hilbert space [21,32,43], but for the sake of brevity, let us assume from now on that A is a matrix of dimension $N < \infty$ and $\|\cdot\|$ is the 2-norm.

Consider, as a function of $z \in \mathbb{C}$, the norm of the resolvent $(zI - A)^{-1}$. When z is an eigenvalue of A, $\|(zI - A)^{-1}\|$ can be thought of as infinite, and we shall use this convention. Otherwise, it is finite. How large? If A is normal, the answer is simple:

$$\|(zI - A)^{-1}\| = \frac{1}{\operatorname{dist}(z, \Lambda(A))}. \tag{5}$$

(Here $\Lambda(A)$ again denotes the spectrum of A, and $\operatorname{dist}(z, S)$ is the usual distance from the point z to the set S.) Thus in the normal case, the surface $\|(zI - A)^{-1}\|$ is determined entirely by the eigenvalues like a tent hanging from its poles. In the non-normal case, however, (5) is only a lower bound and the shape of the surface cannot be inferred from the eigenvalues. As our examples will show, $\|(zI - A)^{-1}\|$ may easily attain values as great as 10^{10} or 10^{20} even when z is far from $\Lambda(A)$.

With this in mind it is natural to define the ϵ-**pseudospectrum** of A, for each $\epsilon \geq 0$, by

$$\Lambda_\epsilon(A) = \{z \in \mathbb{C} : \|(zI - A)^{-1}\| \geq \epsilon^{-1}\}. \tag{6}$$

The ϵ-pseudospectra of A are closed, strictly nested sets with $\Lambda_0(A) = \Lambda(A)$. If A is normal, (5) implies that $\Lambda_\epsilon(A)$ is equal to the union of the closed ϵ-balls about the eigenvalues of A. In general, it may be much larger.

The norm of $(zI - A)^{-1}$ is its largest singular value, i.e., the inverse of the smallest singular value of $zI - A$. Therefore an equivalent definition of the pseudospectrum is:

$$\Lambda_\epsilon(A) = \{z \in \mathbb{C} : \sigma_N(zI - A) \leq \epsilon\}. \tag{7}$$

Another more interesting equivalent definition can be stated in terms of perturbations of A:

$$\Lambda_\epsilon(A) = \{z \in \mathbb{C} : z \text{ is an eigenvalue of } A + E \text{ for some } E \text{ with } \|E\| \leq \epsilon\}. \tag{8}$$

In other words, a pseudo-eigenvalue of A is an eigenvalue of a slightly perturbed matrix. (The equivalence of (6) and (8) is easily proved.) This equivalence suggests a new way of regarding perturbations of eigenvalues: they may give information not only about how the properties of a matrix change when its entries change, but about properties it had already. In other words eigenvalue perturbations can be used as a *visualization device*.

Both matrices A' and A'' of (1) are non-normal, and their pseudospectra are accordingly bigger than the ϵ-neighborhoods about their spectra. Figure 2 plots $\Lambda_\epsilon(A')$ and $\Lambda_\epsilon(A'')$ for the values $\epsilon = 0.05, 0.15, 0.25, \ldots, 0.65$. The two pictures are decidedly different, both near the eigenvalues and far away. For $\epsilon = 0.05$, the innermost contour in each plot, $\Lambda_\epsilon(A')$ extends slightly further to the right in the complex plane than $\Lambda_\epsilon(A'')$. This is a reflection of the defective eigenvalue of A', and can be connected with the behavior of $\|e^{tA'}\|$ as $t \to \infty$. But for all of the larger values of ϵ, $\Lambda_\epsilon(A'')$ extends further to the right than $\Lambda_\epsilon(A')$, and this can be connected with the behavior of $\|e^{tA}\|$ for finite t. (Precise estimates can be derived by means of contour integrals; see [32,33,43].) As for the other limit $\epsilon \to \infty$, it turns out that it can be used to derive the field of values of A, and this is what controls behavior of $\|e^{tA}\|$ for $t \to 0$ [21]. The field of values is equal to the limit (or union) as $\epsilon \to \infty$ of the intersection of all half-planes $H \subseteq \mathbf{C}$ for which $H + \Delta_\epsilon \supseteq \Lambda_\epsilon(A)$, where Δ_ϵ denotes the disk about 0 of radius ϵ [43,25].

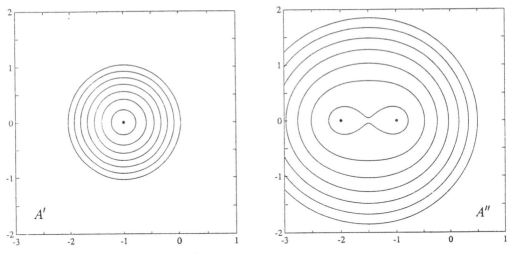

Figure 2. ϵ-pseudospectra of A' and A'' for $\epsilon = 0.05, 0.15, \ldots, 0.65$.

The study of non-normal matrices and operators is a large subject. Some applications of pseudospectral ideas within numerical analysis include:

- convergence of GMRES, CGS, and other nonsymmetric matrix iterations [29],
- design of hybrid iterations that avoid eigenvalue estimates [30],
- convergence of upwind vs. downwind Gauss-Seidel and SOR sweeps,
- backward error analysis of eigenvalue algorithms and polynomial zerofinders [28],
- pseudospectra of Toeplitz matrices [36],

- stiffness of ordinary differential equations [17],
- the Kreiss matrix theorem and generalizations [35],
- Lax-stability of the method of lines [23,24,34,35],
- stability of spectral methods [32,34,41].

Actual or potential applications of pseudospectra in other fields, usually involving operators instead of matrices, include:

- simple differential operators [32],
- convection-diffusion problems [31,32],
- the Fadle-Papcovitch operator (= a biharmonic operator on a half-strip) [39],
- Wiener-Hopf integral operators [1,32],
- the Orr-Sommerfeld operator for Poiseuille flow [33],
- Alfven waves in magnetohydrodynamics [22].

I am in the process of writing a book that will discuss most of these topics [43].

The remainder of this paper has a more limited ambition: to present graphical *examples* of pseudospectra of matrices. The accompanying discussion will give some indication of how pseudospectra may be useful in these application areas, but for full explanations, see the references.

The notion of the pseudospectrum is not new. Under other names, it has been defined previously by Varah [45], Demmel [5,6], Wilkinson [49], Godunov et al. [11], and undoubtedly others. Contour plots as in Figure 2 have appeared in [6] and [11], as well as in the publications [32,33,34] by myself and my colleagues. The recent work of Chatelin and her colleagues is also closely related [4]. Nevertheless this geometric side of linear algebra has received scant attention, and the emphasis has been on how eigenvalues change under perturbations rather than on the exploitation of information that goes beyond eigenvalues. In an era when eigenvalue and singular value computations are routine and graphics tools are universally available, this situation deserves to change.

3 Examples

Our examples consist of thirteen matrices of dimension $N = 32$, all presented in the same format. First the matrix is defined, usually by displaying its 6×6 analogue. Then two plots are presented together on a page. At the head of the page is listed $\kappa(V)$, the condition number of the normalized matrix of eigenvectors.

The upper plot shows 3200 dots representing eigenvalues of 100 perturbed matrices $A + E$, where each E is a random complex matrix with $\|E\| = 10^{-3}$. The entries of E are independent samples from a complex normal distribution with mean 0. To generate E, first a dense matrix $\tilde{E}$ is constructed whose entries are independent samples from the complex normal distribution of mean 0 and standard deviation 1. Then the norm $\|\tilde{E}\|$ is computed and we set $E := 10^{-3} \tilde{E}/\|\tilde{E}\|$.

The lower plot depicts the boundaries of $\Lambda_\epsilon(A)$ for $\epsilon = 10^{-2}, 10^{-3}, \ldots, 10^{-8}$, that is, the level curves $\|(zI - A)^{-1}\| = 10^2, 10^3, \ldots, 10^8$. To determine these curves, the smallest singular value $\sigma_{32}(z_{ij}I - A)$ is computed for each value z_{ij} on a grid of dimensions on the order of 100×100. This takes two or three minutes on a Cray-2. The resulting array of data is then fed to the Matlab contour plotter. The dashed curve on the same plot (sometimes off-scale and hence invisible) is the boundary of the field of values of A computed by the standard algorithm described, for example, in [19]. The thick solid dots are the eigenvalues.

In summary, the upper plot in each example represents a cheap and vivid approximation to the 10^{-3}-pseudospectrum of A. The lower plot depicts the whole surface $\|(zI - A)^{-1}\|$ instead of just a single slice.

Example 1. Jordan block [5,14,48]

There is only one place to begin: with the canonical non-normal matrix, the Jordan block

$$
A_1 = \begin{pmatrix}
0 & 1 & & & & \\
 & 0 & 1 & & & \\
 & & 0 & 1 & & \\
 & & & 0 & 1 & \\
 & & & & 0 & 1 \\
 & & & & & 0
\end{pmatrix}.
$$

(Remember that despite appearances, the matrix under discussion is actually of dimension 32.) This is the best known matrix whose eigenvalues are sensitive to perturbations. The solitary eigenvalue 0 of A_1 is defective with multiplicity 32, and a perturbation of ϵ in the lower-left corner changes it into 32 distinct eigenvalues of magnitude $\epsilon^{1/32} \approx 1$. Consequently, the upper plot reveals a rather elegant halo of dots at a radius slightly less than $(10^{-3})^{1/32} \approx 0.8$. In the lower plot we see concentric circles. By construction, all of the dots in the upper plot lie on or inside the second-largest of these circles, i.e., the boundary of $\Lambda_{10^{-3}}(A_1)$.

It is striking that most of the dots in the upper plot lie near the outside edge. For this example, being defective, the density of dots at the origin is exactly 0, but even in nondefective cases the same edge-clustering phenomenon is commonly observed. It is *not* a result of taking random matrices E with $\|E\| = 10^{-3}$ instead of $\|E\| \leq 10^{-3}$, but simply of choosing examples A that are interesting—that is, whose eigenvalues are sensitive to perturbations.

The fact that the eigenvalues of a Jordan block are highly sensitive to perturbations is widely familiar to numerical analysts. This is not entirely a good thing, for it is the *only* familiar example of this kind of sensitivity, and yet Jordan blocks are by no means representative of non-normal matrices. In general, the pseudospectra of a defective matrix need not look like concentric disks. Conversely, equally interesting pseudospectra can arise for matrices that are nondefective. In fact from the point of view of pseudospectra, Jordan blocks have no special significance whatever. Though it is true that every matrix is similar to a direct sum of Jordan blocks, the similarity transformation involved is in general non-unitary and does not preserve $\Lambda_\epsilon(A)$. By a unitary similarity transformation one can make any matrix triangular (the Schur form), but not bidiagonal.

Example 2. Limaçon [2,32,42]

The first of the points just made can be illustrated by the "super Jordan block"

$$A_2 = \begin{pmatrix} 0 & 1 & 1 & & & \\ & 0 & 1 & 1 & & \\ & & 0 & 1 & 1 & \\ & & & 0 & 1 & 1 \\ & & & & 0 & 1 \\ & & & & & 0 \end{pmatrix}.$$

Mathematically, A_2 is similar to A_1. However, the similarity transformation in question has condition number on the order of 10^{20}. This explains how it is possible that the figures are so different from those of Example 1. The pseudospectra are now bounded approximately by curves known as *limaçons*.

Notice the phenomenon of self-intersection in the curve of dots in the upper plot, which has no counterpart in the lower plot. Like the edge-clustering phenomenon, this is a matter of probability densities that vary greatly with z. Another kind of clustering phenomenon appears if one takes random matrices E that are real instead of complex (try it!). All of these effects are interesting, but for most applications, I believe they are unimportant.

These first two examples are both Toeplitz matrices, i.e., constant along diagonals. If A is an $N \times N$ Toeplitz matrix with entries $a_{1-N}, \ldots, a_0, \ldots, a_{N-1}$, the *symbol* of A is the Laurent polynomial

$$f(z) = \sum_{j=1-N}^{N-1} a_j z^j.$$

For Examples 1 and 2 the symbols are $f_1(z) = z$ and $f_2(z) = z + z^2$. In general, it can be shown that the pseudospectra of a Toeplitz matrix approximate the region in the complex plane bounded by the image of the unit circle under the symbol. More precisely, $\Lambda_\epsilon(A)$ approximates the set of points $z \in \mathbb{C}$ that are enclosed by $f(|z|=1)$ with nonzero winding number. As $N \to \infty$ and $\epsilon \to 0$ this approximation becomes an equality, while for finite N and ϵ, more careful estimates can be derived that involve the image under f of various curves $|z| = r$ with $r = O(\epsilon^{\pm 1/N})$. See [36]. The idea underlying such results is that ϵ-*pseudo-eigenvectors* for small ϵ can be constructed in the form $(1, z, z^2, \ldots, z^{N-1})^T$ when A is triangular, or as linear combinations of such vectors in the nontriangular case.

Example 3. Grcar matrix [15,30]

For a non-triangular Toeplitz example, consider the matrix

$$A_3 = \begin{pmatrix} 1 & 1 & 1 & 1 & & \\ -1 & 1 & 1 & 1 & 1 & \\ & -1 & 1 & 1 & 1 & 1 \\ & & -1 & 1 & 1 & 1 \\ & & & -1 & 1 & 1 \\ & & & & -1 & 1 \end{pmatrix}$$

investigated by Grcar, with symbol

$$f_3(z) = -z^{-1} + 1 + z + z^2 + z^3.$$

The pseudospectra now look like two pearls connected by a chain. All of the eigenvalues of A_3 are sensitive to perturbations, with condition numbers increasing exponentially with N (this is proved in [36]), but the eigenvalues at the end are especially sensitive.

This matrix is nondefective: its eigenvalues are distinct and $\kappa(V)$ is finite, though large. This illustrates the second point made in Example 1, that pseudospectra can be interesting even for diagonalizable matrices. From now on, all but two of our examples are diagonalizable, and all but one have eigenvalues even more sensitive than those of A_3.

An interesting feature of the Grcar example is that the region where the matrix "lives" surrounds the origin but does not contain it. For example, the field of values contains the origin, but it is clear from the pseudospectral contours that not much is happening there; in fact $\sigma_{32}(A)$ has the rather sizeable value 0.951. These properties become important in applications to iterative solution of nonsymmetric systems $Ax = b$, where the origin acquires a special significance. Many methods for such problems make use of an estimated field of values or the convex hull of a set of estimated eigenvalues [26], but for an example like this one, such methods cannot do very well. For better results one must consider pseudospectra or some other non-convex sets [42], or better yet, avoid the complex plane entirely [29].

Additional examples of pseudospectra of Toeplitz matrices are presented in [2], [36] and [46].

Example 4. Wilkinson matrix [47,48]

Many interesting matrices, though not Toeplitz, are approximately Toeplitz in one sense or another. An example is the "Wilkinson matrix"

$$A_4 = \begin{pmatrix} \frac{1}{N} & 1 & & & & \\ & \frac{2}{N} & 1 & & & \\ & & \frac{3}{N} & 1 & & \\ & & & \frac{4}{N} & 1 & \\ & & & & \ddots & 1 \\ & & & & & 1 \end{pmatrix}$$

(Wilkinson himself took $N = 20$ and scaled up the entries by the factor N). Here the diagonal element is not constant but varies smoothly. To a pure mathematician or a careless applied mathematician this may suggest that A_4, being diagonalizable, will behave more or less as if it is diagonal. The truth is that with $\kappa(V)$ greater than 10^{22}, this matrix behaves more like a Jordan block in most respects than like any diagonal matrix. This is certainly the impression suggested by the plots. The pseudospectra approximate ovals which one may think of as superpositions of disks with centers $1/N, 2/N, \ldots, 1$. The fact that the eigenvalues happen to lie exactly where they do inside these ovals is analogous to the difference that distinguishes the carbon isotopes C_{12} and C_{13}: important perhaps in principle, and detectable by specialized experiments, but so deeply shielded by the surrounding electrons as to have negligible effect on the chemistry.

Example 5. Frank matrix [14,48]

The idea of condition numbers become famous in connection with *backward error analysis* of the effects introduced by machine arithmetic. In particular, suppose we want to compute the eigenvalues of a matrix A. Then the standard conclusion of backward error analysis has the following interpretation in terms of pseudospectra: unless A is triangular or has some other special structure, one cannot in general expect to do better on a computer than obtain a set of ϵ-pseudo-eigenvalues for some ϵ on the order of machine precision.

Our next example, the *Frank matrix,* is a classic example of a matrix whose eigenvalues are ill-conditioned and hence difficult to compute:

$$
A_5 = \begin{pmatrix}
6 & 5 & 4 & 3 & 2 & 1 \\
5 & 5 & 4 & 3 & 2 & 1 \\
  & 4 & 4 & 3 & 2 & 1 \\
  &   & 3 & 3 & 2 & 1 \\
  &   &   & 2 & 2 & 1 \\
  &   &   &   & 1 & 1
\end{pmatrix}.
$$

(With $N = 32$ the elements take the values $1, \ldots, 32$.) By a similarity transformation it can be shown that the eigenvalues of A_5 are all real, but this is another similarity transformation whose condition number grows exponentially with N. As Wilkinson pointed out in [48], the lower eigenvalues have huge condition numbers and cannot be computed accurately except in high precision. In other words, the Frank matrix can be described as an example that illustrates that the pseudospectra of a matrix may be much bigger than the spectrum. The plots reveal this.

The philosophy of backward error analysis might be summarized as follows: *if the answer is highly sensitive to perturbations, you have asked a hard question.* However, the main reason for studying pseudospectra is a deeper principle: *if the answer is highly sensitive to perturbations, you have probably asked the <u>wrong</u> question!* When the eigenvalues of a highly non-normal matrix are troublesome to work with, they are probably irrelevant anyway to whatever purpose one genuinely cares about. If the Frank matrix arose in an application, the last thing one would properly want to do is diagonalize it.

Example 6. Kahan matrix [13,20]

Another example of historical interest to numerical analysts is the matrix

$$A_6 = \begin{pmatrix} 1 & -c & -c & -c & -c & -c \\ & s & -sc & -sc & -sc & -sc \\ & & s^2 & -s^2c & -s^2c & -s^2c \\ & & & s^3 & -s^3c & -s^3c \\ & & & & s^4 & -s^4c \\ & & & & & s^5 \end{pmatrix}$$

with $s^{N-1} = 0.1$ and $c = \sqrt{1-s^2}$, which was devised by Kahan in the 1960s to illustrate that QR factorization with column pivoting is not a fail-safe method for determining the rank of a matrix. Rank determination is related to the question of *distance to singularity* of a matrix A [18]: how large a perturbation E is required to make A singular? In the 2-norm the answer is $\sigma_N(A)$, or equivalently, the smallest value of ϵ for which $0 \in \Lambda_\epsilon(A)$.

For the present example, the eigenvalues lie in the interval $[0.1,1]$, which suggests that the matrix may be well-conditioned and hence unambiguously of full rank. But the plots tell a different story. In fact we have $\sigma_{32} = 1.04 \times 10^{-5}$, as can be seen in the plot by noting that the origin lies more or less on the curve $\|(zI - A)^{-1}\| = 10^{-5}$ (check it with a ruler!). This number shrinks at a rate $e^{-C\sqrt{N}}$ as $N \to \infty$.

Geometrically speaking, the point of Kahan's example is that the pseudospectra are *lopsided*, extending substantially to the left of the origin. It is easy to construct matrices with lopsided pseudospectra, but Kahan's ingenuity consisted in finding one that is triangular and whose columns satisfy the requisite pivoting condition.

Example 7. Demmel matrix [6,18]

Another "matrix nearness problem" is that of *distance to instability:* how large a perturbation E is required to move at least one eigenvalue of A into the right half-plane? Our next example,

$$A_7 = -\begin{pmatrix} 1 & B & B^2 & B^3 & B^4 & B^5 \\ & 1 & B & B^2 & B^3 & B^4 \\ & & 1 & B & B^2 & B^3 \\ & & & 1 & B & B^2 \\ & & & & 1 & B \\ & & & & & 1 \end{pmatrix},$$

with $B^{N-1} = 10^8$, was devised by Demmel in order to disprove a conjecture due to Van Loan [44]. The eigenvalues of A_7 are all -1, but under perturbations, they move away

244

from the real axis and first hit the right half-plane at a point with sizable imaginary part. This is evident in the plot. The Van Loan conjecture amounted to the assertion that the point of impact must be real.

In a sense Demmel's example is opposite to Kahan's: its purpose is to make $\sigma_N(A)$ as large as possible, not as small as possible. Certainly the plots are very different. In Example 7 the origin lies in a hole in the pseudospectra, entirely surrounded by regions of $\Lambda_\epsilon(A)$ for smaller ϵ. The reader may find these pictures a bit hard to interpret, for A has entries as large as 10^8, and some of the action, including many of the 3200 dots, is off-scale. The contour $\epsilon = 10^{-8}$ is the large oval in the left half-plane that encloses the eigenvalue -1, and the contour $\epsilon = 10^{-2}$ is the smallest circle in the right half-plane. The contours for other values of ϵ lie in between.

A plot like the lower one of Example 7 (with $N = 3$ and $B = 100$) appears in Demmel's 1987 paper [6]. This is the earliest plot I have seen in print of computed pseudospectra.

Example 8. Matrix of Lenferink and Spijker [24]

Our next matrix is tridiagonal:

$$
A_8 = \begin{pmatrix}
-5 & 2 \\
\frac{1}{2} & -7 & 3 \\
& \frac{1}{3} & -9 & 4 \\
& & \frac{1}{4} & -11 & 5 \\
& & & \frac{1}{5} & -13 & 6 \\
& & & & \frac{1}{6} & -15
\end{pmatrix}.
$$

This elegant example was devised by Lenferink and Spijker to illustrate a point in the theory of numerical stability for discrete approximations of partial differential equations. In that field eigenvalues are often used for heuristic stability analysis, but they cannot be relied upon. Highly non-normal examples require a more careful analysis based on other ideas such as circle conditions [24], a generalized numerical range [24,25], or pseudospectra [34,35], and we shall comment further on such applications with Example 11.

It is obvious that A_8 can be symmetrized by the diagonal matrix $D = \mathrm{diag}(1, 2!, 3!, \ldots, N!)^T$, yielding a mathematically similar matrix in which all of the off-diagonal elements are replaced by 1. Therefore the spectrum of A_8 is real. What is interesting is the position-dependence of the similarity transformation. Near the upper-left corner, to speak loosely, A_8 is "close to normal," but the departure from normality grows steadily as one goes down the matrix. Each eigenvalue is more sensitive than the last, and in the complex plane this shows up as a family of pseudospectra shaped like wedges about the negative real axis.

Some of our examples become trivial in the limit $N \to \infty$. This is true in particular of the Toeplitz matrices, for which the pseudospectra for finite N approach the spectrum of the corresponding Toeplitz operator with $N = \infty$, padded by a border of width $O(\epsilon)$ [36]. The Lenferink-Spijker example is different. Computations indicate that the operator A_8 of

245

dimension $N = \infty$ has a discrete negative real spectrum, like its finite-dimensional sections, with pseudospectra spreading approximately in unbounded wedges of equal angles about that axis. For any finite ϵ, no matter how small, Λ_ϵ is approximately an infinite wedge, not a subset of the real axis. All of this structure is invisible if one looks just at the spectrum. It is invisible also, incidentally, to the device for analyzing families of finite-dimensional non-normal matrices known as the "spectrum of the family" [12,23].

Example 9. Companion matrix [27]

Example 9 is a companion matrix:

$$A_9 = \begin{pmatrix} 0 & 1 & & & & \\ & 0 & 1 & & & \\ & & 0 & 1 & & \\ & & & 0 & 1 & \\ & & & & 0 & 1 \\ -c_0 & -c_1 & -c_2 & -c_3 & -c_4 & -c_5 \end{pmatrix},$$

with c_j chosen so that the polynomial

$$p(z) = c_0 + c_1 z + \cdots + c_{N-1} z^{N-1} + z^N$$

has zeros equally spaced in $[-2, 2]$: $\lambda_j = -2 + 4(j-1)/(N-1)$. The same numbers $\{\lambda_j\}$ are the eigenvalues of A_9; a proof consists of exhibiting the eigenvectors $(1, \lambda_j, \lambda_j^2, \ldots, \lambda_j^{N-1})^T$.

From the plots and the value $\kappa(V) \approx 10^9$ it is evident that companion matrices may be far from normal. The eigenvalues are highly sensitive to perturbations, and there is interesting structure in the pseudospectra. So far as I am aware, little is known about this structure. This is unfortunate, for such a knowledge would have application to the analysis of a familiar but poorly understood algorithm: computing zeros of polynomials by solving matrix eigenvalue problems [27]. In the language of the present paper, the essential question in evaluating this algorithm is: what is the relationship between the pseudospectra of a companion matrix and the sets of *pseudozeros* of the corresponding polynomial, that is, the sets of complex numbers obtainable as zeros of slightly perturbed polynomials? If the pseudozero sets are much smaller than the pseudospectra, the algorithm is unstable, but to the extent that they can be made comparable in size, the algorithm is stable.

It was Wilkinson who made the possible ill-conditioning of polynomial zerofinding problems famous, but he did not view the problem geometrically [47]. For a geometrical study leading to a surprisingly simple algebraic characterization of pseudozero sets in the ∞-norm, see [28]. Pseudozero sets may also be interpreted as *structured pseudospectra* of the companion matrix, defined as in (8) except with perturbations permitted only in the bottom row.

One thing apparent in the plots is that in contrast to the Wilkinson matrix A_4, whose eigenvalues are also equally spaced and whose behavior might consequently have been expected to be analogous, the pseudospectra of A_9 exhibit a marked scale-dependence: the behavior is quite different inside and outside the unit circle. This is a consequence of having 1 in the super-diagonal. By a diagonal similarity transformation, this number 1 can be replaced by any $\alpha \neq 0$, if c_j is adjusted by a factor α^{N-j}; this transformation is essentially the scale change $z \mapsto \alpha z$ applied to the polynomial $p(z)$. More generally, the technique of matrix "balancing" employed, e.g., in Eispack and Matlab introduces an arbitrary diagonal similarity transformation. Little is known about how close such balancing operations may bring the pseudospectra of A_9 to the pseudozero sets of $p(z)$ [27], and this is a topic ripe for further research.

Example 10. Gauss-Seidel iteration matrix [8,9,36]

With the next example we turn to one of the icons of classical numerical analysis: the Gauss-Seidel iteration matrix $A_{10} = -(L+D)^{-1}U$ obtained from the symmetric tridiagonal matrix

$$
B = L+D+U = \begin{pmatrix}
-2 & 1 & & & & \\
1 & -2 & 1 & & & \\
& 1 & -2 & 1 & & \\
& & 1 & -2 & 1 & \\
& & & 1 & -2 & 1 \\
& & & & 1 & -2
\end{pmatrix}.
$$

Here L, D and U denote the lower-triangular, diagonal, and upper-triangular parts of B, respectively. Though B is symmetric, A_{10} is not. It is a highly non-normal matrix that is lower-Hessenberg and Toeplitz, except for zeros in the first column:

$$
A_{10} = \begin{pmatrix}
0 & \frac{1}{2} & & & & \\
0 & \frac{1}{4} & \frac{1}{2} & & & \\
0 & \frac{1}{8} & \frac{1}{4} & \frac{1}{2} & & \\
0 & \frac{1}{16} & \frac{1}{8} & \frac{1}{4} & \frac{1}{2} & \\
0 & \frac{1}{32} & \frac{1}{16} & \frac{1}{8} & \frac{1}{4} & \frac{1}{2} \\
0 & \frac{1}{64} & \frac{1}{32} & \frac{1}{16} & \frac{1}{8} & \frac{1}{4}
\end{pmatrix}.
$$

The eigenvalues of A_{10} were determined analytically by Frankel in 1950 [9] and can also be derived from the more general theory of David Young, to be found in many textbooks.

The plots reveal pseudospectra in the shape of snowshoes. The largest eigenvalues of A_{10} are insensitive to perturbations, but the smaller eigenvalues are so sensitive as to be meaningless for practical purposes. Fortunately, they are never used for practical purposes! It is the largest eigenvalue that determines the convergence rate, conventionally speaking, and consequently, the non-normality of the Gauss-Seidel iteration matrix has been ignored over the years with no ill effects.

If B is replaced by a nonsymmetric matrix, however, such as might arise in the modeling of convection-diffusion equations, the situation changes. Now even the dominant eigenvalue of A become highly sensitive to perturbations, and predictions about convergence based on eigenvalues may be misleading. In particular one encounters the curious anomaly that the spectral radius, hence the classical convergence rate, is unaffected by whether the Gauss-Seidel sweeps run "upwind" or "downwind," even though in actuality, the direction of sweep makes a great difference to convergence [16]. Such upwind-downwind anomalies become less surprising when one considers the pseudospectra for the two cases, which are utterly different. Realistic convergence bounds can be obtained through the use of the inequality [43]

$$\|A^n\| \leq \epsilon^{-1} \rho_\epsilon(A)^{n+1}, \tag{9}$$

valid for any $\epsilon > 0$, where $\rho_\epsilon(A)$ denotes the ϵ-*pseudospectral radius* of A defined by $\rho_\epsilon(A) = \sup_{z \in \Lambda_\epsilon(A)} |z|$ (compare (3)). I hope to discuss these matters in a future paper.

Example 11. Chebyshev spectral [3,34,41]

The next example comes from the field of spectral methods for the numerical solution of partial differential equations, a source of many fascinating matrices. Let $x_0, \ldots, x_N$ denote the Gauss-Lobatto-Chebyshev points $x_j = \cos(j\pi/N)$, $0 \leq j \leq N$. Given data $y_1, \ldots, y_N$ at the points $x_1, \ldots, x_N$, let f_j be the quantity obtained by (1) interpolating the values (x_j, y_j), together with the boundary condition $(x_0, 0)$, by a polynomial $p_N(x)$ of degree N; (2) setting $f_j := p_N'(x_j)$, $1 \leq j \leq N$. Clearly $\{y_j\} \mapsto \{f_j\}$ defines a linear map from $\mathbf{R}^N$ to $\mathbf{R}^N$, hence a matrix which we denote A_{11}:

$$A_{11} = N^{-2} \times \text{Chebyshev spectral differentiation matrix with b.c. } u(1) = 0.$$

The entries of A_{11} can be expressed by explicit formulas [3]. A Matlab program to generate A_{11} can be derived either from these formulas or via the FFT; see [19].

The plots for A_{11} reveal a remarkable structure. The most striking feature is the set of nearly perfectly straight lines to the left of the origin. This cannot be an accident, and in fact, the explanation is that the differential operator that this matrix is approximating, namely the first derivative operator on $[-1, 1]$ with boundary condition $u(1) = 0$, has pseudospectra exactly in the form of half-planes. The resolvent contours $\|(zI - A)^{-1}\| = \epsilon^{-1}$ for that operator are all vertical lines in the complex plane, and ϵ^{-1} grows exponentially as $\mathrm{Re}\, z \to -\infty$.

Finite difference discretizations yield less dramatic agreement of matrix pseudospectra with those of the underlying operator. But it is likely that a wide class of discretizations, spectral or finite-difference, satisfy at least the identity

$$\lim_{N \to \infty} \Lambda_\epsilon(\mathcal{L}_N) \to \Lambda_\epsilon(\mathcal{L}) \qquad (\forall \epsilon > 0) \tag{10}$$

if this limit is defined to mean, for example, pointwise convergence of $\|(zI - \mathcal{L}_N)^{-1}\|$ to $\|(zI - \mathcal{L})^{-1}\|$ as $N \to \infty$, or uniform convergence on compact subsets. See [32] and [36]. Convergence of this kind leaves open the possibility of spurious eigenvalues that have no connection with the differential operator, so long as they move to ∞ as $N \to \infty$. Four of these "outlying eigenvalues" are visible in the plots for A_{11}. They are common but not universal in spectral discretization matrices [34].

It is essential to come to grips with the non-normality of spectral differentiation matrices if one is to make correct predictions about their numerical stability. Typically a matrix like A_{11} effects the space discretization of a time-dependent partial differential equation, while the time-discretization is handled by an o.d.e. formula with time step Δt; this is the *method of lines*. If the matrix were normal, it would be enough to fit its eigenvalues in the stability region of the o.d.e. formula to ensure Lax-stability of the overall computation. In the non-normal case, however, the restriction on Δt suggested by eigenvalue analysis can be too generous by a factor as large as N, the number of points in the grid [34,41]. It has been shown that an appropriate stability criterion in the non-normal case is that all points of the ϵ-pseudospectrum must lie within a distance $O(\epsilon)$ of the stability region as $\epsilon \to 0$ [34,35]. This result can be interpreted as a transplantation of the Kreiss matrix theorem [37] from the unit disk (for powers of matrices) to an arbitrary stability region (for more general recurrences).

Example 12. Random [7,10]

Our penultimate example is a simple one:

$$A_{12} = \text{random matrix.}$$

To be precise, A_{12} is an $N \times N$ matrix (with $N = 32$ as always) whose entries are random samples from the complex normal distribution of mean 0 and standard deviation $N^{-1/2}$. The standard deviation is normalized in this way so that the limit $N \to \infty$ will be well-behaved. In fact one has $\rho(A_{12}) \approx 1$, $\|A_{12}\| \approx 2$, where $\rho(A_{12})$ denotes the spectral radius, and these approximations can be shown to become equalities in various probabilistic senses as $N \to \infty$.

To clarify, A_{12} is a particular, fixed matrix constructed as described above. Another random matrix would look different in detail but probably not very different in general features.

The upper plot for A_{12} is unlike all those presented heretofore: it reveals just 32 dots instead of 3200! Of course each of these dots is a superposition of 100 copies, which look the same to this plotting resolution. The eigenvalues of A_{12} are insensitive to perturbations in comparison with the other examples we have presented. Still, the lower plot reveals that the eigenvalues of smaller modulus are somewhat sensitive. So far as I know, no theorems have been established to make these observations precise.

There is a lesson to be drawn from this example. If the matrices and operators that we cared about arose at random, it might be appropriate to say that eigenvalue analysis failed only in pathological cases of little importance. However, they do not arise at random. Highly non-normal systems are of special interest in the sciences, and we have a special interest in understanding them.

Example 13. Random upper-triangular [45]

Randomness alone does not ensure closeness to normality. For a final example we define A_{13} exactly like A_{12} except that all entries below the diagonal are set to zero:

$$A_{13} = \text{upper-triangular random matrix.}$$

The pictures change dramatically. Suddenly we have an exponential degree of non-normality again, to judge by the cloud of dots in the upper plot and the approximately circular contours, including all values down to 10^{-8}, in the lower plot. The condition number $\kappa(V)$ has increased by ten orders of magnitude.

I consider this example a particularly fascinating illustration of the subtle relationship between spectra and pseudospectra. Of course A_{13} has nonzero eigenvalues; they are the diagonal entries of the matrix, a set of N random numbers from the complex normal distribution of standard deviation $N^{-1/2}$. On the other hand the pictures suggest that the significance of these eigenvalues is open to question. The eigenvalue to the lower-right is a clear outlier, insensitive to perturbations; it will "behave like an eigenvalue" by any measure. The remaining eigenvalues become less and less sharply defined as one moves in towards the origin. It is impossible to draw a sharp line between meaningful eigenvalues and meaningless ones.

If A_{13} is replaced by a *strictly* upper-triangular random matrix, with zeros on the diagonal as well as the subdiagonals, the pictures change modestly. The curves become more nearly circular, but the scale $O(N^{-1/2})$ remains the same. The spectrum vanishes to the origin, but the pseudospectra remain.

4 Conclusion

I hope the examples in this paper have convinced the reader that there is a geometrical aspect to non-normality, and that it may be beautiful. The idea that it is also informative has been asserted but not argued in depth. For discussions of this point in various applications see [29,30,34,35,33] and the book [43], to appear, that will synthesize these and other developments.

My advice in practice is simple: if you find yourself computing eigenvalues of non-normal matrices, try perturbing the entries by a few percent and see what happens! If the effect on the eigenvalues is negligible, it is probably safe to forget about non-normality. If the effect is considerable, the time has come to be more careful.

Matrices are fascinating, but operators can be even more so [Reddy91]. That is where some of the most remarkable effects of non-normality appear, and in a few years, when I have studied more examples and found better ways to compute pseudospectra, I hope to be able to write a paper like this one with $N = \infty$.

Acknowledgments

The work described here was brought to its present form during sabbatical visits in 1990–91 to the University of New South Wales (Sydney), the Université Pierre et Marie Curie (Paris), and Oxford University. I am grateful for the support at these institutions of Ian Sloan, Philippe Ciarlet and Yvon Maday, and K. W. Morton. Additional support has been provided by an NSF Presidential Young Investigator award. Computations were carried out on the Cray-2 at MIT and with the invaluable aid of Matlab. The work of Godunov, et al. [11] was brought to my attention in May 1991 by Michael Eiermann at the University of Karlsruhe, and a German translation was graciously supplied by his colleague Elvira Winterfeld. Finally, my special thanks go to Satish Reddy for his many contributions to the theory of pseudospectra during several years of collaboration.

References

[1] P. M. Anselone and I. H. Sloan, *Spectral approximations for Wiener-Hopf operators*, J. Int. Eqs. Applics. 2 (1990), 237–261.

[2] R. M. Beam and R. F. Warming, *The asymptotic eigenvalue spectra of banded Toeplitz and quasi-Toeplitz matrices*, to appear.

[3] C. Canuto, M. Y. Hussaini, A. Quarteroni, and T. A. Zang, *Spectral Methods in Fluid Dynamics*, Springer-Verlag, New York, 1988.

[4] F. Chatelin, *Resolution Approchée D'Equations sur Ordinateur*, Lecture notes, Lab. de Statistique Théorique et Appliquée, Univ. Pierre et Marie Curie, Paris, 1989.

[5] J. W. Demmel, *A Numerical Analyst's Jordan Canonical Form*, PhD thesis, U. Calif. Berkeley, 1983.

[6] J. W. Demmel, *A counterexample for two conjectures about stability*, IEEE Trans. Aut. Control AC-32 (1987), 340–342.

[7] A. Edelman, *Eigenvalues and Condition Numbers of Random Matrices*, PhD thesis, MIT, 1989.

[8] G. Fairweather, *On the eigenvalues and eigenvectors of a class of Hessenberg matrices*, SIAM Review 13 (1971), 220–221.

[9] S. Frankel, *Convergence rates of iterative treatments of partial differential equations*, Math. Comp. 4 (1950), 65–75.

[10] V. L. Girko, *Theory of Random Determinants*, Kluwer Academic Press, Boston, 1990.

[11] S. K. Godunov, O. P. Kiriljuk, and W. I. Kostin, *Spectral portraits of matrices*, Preprint #3, Inst. of Math., Acad. Sci. USSR, Novosibirsk, 1990 (Russian; see Acknowledgements).

[12] S. K. Godunov and V. S. Ryabenki, *Theory of Difference Schemes*, North-Holland, Amsterdam, 1964.

[13] G. H. Golub and C. F. Van Loan, *Matrix Computations,* 2nd ed., Johns Hopkins University Press, Baltimore, 1989.

[14] G. H. Golub and J. H. Wilkinson, *Ill-conditioned eigensystems and the computation of the Jordan canonical form,* SIAM Review 18 (1976), 578–619.

[15] J. Grcar, *Operator coefficient methods for linear equations,* Sandia National Lab. Rep. SAND89-8691, November, 1989.

[16] H. Han, V. P. Il'in, R. B. Kellogg, and W. Yuan, *Analysis of flow directed iterations,* Tech. Note BN-1109, Inst. Phys. Sci. and Tech., U. Maryland, College Park, MD, February 1990.

[17] D. J. Higham and L. N. Trefethen, *Stiffness of ODEs,* in preparation.

[18] N. J. Higham, *Matrix nearness problems and applications,* in M. J. C. Gover and S. Barnett, eds., Applications of Matrix Theory, Oxford U. Press, 1989.

[19] N. J. Higham, *A collection of test matrices in MATLAB,* ACM Trans. Math. Soft., to appear.

[20] W. Kahan, *Numerical linear algebra,* Canad. Math. Bull. 9 (1966), 757–801.

[21] T. Kato, *Perturbation Theory for Linear Operators,* Springer-Verlag, New York, 1976.

[22] W. Kerner, *Large-scale complex eigenvalue problems,* J. Comp. Phys. 85 (1989), 1–85.

[23] V. Lakshimkantham and D. Trigiante, *Theory of Difference Equations: Numerical Methods and Applications,* Academic Press, New York, 1988.

[24] H. W. J. Lenferink and M. N. Spijker, *On the use of stability regions in the numerical analysis of initial value problems,* Rep. TW-89-07, Dept. Math. and Comp. Sci., U. Leiden, 1989.

[25] H. W. J. Lenferink and M. N. Spijker, *A generalization of the numerical range of a matrix,* Lin. Alg. Applics. 140 (1990), 251–266.

[26] T. A. Manteuffel, *Adaptive procedure for estimating parameters for the nonsymmetric Tchebychev iteration,* Numer. Math. 31 (1978), 183–208.

[27] C. Moler, *ROOTS—of polynomials, that is,* Mathworks Newsletter (MathWorks, Inc.) 5 (1991), 8–9.

[28] R. G. Mosier, *Root neighborhoods of a polynomial,* Math. Comput. 47 (1986), 265–273.

[29] N. M. Nachtigal, S. C. Reddy, and L. N. Trefethen, *How fast are nonsymmetric matrix iterations?,* SIAM J. Matrix Anal. Applics., to appear.

[30] N. M. Nachtigal, L. Reichel, and L. N. Trefethen, *A hybrid GMRES algorithm for nonsymmetric linear systems,* SIAM J. Matrix Anal. Applics., to appear.

[31] D. Pathria, S. C. Reddy, and L. N. Trefethen, *Eigenvalues and pseudo-eigenvalues of convection-diffusion operators,* in preparation.

[32] S. C. Reddy, *Pseudospectra of operators and discretization matrices and an application to stability of the method of lines,* PhD thesis, MIT, 1991.

[33] S. C. Reddy, P. Schmid and D. Henningson, *Pseudospectra of the Orr-Sommerfeld operator,* submitted to SIAM J. Appl. Math.

[34] S. C. Reddy and L. N. Trefethen, *Lax-stability of fully discrete spectral methods via stability regions and pseudo-eigenvalues,* Comp. Meth. Appl. Mech. Engr. 80 (1990), 147–164.

[35] S. C. Reddy and L. N. Trefethen, *Stability of the method of lines,* submitted to Numer. Math.

[36] L. Reichel and L. N. Trefethen, *Eigenvalues and pseudo-eigenvalues of Toeplitz matrices,* Lin. Alg. Applics., to appear.

[37] R. D. Richtmyer and K. W. Morton, *Difference Methods for Initial-Value Problems,* Wiley-Interscience, New York, 1967.

[38] F. Riesz and B. Sz-Nagy, *Functional Analysis,* Frederick Ungar, New York, 1955.

[39] D. A. Spence, *A class of biharmonic end-strip problems arising in elasticity and Stokes flow,* IMA J. Numer. Anal. 30 (1989), 107–139.

[40] G. W. Stewart and Sun, *Matrix Perturbation Theory,* Academic Press, 1990.

[41] L. N. Trefethen, *Lax-stability vs. eigenvalue stability of spectral methods,* In: K. W. Morton and M. J. Baines (eds.), Numerical Methods for Fluid Dynamics III, 237–253. Oxford: Clarendon Press 1988.

[42] L. N. Trefethen, *Approximation theory and numerical linear algebra,* in J. C. Mason and M. G. Cox, eds., Algorithms for Approximation II, Chapman and Hall, London, 1990.

[43] L. N. Trefethen, *Non-Normal Matrices and Pseudospectra,* book to appear.

[44] C. Van Loan, *How near is a stable matrix to an unstable matrix?,* in R. Brualdi, et al., eds., Contemporary Mathematics, v. 47, Amer. Math. Soc., 1985.

[45] J. M. Varah, *On the separation of two matrices,* SIAM J. Numer. Anal. 16 (1979), 216–222.

[46] R. F. Warming and R. M. Beam, *An eigenvalue analysis of finite-difference approximations for hyperbolic IBVPs II: the auxiliary Dirichlet problem,* in Proc. Third Intl. Conf. Hyperbolic Problems.

[47] J. H. Wilkinson, *Rounding Errors in Algebraic Processes,* Prentice-Hall, Englewood Cliffs, NJ, 1963.

[48] J. H. Wilkinson, *The Algebraic Eigenvalue Problem,* Clarendon Press, Oxford, 1965.

[49] J. H. Wilkinson, *Sensitivity of Eigenvalues II,* Utilitas Math. 30 (1986), 243–286.

Department of Computer Science
Cornell University
Ithaca, NY 14853 USA
lnt@cs.cornell.edu

Example 1. Jordan block $\kappa(V) = \infty$

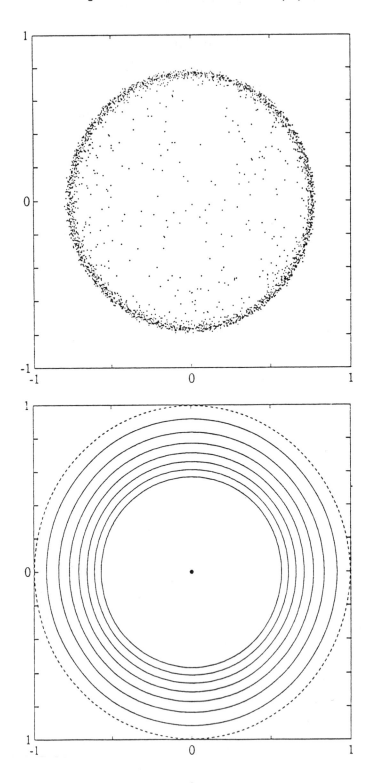

Example 2. Limaçon $\kappa(V) = \infty$

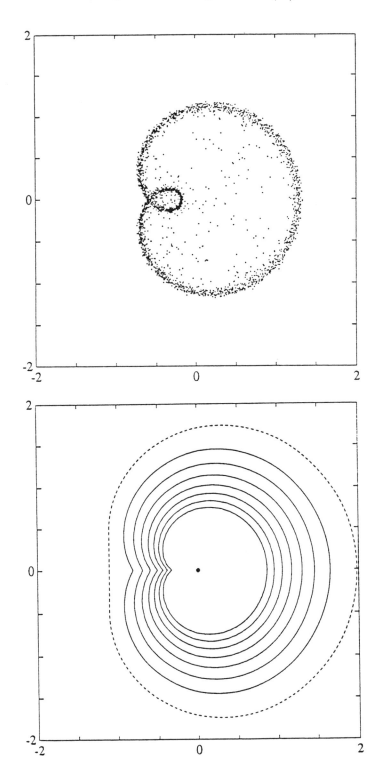

Example 3. Grcar matrix $\kappa(V) = 9.80 \times 10^4$

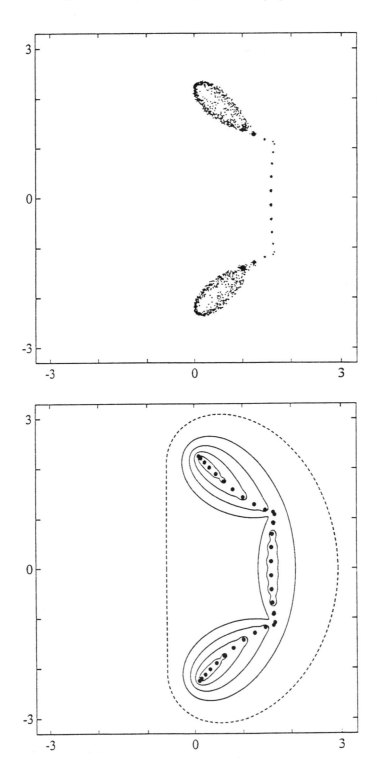

Example 4. Wilkinson matrix $\quad \kappa(V) = 2.62 \times 10^{22}$

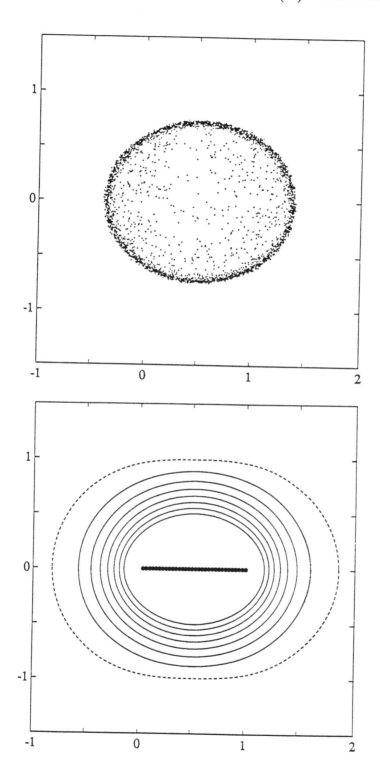

Example 5. Frank matrix $\kappa(V) = 7.81 \times 10^{11}$

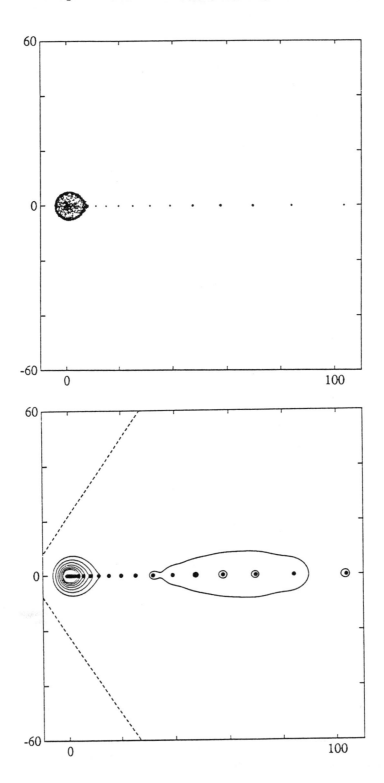

Example 6. Kahan matrix $\quad \kappa(V) = 6.84 \times 10^8$

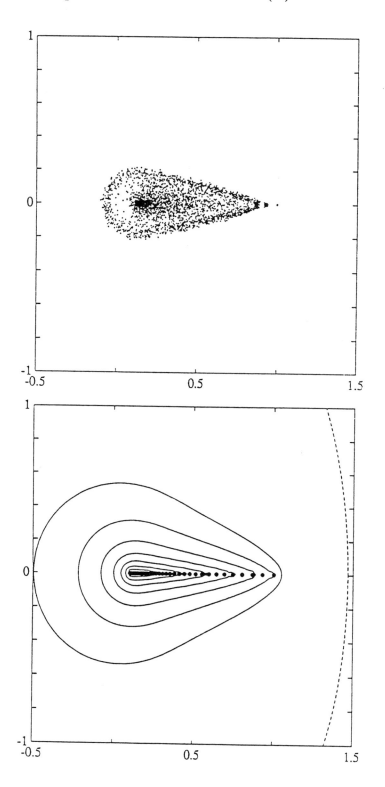

Example 7. Demmel matrix $\quad \kappa(V) = \infty$

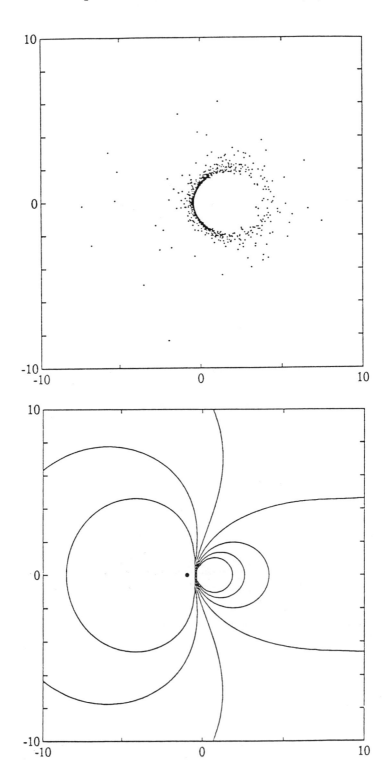

Example 8. Matrix of Lenferink and Spijker $\kappa(V) = 1.75 \times 10^9$

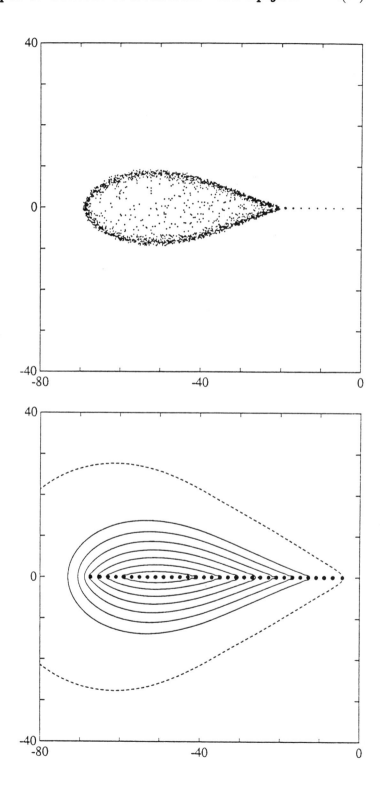

Example 9. Companion matrix $\kappa(V) = 1.55 \times 10^9$

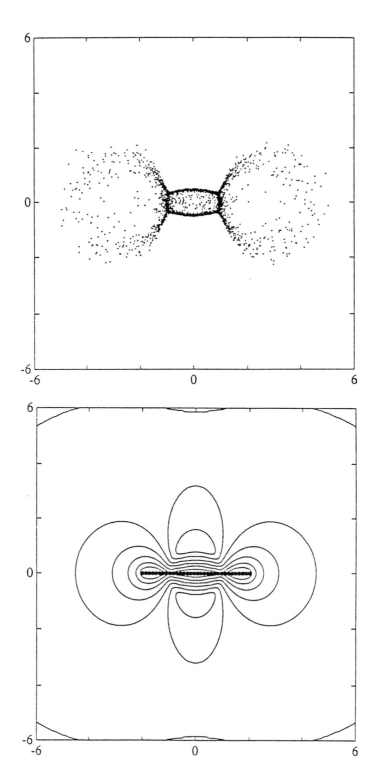

Example 10. Gauss-Seidel $\kappa(V) = \infty$

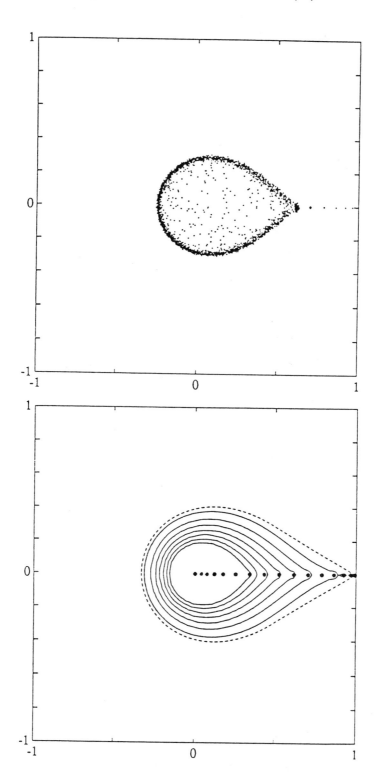

Example 11. Chebyshev spectral $\kappa(V) = 2.69 \times 10^{14}$

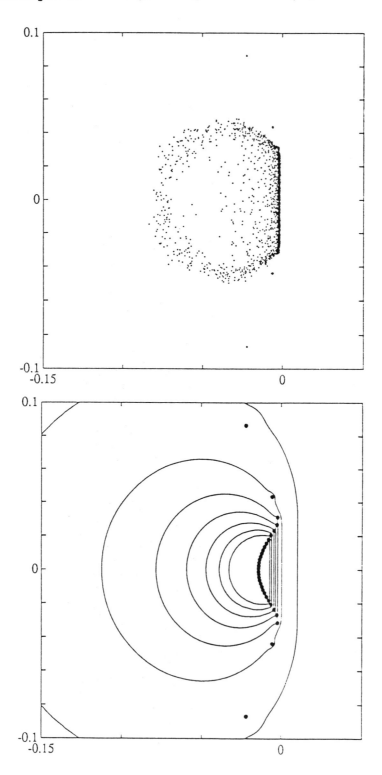

Example 12. Random $\kappa(V) = 25.80$

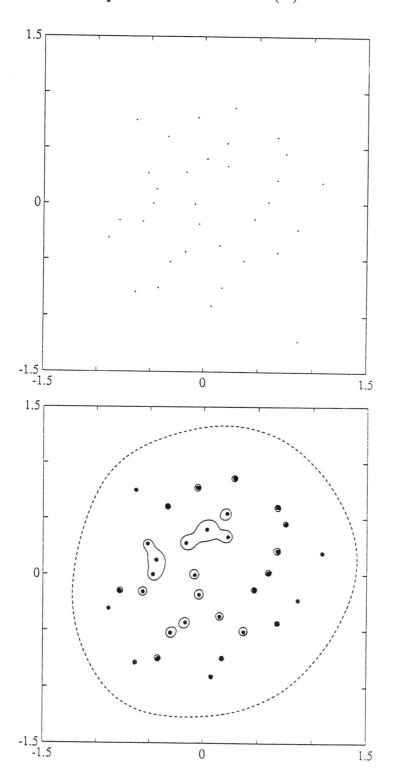

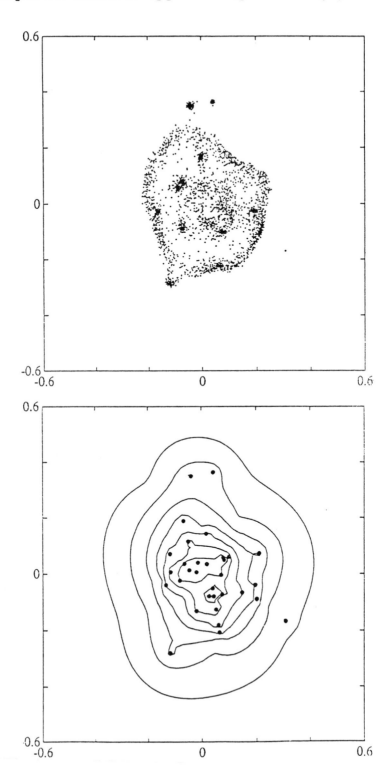

J G VERWER AND R A TROMBERT
Analysis of an adaptive finite-difference method for time dependent PDEs

1. Introduction

Standard numerical methods integrate on a fixed spatial grid, a priori chosen for the whole temporal integration interval. For solutions exhibiting strong local phenomena, like steep moving fronts, emerging layers, etc., this may necessitate a very fine grid covering the entire spatial domain for all values of time. Needless to say that this then may involve unacceptably high computational costs, notably for multi-space dimensional problems. In such cases adaptive-grid methods may provide a remedy. An adaptive-grid method refines the space grid only locally, thus striving for a substantial reduction in number of grid points and computer costs.

Two main categories of adaptive methods are distinguished, viz dynamic and static methods. Dynamic (in time) methods adapt the grid in a continuous-time manner, like classical Lagrangian methods. As examples we mention the moving-finite-element method of Miller [15] and the moving-finite-difference method discussed in Verwer et al. [19]. These methods usually not only provide a local spatial refinement, but also allow larger time steps due to their Lagrangian nature. As a rule, the dynamic approach is feasible for 1D problems, but much less so in 2D and 3D. The second main category of methods, the static (in time) methods, adapt the grid only at discrete times. Static methods are in principle well feasible in any space dimension, but they do not soften temporal solution behaviour like in the dynamic approach, since the actual time stepping is carried out on a non-moving grid.

We are interested in the development of adaptive-grid techniques for time-dependent PDE problems which are suitable for general implementation and are, as much as possible, as feasible as fixed-single-grid methods. The present paper is devoted to a static technique called local uniform grid refinement (LUGR). Our main purpose here is to discuss this technique and to show how it can be combined with standard integration methods of the linear multistep and Runge-Kutta type. Following the method-of-lines approach, we start our discussion with the implicit Euler method. Later in the paper we show how the results obtained on the implicit Euler LUGR method can be extended to general Runge-Kutta and linear multistep methods.

Throughout the paper the discussion focusses on the local refinement analysis problem. Further, for the purpose of demonstration, our examples are restricted to Cartesian grids but we emphasize that the main results of our error analysis, as presented in Section 4, are also valid for other types of grids, such as triangular grids for two-space dimensional problems.

2. The LUGR technique

Although its mathematical elaboration is complicated, in essence the LUGR technique is simple and straightforward. Therefore we begin with an outline of the technique and the grid structure involved (cf. [17, 18] and [16]). Let $\Omega \times [0,T]$ be the space-time domain (in any dimension) with boundary $\partial\Omega$ parallel to the co-ordinate axes. Let ω_1 be a coarse, uniform grid covering Ω. This grid is called the base grid. Starting at ω_1, one base time step with the implicit Euler LUGR method consists of repeated integrations on nested, finer-and-finer local subgrids, possessing internal nonphysical boundaries (grid interfaces). Each of the single integrations spans the same step interval. Hence on each subgrid a new initial-boundary value problem is solved, over the base time step interval in use. Required initial values are defined by interpolation from the next coarser subgrid or taken from a possibly existing subgrid from the previous time step interval. Boundary values required at internal boundaries are also interpolated from the next coarser subgrid. Internal boundaries are treated as Dirichlet boundaries. The generation of the nested subgrids is continued up to a level considered fine enough for resolving the fine scale structure at hand. Having completed the integration on the finest level, the process

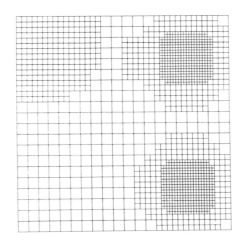

Figure 1. Example of a 2D composite grid for a particular point of time.

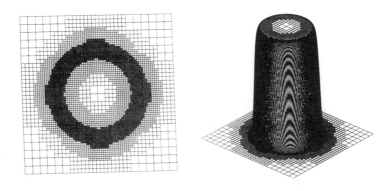

Figure 2. Circular wave front computed on a circular 4-level grid.

is repeated for the next base time step interval until the physical end time t = T is reached.

Thus, for each base time step the computation starts at the base grid, while using the most accurate solution available, since fine grid solution values are always injected in coinciding coarse grid points. Moreover, grid points already living at a certain level of refinement are used for step continuation (all fine subgrids are kept in storage). In conclusion, each base step consists of the following operations:

     1. *Integrate on coarse base grid.*
     2. *Determine new fine subgrid at forward time.*
     3. *Interpolate internal boundary values at forward time.*
     4. *Interpolate new internal boundary values at backward time.*
     5. *Interpolate new initial values at backward time.*
     6. *Integrate on subgrid, using the same steplength.*
     7. *Inject fine grid values at coinciding coarse grid points.*
     8. *Goto 2 until the desired number of refinement levels is reached.*

Figure 1 shows an example of a composite LUGR grid in 2D for a certain point of time. Note that we allow disjunct subgrids and that these need not be a rectangle. Further one sees that the actual refinement is cellular and carried out by bisection of sides of coarse grid cells. An important point to notice is that the repeated use of all varying subgrids, from coarse to fine, is essential in our method. This way we generate the required boundary conditions at the internal boundaries and keep the subgrids uniform. This approach is necessarily a bit wasteful in situations where sharp transitions move very slowly, e.g. when approaching steady state. On the other hand, the workload on the coarser grids will normally be small and we consider the use of uniform grids attractive. Uniform grids allow an efficient use of vector based algorithms and finite-difference approximations on uniform grids are more accurate and faster to compute than on nonuniform grids. In this respect the current approach is to be contrasted with pointwise refinement leading to truly nonuniform grids. Pointwise refinement techniques also require a more complex and expensive data structure.

Previous work on LUGR methods has been published by Berger and Oliger [4], Gropp [10,11], and others. In [4], and also in Arney and Flaherty [3], LUGR methods are examined which are based on noncellular refinement and truly rectangular local subgrids which may rotate and overlap to align with an evolving fine scale structure. We avoid these difficulties. Our local subgrids may not overlap and internal boundaries are always parallel to co-ordinate axes. On the other hand, our subgrids need not be a rectangle. Figure 2 shows how a circular wave front is handled with our grid structure.

3. Mathematical formulation

We will now give a formulation of the implicit Euler LUGR method which enables us to set up a general analysis of the interplay between spatial discretization and interpolation errors.

3.1. The semi-discrete problem

Consider a well-posed real abstract Cauchy problem

$$u_t = L(\underline{x},t,u), \quad (\underline{x},t) \in \Omega \times (0,T], \quad u(\underline{x},t,0) = u_0(\underline{x}), \tag{3.1}$$

where L is a d-space dimensional PDE operator, of at most second order. L is supposed to be provided with appropriate boundary conditions for t>0 such that the true solution u(x,t) uniquely exists and is as often differentiable on $(\Omega \cup \partial\Omega) \times [0,T]$ as the numerical analysis requires. Recall, however, that we aim at non-smooth solutions, like steep travelling fronts. Thus, non-smoothness means here the occurrence of rapid transitions in the space-time domain, with a sufficient degree of differentiability. The function u(x,t) may be vector-valued and the number of space dimensions may be arbitrary.

LUGR methods use local subgrids of varying size in time and thus generate approximation vectors of a varying dimension. This varying dimension complicates the error analysis. For the sake of analysis, we circumvent this problem by expanding the fine local subgrids over the entire domain Ω. Integration then takes place on part of the expanded grid and interpolation on its complement. We stipulate that this grid expansion does not take place in the application of the method, merely in the formulation.

Let ω_k, $1 \le k \le r$, be uniform space grids, with the uniformity meant directionwise. The integer r is the number of refinement levels. Each grid covers the whole of Ω (grid expansion) and has no points on $\partial\Omega$. Given the base grid ω_1, ω_2 is constructed by bisecting all sides of all cells of ω_1, etc., so that the grids are naturally nested. Because $\partial\Omega$ is locally parallel to the co-ordinate axes, it is always possible to construct such a set of grids. To the PDE problem (3.1) we now associate on each ω_k a real Cauchy problem for an explicit ODE system

$$\frac{d}{dt} U_k(t) = F_k(t, U_k(t)), \qquad 0 < t \le T, \quad U_k(0) = U_k^0, \tag{3.2}$$

obtained by an appropriate spatial discretization of (3.1) on ω_k. Hence U_k and F_k are vectors representing grid functions on ω_k. It is assumed that the boundary conditions have been worked into (3.2) by elimination of variables on $\partial\Omega$.

Let d_k be the length of U_k. In the remainder we let S_k with $\dim(S_k) = d_k$ denote the vector space of all grid functions on ω_k. Specific elements of S_k are U_k, F_k and u_k, where $u_k = u_k(t)$ represents the natural restriction of the true PDE solution u(x,t) to ω_k. In the space S_k the initial value problems (3.1), (3.2) are related by the local space (discretization) error

$$\alpha_k(t) = \frac{d}{dt} u_k(t) - F_k(t, u_k(t)), \qquad 0 \le t \le T. \tag{3.3}$$

In particular, the continuous time grid functions u_k and α_k are supposed to be sufficiently often differentiable in t, while α_k has the order of consistency of the spatial finite-difference formulas employed.

3.2. **The implicit Euler LUGR formula**
The base time step from t_{n-1} to t_n is formulated by

$$U_1^n = R_{r1} U_r^{n-1} + \tau F_1(t_n, U_1^n), \tag{3.4a}$$

$$U_k^n = D_k^n [R_{rk} U_r^{n-1} + \tau F_k(t_n, U_k^n)] + (I_k - D_k^n)[P_{k-1k} U_{k-1}^n + b_k^n], \tag{3.4b}$$

where $k = 2, 3, \ldots, r$ and τ is the stepsize in time. $U_k^n \in S_k$ is the approximation to $u_k(t_n)$ at ω_k. $I_k: S_k \to S_k$ is the unit matrix. $D_k^n: S_k \to S_k$ is a diagonal matrix with entries $(D_k^n)_{ii}$ either unity or zero. $R_{rk}: S_r \to S_k$ is the natural restriction operator from ω_r to ω_k, $R_{rr} = I_r$. $P_{k-1k}: S_{k-1} \to S_k$ is an interpolation operator from ω_{k-1} to ω_k and $b_k^n \in S_k$ contains time-dependent terms emanating from the boundary $\partial\Omega$.

Inspection of (3.4b) reveals that the diagonal matrices D_k^n are used to define the integration domains, which are just the local subgrids without grid expansion. Specifically, if at a certain node integration takes place, then $(D_k^n)_{ii} = 1$, while at all remaining nodes where interpolation is carried out, $(D_k^n)_{ii} = 0$. The actual selection of integration and interpolation nodes is made by the regridding strategy which is discussed later. The nesting property of the integration domains is also induced by this strategy and cannot be recovered from the above formulation, as it is hidden in the actual definition of D_k^n.

The interpolation step can be written as

$$(I_k - D_k^n) U_k^n = (I_k - D_k^n) [P_{k-1k} U_{k-1}^n + b_k^n]. \tag{3.5}$$

The gridfunction b_k^n plays an auxiliary role here, but we need to include it due to the fact that boundary conditions on $\partial\Omega$ have been worked into the semi-discrete system. For the analysis it plays no role, whatsoever. Noteworthy is that the choice of interpolation is still free. The integration step on level $k \geq 1$ can be represented by $(D_1^n = I_1)$

$$D_k^n U_k^n = D_k^n [R_{rk} U_r^{n-1} + \tau F_k(t_n, U_k^n)]. \tag{3.6}$$

Values at or beyond internal boundaries needed in $F_k(t_n, U_k^n)$ are defined by (3.5). Thus (3.4) automatically prescribes values at internal boundaries. Recall that for nodes at and beyond internal boundaries, the associated entry of D_k^n is zero, by definition. Also note that (3.6) is coupled with (3.5) through these internal boundaries. Finally, due to the injection, at each grid level the fine grid solution $D_k^n R_{rk} U_r^{n-1}$ is used as initial function.

An important point to notice is that (3.4) - (3.6) live in the space S_k, due to the grid expansion. This means that the interpolation is considered to take place on the whole of ω_k. However, in actual application we interpolate only over the current integration domain. This deviation is justified as our regridding strategy is such that as far as the choice of the next finer integration domain is concerned, it makes no difference whether we interpolate over the whole of ω_k or restrict it to the current domain. Our strategy, together with this issue of restricted interpolation, is discussed at length in [17]. In the next section we will briefly review the main underlying ideas.

Finally, in (3.4) the number of levels, r, is a priori fixed independent of time. This fixed-level mode of operation may be inefficient and it is obvious that the method should be capable of working in a variable-level mode of operation. For general Runge-Kutta methods we discuss this in [18]. In this paper we restrict ourselves to fixed r, for reasons of space.

4. The regridding strategy

A strategy should fulfil two basic accuracy requirements. It should induce a sufficient local refinement in regions where the spatial errors are larger than elsewhere, and it should involve automatic control of the inevitable interpolation errors. This second requirement is often neglected, but is of significant importance. The reason is that if we regrid at each base time step, we interpolate at each base time step. Thus the interpolation errors can accumulate linearly with the base time steps, so that reducing τ may result in error growth, rather than in error decay. Although less, this threat remains if we would not regrid at every base time step, but per certain number > 1 of such steps.

In [17] we have developed a strategy which meets both requirements. We demand that the refinement is such that the spatial accuracy on the composite final grid is comparable to the spatial accuracy on ω_r if integration takes place on the whole of ω_r. As far as accuracy is concerned, this is the maximum we can ask for. In addition, this way we force the interpolation error to remain negligible when compared with the common spatial error on ω_r. This means that the accumulation of the interpolation error is automatically controlled. In this section we will outline the analysis upon which our regridding strategy is based.

In (3.4) the matrices D_k^n determine the integration domains. Two extreme choices are I_k and 0, which imply, respectively, integration or interpolation over the entire domain Ω. In our analysis the diagonal matrices D_k^n act as control matrices used to satisfy the above mentioned accuracy demand. Introduce the global error

$$e_k^n = u_k^n - U_k^n, \qquad u_k^n = u_k(t_n), \tag{4.1}$$

and the perturbed scheme

$$u_1^n = R_{r1}u_r^{n-1} + \tau F_1(t_n, u_1^n) + \delta_1^n, \tag{4.2a}$$

$$u_k^n = D_k^n [R_{rk}u_r^{n-1} + \tau F_k(t_n, u_k^n)] + (I_k - D_k^n) [P_{k-1k}u_{k-1}^n + b_k^n] + \delta_k^n, \tag{4.2b}$$

where δ_k^n is the defect left by substituting u_k^n into (3.4). Let $D_1^n = I_1$. As $u_k^{n-1} = R_{rk}u_r^{n-1}$, we then can write

$$\delta_k^n = D_k^n (\tau \alpha_k^n + \beta_k^n) + (I_k - D_k^n) \gamma_k^n, \qquad k = 1, \dots, r, \tag{4.3}$$

where

$$\alpha_k(t) = \frac{d}{dt}u_k(t) - F_k(t, u_k(t)) \qquad \text{(local space error)} \tag{4.4a}$$

$$\beta_k(t) = u_k(t) - u_k(t - \tau) - \tau \frac{d}{dt}u_k(t) \qquad \text{(local time error)} \tag{4.4b}$$

$$\gamma_k(t) = u_k(t) - P_{k-1k} u_{k-1}(t) - b_k(t) \qquad \text{(interpolation error)} \tag{4.4c}$$

Naturally, δ_k^n is composed of the three common local errors encountered in the space S_k. We omit here to discuss the asymptotics on these errors, as this is not relevant for the remainder.

272

Next subtract (3.4) from (4.2). This yields

$$Z_1^n e_1^n = R_{r1} e_r^{n-1} + \delta_1^n, \tag{4.5a}$$

$$Z_k^n e_k^n = D_k^n R_{rk} e_r^{n-1} + (I_k - D_k^n) P_{k-1k} e_{k-1}^n + \delta_k^n, \qquad k = 2, \ldots, r, \tag{4.5b}$$

where

$$Z_k^n = I_k - \tau D_k^n \int_0^1 \frac{\partial}{\partial U} F(t_n, \theta u_k^n + (1-\theta) U_k^n) \, d\theta \tag{4.6}$$

is the integrated Jacobian matrix resulting from applying the mean value theorem for vector functions. Note that so far we consider the general nonlinear, semi-discrete problem (3.2). In the remainder it is now tacitly assumed that the matrix (4.6) is regular, which is a most natural condition satisfied in any realistic application. We also introduce the interpolation operator X_k^n : $S_{k-1} \to S_k$, given by

$$X_k^n = (Z_k^n)^{-1} (I_k - D_k^n) P_{k-1k}. \tag{4.7}$$

The global errors e_k^n then can be shown to satisfy an inner-outer recurrence of the one-step type

$$e_k^n = G_k^n e_r^{n-1} + \psi_k^n, \qquad n = 1, 2, \ldots ; k = 1, \ldots, r, \tag{4.8}$$

where the step operator G_k^n and the local error ψ_k^n are also recursively defined:

$$G_1^n = (Z_1^n)^{-1} R_{r1}, \qquad G_k^n = X_k^n G_{k-1}^n + (Z_k^n)^{-1} D_k^n R_{rk}, \qquad k = 2, \ldots, r, \tag{4.9}$$

$$\psi_1^n = (Z_1^n)^{-1} \delta_1^n, \qquad \psi_k^n = X_k^n \psi_{k-1}^n + (Z_k^n)^{-1} \delta_k^n, \qquad k = 2, \ldots, r. \tag{4.10}$$

The occurrence of the backward finest level error e_r^{n-1} in (4.8) emanates from the injection. We are primarily interested in these finest level errors. G_k^n represents the step operator that advances e_r^{n-1} to the forward level k and ψ_k^n denotes the full local error at this level. Note that both G_k^n and ψ_k^n depend on quantities living on all grids ω_j, $1 \le j \le k$. Specifically, G_k^n is composed of the involved integration, interpolation and restriction operators, while the local error ψ_k^n is composed of the three types of local errors encountered.

As outlined above, a regridding strategy should find a proper balance between the two types of spatial errors and at the same time control the accumulation of interpolation errors for evolving time. The following result can be shown. First we split ψ_k^n into its temporal and spatial part. That is, we write

$$\psi_k^n = \psi_{kt}^n + \psi_{ks}^n. \tag{4.11}$$

From (4.3), (4.9) and (4.10) we then can deduce that $\psi_{kt}^n = G_k^n \beta_r^n$ and

$$\psi_{1s}^n = (Z_1^n)^{-1} \tau \alpha_1^n, \qquad \psi_{ks}^n = (Z_k^n)^{-1} [\tau D_k^n \alpha_k^n + (I_k - D_k^n)\rho_k^n], \qquad k = 2, \ldots, r, \qquad (4.12)$$

where the new grid function ρ_k^n satisfies

$$\rho_1^n = 0, \qquad \rho_k^n = \gamma_k^n + P_{k-1k} \psi_{k-1s}^n, \qquad k = 2, \ldots, r. \qquad (4.13)$$

Because our main interest lies in the local refinement analysis, we now focuss on the spatial local error expressions (4.12), (4.13) which contain all possible local contributions from the spatial discretization and interpolation. Specifically, in (4.12) we have separated the common local space error

$$\tau D_k^n \alpha_k^n, \qquad (4.14)$$

restricted to the level-k integration domain, from all other space error contributions living outside this integration domain. These are collected in

$$(I_k - D_k^n) \rho_k^n. \qquad (4.15)$$

From (4.13) we see that ρ_k^n contains the level-k interpolation error γ_k^n plus the spatial local error $P_{k-1k} \psi_{k-1s}^n$ prolongated from the next coarser grid. This separation of spatial discretization and interpolation errors is very useful, since it enables us to constraint the diagonal matrices D_k^n in such a way that the space discretization error contribution (4.14) dominates its parasitic counterpart (4.15), which is entirely due to interpolation.

Let $\| \cdot \|$ be the maximum norm in the space S_k. The definition of the control matrices D_k^n is then accomplished through the so-called *refinement condition*. This condition reads

$$\| (Z_k^n)^{-1} \tau D_k^n \alpha_k^n \| \geq \frac{1}{c} \| (Z_k^n)^{-1} (I_k - D_k^n)\rho_k^n \|, \qquad k = 2, \ldots, r, \qquad (4.16)$$

where $c > 0$ is a control parameter to be specified. Substitution into (4.8) and taking norm bounds, gives for $k = r$ the global error inequality

$$\| e_r^n \| \leq \| G_r^n e_r^{n-1} \| + \| G_r^n \beta_r^n \| + (1+c) \| (Z_r^n)^{-1} \tau D_r^n \alpha_r^n \|. \qquad (4.17)$$

The crucial observation is now that by imposing (4.16), the parasitic interpolation error (4.15) has been virtually removed. Apart from the factor $1+c$, the overall spatial error at the finest grid level is bounded by the common spatial error bound $\| (Z_r^n)^{-1} \| \| \tau \alpha_r^n \|$, which is also found on ω_k without any adaptation at all. This result is in agreement with the two requirements on the regridding strategy mentioned above.

For practical use the refinement condition (4.16) needs to be brought into a workable form. Needless to say that this involves various aspects of implementation, like estimation of local space and interpolation errors. We conclude this section by briefly outlining how we have dealt with (4.16). The main point is that we have replaced it by the related condition

$$\| (I_k - D_k^n)(\gamma_k^n + P_{k-1k}(Z_{k-1}^n)^{-1} \tau D_{k-1}^n \alpha_{k-1}^n) \| \leq \frac{c}{r-1} \| (Z_r^n)^{-1} \tau D_r^n \alpha_r^n \|, \qquad (4.18)$$

where $k = 2, \ldots, r$, which is more feasible for implementation. If implicit Euler is contractive and the interpolation stable, i.e., $\| (Z_k^n)^{-1} \| \leq 1$ and $\| P_{k-1k} \| = 1$, then (4.18) can be proved to be stronger than (4.16). If implicit Euler is not truly contractive, but merely stable, then the bound 1 needs to be replaced by (unknown) stability bounds for $(Z_k^n)^{-1}$ which also enter (4.18). However, these are close to 1 so that also in this case (4.18) remains valid, approximately. We also have the experience that the stability condition on the interpolation is not crucial. Note that this condition is somewhat restrictive. For example, it holds for standard linear interpolants, but not for higher order Lagrangian ones.

In actual application (4.18) thus determines the integration domains within each base time step. This goes as follows. Suppose that at level-(k-1) a solution has been computed. Condition (4.18) is then checked in a flagging procedure which scans all level-k points lying within the level-(k-1) integration domain (recall the nesting property). Specifically, at these level-k points we check the inequality

$$|(\lambda_k^n)_i| > \frac{c}{r-1} \| (Z_r^n)^{-1} \tau D_r^n \alpha_r^n \|, \qquad \lambda_k^n = \gamma_k^n + P_{k-1k}(Z_{k-1}^n)^{-1} D_{k-1}^n \alpha_{k-1}^n. \qquad (4.19)$$

If (4.19) is true, then it is decided that the point will remain within the integration domain. Otherwise we put $(D_k^n)_{ii} = 0$, saying that the grid point will lie outside this domain. This way the refinement conditions (4.16) and (4.18) are satisfied.

Inspection of (4.19) reveals that the integration is redone at a grid point if the sum of the interpolation and prolongated spatial discretization error is larger than the maximum spatial discretization error on the finest grid ω_r. This requirement is quite restrictive for the interpolation, because the discretization errors are multiplied with τ and the interpolation error is not. This actually means that for decreasing τ the interpolation becomes more and more dominant. In other words, the smaller τ, the more points will be flagged into the new integration domain, so as to keep the interpolation error sufficiently small. This is what really should happen, because when going to a higher level within the current base time step, we never return to a point where the solution has been interpolated. This means that the error will be carried along to the next base time step. Our strategy only allows this if the interpolation error is so small that the final spatial accuracy is not affected, according to the global error bound (4.17). Naturally, these observations suggest to use higher order interpolation. In application the use of higher order interpolation indeed leads to smaller integration domains than found with linear interpolation. We stress, however, that simple linear interpolation can be used.

The control parameter c plays a minor role. We usually take $c \approx 1$. Note that the larger c, the easier it will be to satisfy the refinement condition in the flagging procedure. Thus a 'large' value for c will lead to 'small' integration domains. At first sight this seems attractive. However, according to (4.17), a 'large' value for c will most likely result in 'large' spatial errors. Apparently, a 'large' value for c is not in accordance with our two regridding requirements.

Finally, to save space, we omit a discussion on the stability properties of the implicit Euler LUGR method. However, as a rule stability is not affected by interpolation. If the semi-discrete PDE system (3.2) is dissipative in the maximum norm, and the interpolation is stable, one can even prove that the maximum norm contractivity of implicit Euler is maintained ($\| G_r^n \| \le 1$). We also recall that our analysis takes place in the space S_k (grid expansion), but that the interpolation and the subsequent scanning of grid points in the flagging procedure is restricted to the current integration domain. In [17] it is shown that the deviation caused by the restricted interpolation is allowed.

5. Numerical illustration

We will numerically illustrate the outcome of imposing the refinement condition for a parabolic model problem. See [17] for details on estimation, implementation and strategy aspects. The model is hypothetical and due to [1]:

$$u_t = u_{xx} + u_{yy} + f(x,y,t), \quad 0 < x,y < 1, \quad t > 0. \tag{5.1}$$

The initial function, the Dirichlet boundary conditions, and the source term f are such that

$$u(x,y,t) = \exp(-80[(x - .25(2 + \sin(\pi t)))^2 + (y - .25(2 + \cos(\pi t)))^2]). \tag{5.2}$$

This solution is a cone that is initially centered at (1/2,3/4) and symmetrically rotates around (1/2,1/2) in the clockwise direction. The speed of rotation is constant with period 2. This problem is not a very difficult one in the sense that the spatial gradients are not very large, that is, the cone is not very steep. However, the problem is suitable to subdue the LUGR method to a convergence test and to check the refinement condition. The spatial discretization is based on standard 2-nd order differencing (implicit Euler is then contractive).

We have carried out two identical experiments. In the first linear interpolation has been used and in the second 4-th order Lagrangian. In both the solution is computed two times over the interval $0 \le t \le 2$ using a constant τ. In the first computation $r = 3$ and in the second $r = 4$, using a uniform 10 x 10 base grid. When going from $r = 3$ to $r = 4$, τ is decreased with the factor 2^2, since the smallest meshwidth is halved and the Euler method is of order one only. Then, in line with our analysis, the maximal global error should also decrease with this factor 2^2, approximately.

Table 1 shows the maxima of global errors restricted to the finest available subgrid. Inspection of this table clearly reveals the 2-nd order (this conclusion also follows if we would examine the global errors at the entire composite grid). Note the striking correspondence with the single fine-grid error. We conclude that the interpolation error is well controlled and that the spatial accuracy of the single finest grid is maintained. At this point we should emphasize that in spite of the relatively large values for τ, the spatial error dominates the global errors shown in the table. In other words, conclusions on the spatial error behaviour induced by the local refinement algorithm can be drawn from these results. The threshold factor $c = 1$ has apparently no influence on the error. We owe this to the fact that the refinement condition has been derived from errors bounds. Furthermore, buffering in the flagging procedure may also have some interfering effect.

Table 1. Results for model problem (5.1)-(5.2).

τ	r	interpolation	single grid	error at t = 2
1/8	3	linear		0.01369
		4-th order		0.01376
			40 x 40	0.01389
1/32	4	linear		0.00340
		4-th order		0.00359
			80 x 80	0.00347

As anticipated by our strategy, the choice of interpolant has no notable influence on the error. The use of the two different interpolants is expressed in the slightly different integration domains shown in Figure 3. As expected, at the higher levels linear interpolation gives rise to somewhat larger domains. This means that the use of linear interpolation is more expensive. For both interpolants the moving domains accurately reflect the symmetric rotation of the cone, which once again nicely illustrates the reliability of the implemented refinement condition with the various estimators.

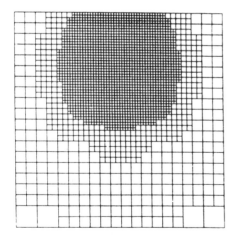 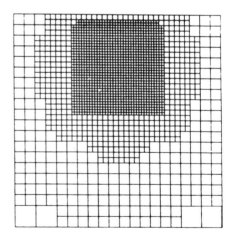

Figure 3. Problem (5.1)-(5.2). Composite grids of the r = 4 runs at t = 2. The left picture corresponds with linear interpolation, the right one with 4-th order Lagrangian.

6. Runge-Kutta and linear multistep methods

Implicit Euler has excellent stability properties, but in connection with accuracy its order one is often considered to be too restrictive. Higher order methods are found in the familiar Runge-Kutta and linear multistep families. Given such a method, the question arises how to extend the implicit Euler LUGR technique, together with the analysis, so as to obtain a similar control of interpolation errors and maintenance of fine-grid spatial accuracy as found for (3.4).

6.1. Runge-Kutta methods

Suppressing index k in the ODE formula (3.2), we denote the general s-stage RK scheme by

$$U^{(i)} = U^{n-1} + \tau \sum_{j=1}^{s} a_{ij} F(t_{n-1} + c_j\tau, U^{(j)}) \quad (1 \leq i \leq s+1), \quad U^n = U^{(s+1)}. \tag{6.1}$$

Note that bracketed upperscripts are used for the intermediate stage values. In contravention of the usual notation, we use one and the same expression for the intermediate stages and the (s+1)-st output stage. The fact that $U^{(s+1)}$ is the approximation at $t = t_n$ is clear from the context. We adhere to the convention $c_i = a_{i1} + ... + a_{is}$. Hence $U^{(i)}$ is an approximation at the intermediate point $t_{n-1} + c_i\tau$. Note that class (6.1) is supposed to represent the general RK class, containing all implicit and explicit methods.

The general Runge-Kutta LUGR formula now reads

$$U_1^{(i)} = R_{r1}U_r^{n-1} + \tau \sum_{j=1}^{s} a_{ij} F_1(t_{n-1} + c_j\tau, U_1^{(j)}) \quad (1 \leq i \leq s+1) \tag{6.2a}$$

$$U_k^{(i)} = D_k^n [R_{rk}U_r^{n-1} + \tau \sum_{j=1}^{s} a_{ij} F_k(t_{n-1} + c_j\tau, U_k^{(j)})]$$
$$+ (I_k - D_k^n)[P_{k-1k}U_{k-1}^{(i)} + b_k^{(i)}] \quad (1 \leq i \leq s+1), \tag{6.2b}$$

where $k = 2, ... , r$. Hence, $U_k^{(i)} \in S_k$ is the approximation to $u_k(t_{n-1} + c_i\tau)$ at grid ω_k. This formula is similar to the implicit Euler LUGR formula (3.4). In essence there is no difference with the implicit Euler formulation due to the fact that the interpolation step is carried out stagewise. In application the same operations are carried out, the only difference being that the interpolation is done at all stages and a more complicated integration formula is used. However, there exists one small exception for methods having $a_{1j} = 0$, $1 \leq j \leq s$. For such methods, including all explicit ones and the implicit Lobatto IIIA-methods (trapezoidal rule), no integration is carried out at the first stage. Now, due to the fact that the regridding matrix D_k^n depends on the stepnumber n and the level index k, and not on the stage index i, a consequence is that for these methods interpolation is carried out at the first stage, which is redundant since there is no integration at this stage. It is therefore more natural to put $U_k^{(1)} = R_{rk}U_r^{n-1}$, which is what we do. We make no exception for RK methods having this zero coefficient row property at a higher stage, as these are likely to be of little practical interest.

The close correspondence with the implicit Euler formulation is seen very clearly if we rewrite (6.1), (6.2) in the compact Kronecker-product form notation (see, e.g. [8], Section 5.1). For example, if we denote

$$U^n = [U^{(1)T}, ... , U^{(s+1)T}]^T, \quad F(t_n,U^n) = [F^{(1)T}, ... , F^{(s)T}, 0^T]^T, \tag{6.3}$$

where $F^{(i)} = F(t_{n-1} + c_i\tau, U^{(i)})$, and put $a_{is+1} = 0$, then (6.1) then can be rewritten as

$$U^n = e \otimes U^{n-1} + \tau(A \otimes I) \, F(t_n, U^n), \tag{6.4}$$

where $e = [1, \dots, 1]^T$ and A is the coefficient matrix (a_{ij}). This general RK formula can be interpreted as an 'implicit Euler formula' in the augmented vector space. It shall be clear that (6.2) can be rewritten in a similar way, leading to an 'implicit Euler LUGR formula' in the augmented spaces $S_k = S_k^{s+1}$. This interpretation is very helpful for error analysis directed to controlling interpolation errors and maintenance of fine-grid spatial accuracy. Actually, this is one of the principal reasons to write the RK formula in the somewhat unusual form (6.1).

The error analysis we are referring to is given in [18]. There we derive a refinement condition similar to (4.16). Because this condition lives in S_k, it must be elaborated further into a condition in S_k, similar to (4.18), so as to reduce overhead in the actual application. Our paper [18] contains an example of such an elaboration for the strongly A-stable DIRK method

$$\begin{array}{c|ccc}
0 & 0 & 0 & 0 \\
2\theta & \theta & \theta & 0 \\
1 & b_1 & b_2 & \theta \\
\hline
& b_1 & b_2 & \theta
\end{array}
\qquad
\begin{array}{l}
\theta = (3 + \sqrt{3})/6 \\
b_1 = 1.5 - \theta - (4\theta)^{-1} \\
b_2 = -0.5 + (4\theta)^{-1}
\end{array}
\tag{6.5}$$

This method has classical order of consistency 3, stage order 2 [8], and is due to [2,7]. For a linear PDE problem we have proved that this lower stage order causes no order reduction at the internal (Dirichlet) boundaries, provided the accuracy order of the interpolation $\geq$ the order of the PDE operator. Hence, for a parabolic PDE simple linear interpolation is allowed.

We now present results of a numerical example for the DIRK method (6.5). The example is similar as in Section 5 and serves to illustrate the outcome of imposing the regridding condition, i.e., control of interpolation errors and maintenance of fine-grid spatial accuracy (for details see [18]). The equation is again linear and parabolic and given by

$$u_t = u_{xx} + u_{yy} - u_x - u_y + f(x,y,t), \quad 0 < x,y < 1, \quad t > 0. \tag{6.6}$$

Note, however, that the choice of PDE operator is not our main concern here. This is the type of solution to be computed. Following [3], the initial function, the Dirichlet boundary conditions, and the source term f are such that

$$u(x,y,t) = 1 - \tanh(25(x - t) + 5(y - t)). \tag{6.7}$$

This solution is a skew wave, propagating from left to right. The wave starts near the left boundary and arrives at the right boundary at approximately $t = 0.8$. We use the interval $[0, 0.6]$. Symmetric 2-nd order finite-differences are used for the spatial discretization and the interpolation is linear. We give results from 4 integrations using, respectively, 1, 2, 3 and 4 levels. In all 4 cases the base grid is 20 x 20. During an integration τ is fixed, but when a grid level is added, we also halve τ. In view of the 2-nd order in space and the stage order 2 of the DIRK scheme, a gain factor of 4 in the global errors is to be expected. For comparison we also solved the problem on single-fine grids. The stepsizes in time and space are such that the space error dominates the time error, so that we are able to draw valid conclusions on the performance of the LUGR method.

Table 2 shows maxima of global errors restricted to the finest subgrid (examining the composite grid covering the whole of Ω leads to the same conclusions). We see that the LUGR solutions converge according to the theory: the errors decrease with approximately the gain factor 4 for increasing r. We see that the errors are also very close to the single-finest grid errors. Hence, the results are in full accordance with the theory underlying the local refinement. Figure 4 shows grids obtained with r = 2, 3 and 4 at t = 0.3 and t = 0.6. Note that the grids nicely align with the wave front. In particular, this figure illustrates that if τ is decreasing, the local subgrids grow in order to satisfy the regridding condition (see inequality (4.8)). Note that we have used simple linear interpolation.

Table 2. Results for model problem (6.6)-(6.7).

τ	r	single grid	error at t = 0.3	error at t = 0.6
1/10	1	20x20	0.17319	0.17401
1/20	2		0.02728	0.02815
		40x40	0.02789	0.02810
1/40	3		0.00624	0.00716
		80x80	0.00680	0.00684
1/80	4		0.00177	0.00174
		160x160	0.00168	0.00169

6.2. Linear multistep methods

Extending the implicit Euler local refinement analysis to one-step RK methods is complicated by the multi-stage character of these methods. For linear multistep (LM) methods this is much simpler, since there are no intermediate stages. A most popular LM method for time-dependent PDEs is the backward-differentiation (BDF) method. Therefore, to save space, we confine ourselves here to BDF.

Suppressing the level index k in the ODE formula (3.2), we denote the familiar s-step BDF formula by

$$U^n = a_1 U^{n-1} + \dots + a_s U^{n-s} + \tau \theta_s F(t_n, U^n). \tag{6.8}$$

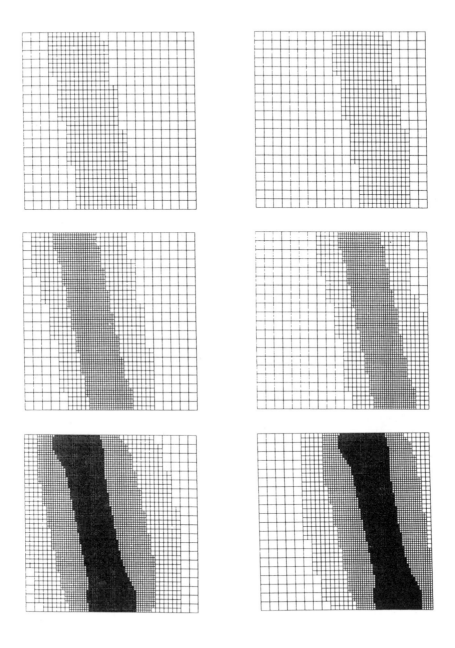

<u>Figure</u> 4. Problem (6.6)-(6.7). Composite grids for r = 2, 3 and 4 at t = 0.3 and t = 0.6.

Due to its close analogy with implicit Euler (s = 1), in the spaces S_k the LUGR formulation for the BDF method immediately follows (see (3.4)):

$$U_1^n = R_{r1} \mathbf{U}_r^{n-1} + \tau\theta_s F_1(t_n, U_1^n), \tag{6.9a}$$

$$U_k^n = D_k^n [R_{rk} \mathbf{U}_r^{n-1} + \tau\theta_s F_k(t_n, U_k^n)] + (I_k - D_k^n) [P_{k-1k} U_{k-1}^n + b_k^n], \quad 2 \le k \le r, \tag{6.9b}$$

where $\mathbf{U}_r^{n-1} = a_1 U_r^{n-1} + ... + a_s U_r^{n-s}$ collects the past solution values at the finest-grid level r. Except for this past-values vector, (6.9) and (3.4) are completely identical, and so is their local refinement analysis. The reason is that we here merely deal with the question of how to balance the local space and interpolation errors committed at the current point of time $t = t_n$. The past values do of course show up in any global error analysis, but not in the local error analysis of which this particular local refinement analysis is part of. In conclusion, for (6.9) the control matrices D_k^n are defined by the same refinement condition as for implicit Euler.

The correspondence with the implicit Euler case relies on the premise of local subgrid expansion leading to formulation in the spaces S_k. In theory we then interpolate over the whole of the grids ω_k which is of course too expensive. As mentioned before, in application we use restricted interpolation (to the current integration domain) and can also justify this (see Section 3). However, for linear multistep methods like BDF, we now encounter the practical difficulty that due to the restricted interpolation not all components of the past-solution vector $\mathbf{U}_r^{n-1}$ required at $t = t_n$ will be available, simply because the integration domains move for evolving time. This difficulty can only be solved for by additional interpolation and injection of missing past solution values, assuming we keep regridding at every base time step. This necessarily involves a so-called flying restart per base time-step. The flying restart possibility has been proposed before by Furzeland [9] (see also Berzins et al. [5]). If we do not regrid at every base time step, then a cold restart is of course also possible. However, delaying regridding has the disadvantage that more points will be needed to resolve moving fine-scale structures. In these respects, LM methods are less amenable for static regridding than the one-step RK methods for which the backward interpolation difficulty does not exist. We note that the approach of delaying the regridding is followed by Bieterman [6].

So far we haven't implemented our LUGR technique in the BDF method, but we plan to do so in the future. The strong potential of the BDF method for the method of lines warrants further investigations.

7. Some remarks on implicitness

In this paper we have focussed on the analysis of the local spatial refinement and refrained from examining efficiency questions. For the actual application one should have sufficient insight into the issues of efficiency and overhead, so as to be able to judge whether the tradeoff between the expected gain in efficiency due to reduction in number of grid points, and loss of efficiency due to extra overhead, is sufficiently large to justify adaptive grids.

In this respect, a subject of major importance is implicitness. When using an explicit time-stepping scheme, the only additional overhead emanates from data structure operations. Because for LUGR methods the data structure is not so complicated, this overhead is readily earned back. As a rule, explicit time-stepping is attractive to combine with LUGR, provided stability renders no problems. In contrast, implicitness leads to overhead costs for solving systems of linear or nonlinear algebraic equations. For LUGR methods implicitness has a still larger impact than for common single-grid methods, because multilevel local refinement results in a new system of linear or nonlinear algebraic equations to solve per level per base time step (we now tacitly assume adaptation at each base time step). Needless to say that the use of direct solvers requiring the Jacobian matrix be given in explicit form then will be expensive. These overhead costs can be reduced by not adapting the grids at each base time step, of course at the expense of accuracy. An attractive alternative may be provided by matrix-free iteration methods. A matrix-free iteration method does not require the Jacobian matrix be given explicitly, as it is based on matrix-vector product operations. Matrix-free iteration methods are, in theory, also directly applicable to nonlinear problems (inexact Newton methods). These methods are accelerated using some form of (problem dependent) preconditioning. Recently quite some research is in progress on using these methods in implicit method-of-lines schemes (see [12-14]). In view of our grid structure, multigrid is of course also a natural approach to combine with our locally uniform, nested grid approach, provided the iteration scheme can be used matrix-free. Our current research code works with the Harwell sparse matrix solver MA28 combined with standard Newton iteration. Although MA28 is well suited to cope with the nonregular band structures we meet, for our application it is rather time consuming for the reasons just discussed.

References

[1] S. Adjerid and J.E. Flaherty (1988): *A local refinement finite element method for two-dimensional parabolic systems*. SIAM J. Sci. Stat. Comput. 9, 792-811.

[2] R. Alt (1973): Thèse de Troisième Cycle, Université de Paris 6.

[3] D.C. Arney and J.E. Flaherty (1989): *An adaptive local mesh refinement method for time-dependent partial differential equations*. Appl. Numer. Math. 5, 257-274.

[4] M.J. Berger and J. Oliger (1984): *Adaptive mesh refinement for hyperbolic partial differential equations*. J. Comput. Phys. 53, 484-512.

[5] M. Berzins, P.M. Dew and R.M. Furzeland (1989): *Developing software for time-dependent problems using the method of lines and differential-algebraic integrators*. Appl. Numer. Math. 5, 375 - 398.

[6] M.B. Bieterman (1983): *A posteriori error estimation and adaptive finite element grids for parabolic equations*. In: Adaptive computational methods for partial differential equations, eds. I. Babuska, J. Chandra and J.E. Flaherty, SIAM Publications, pp. 123 - 143.

[7] M. Crouzeix and P.A. Raviart (1980): *Approximation des problèmes d'evolution. Première partie: étude des méthodes linéaires à pas multiples et des méthodes de Runge-Kutta*. Unpublished Lecture Notes, Université de Rennes, France.

[8] K. Dekker and J.G. Verwer (1984): *Stability of Runge-Kutta methods for stiff nonlinear differential equations*. North-Holland, Amsterdam - New York - Oxford.

[9] R.M. Furzeland (1985): *The construction of adaptive space meshes*. Report TNER.85.022, Shell Research Ltd, Thornton Research Centre, Chester, England.

[10] W.D. Gropp (1987): *Local uniform mesh refinement on vector and parallel processors*.

In: Large Scale Scientific Computing, eds. P. Deuflhard and B. Engquist, Birkhäuser Series Progress in Scientific Computing, Vol. 7, pp. 349-367.

[11] W.D. Gropp (1987): *Local uniform mesh refinement with moving grids*. SIAM J. Sci. Stat. Comput. 8, 292-304.

[12] A.C. Hindmarsh and P.N. Brown (1987): *Reduced storage techniques in the numerical method of lines*. Preprint UCRL - 96261, Lawrence Livermore National Laboratory.

[13] A.C. Hindmarsh and S.P. Nørsett (1988): *KRYSI, an ODE solver combining a semi-implicit Runge-Kutta method and a preconditioned Krylov method*. Report UCID - 21422, Lawrence Livermore National Laboratory.

[14] A.C. Hindmarsh (1989): *Combining the method of lines, stiff integrators, and Krylov methods*. Lecture presented at the SIAM Annual Meeting, San Diego, July 1989.

[15] K. Miller (1986): *Recent results on finite element methods with moving nodes*. In: Accuracy estimates and adaptive refinements in finite element computations, eds. I. Babuska, O.C. Zienkiewicz, J. Gago and E.R. de A. Oliveira, Wiley, pp. 325 - 338.

[16] R.A. Trompert and J.G. Verwer (1989): *A static-regridding method for two-dimensional parabolic partial differential equations*. Report NM-R8923, CWI, Amsterdam (to appear in Appl. Numer. Math.).

[17] R.A. Trompert and J.G. Verwer (1990): *Analysis of the implicit Euler local uniform grid refinement method*. Report NM-R9011, CWI, Amsterdam.

[18] R.A. Trompert and J.G. Verwer (1990): *Runge-Kutta methods and local uniform grid refinement*. Report NM-R9022, CWI, Amsterdam.

[19] J.G. Verwer, J.G. Blom, R.M. Furzeland and P.A. Zegeling (1989): *A moving-grid method for one-dimensional PDEs based on the method of lines*. In: Adaptive methods for partial differential equations, eds. J.E. Flaherty, P.J. Paslow, M.S. Shephard and J.D. Vasilakis, SIAM Publications, pp. 160 - 175.

J.G. Verwer and R.A. Trompert
Center for Mathematics and Computer Science
Department of Numerical Mathematics
Kruislaan 413, 1098 SJ Amsterdam
The Netherlands
email: janv@cwi.nl

Contributed Papers[1]

New Members of the Broyden Family of Formulae.
M. Al–Baali, Ajman University College, P O Box 346, Ajman, United Arab Emirates.

Moving Finite Elements, Approximate Characteristics and a.e. Best Fits with Adjustable Nodes.
M.J. Baines, University of Reading, Mathematics Department, Whiteknights, PO Box 220, Reading, RG6 2AX Berks.

Error Analysis of Update Methods for the Symmetric Eigenvalue Problem .
Jesse L. Barlow , Pennsylvania State University, Computer Science Department, University Park, PA 16802, USA.

Finite Element Approximation of the P–Laplacian.
John W. Barrett and W.B. Liu, Imperial College, Mathematics Department, Queens Gate, London SW7 2BX.

Dynamics of the QR algorithm.
Steve Batterson, Emory University, Department of Math, Atlanta, GA 30322, USA.

Collocation Techniques for Differential/Algebraic Boundary Value Problems.
C.C. Beardah, UMIST, Department of Mathematics, PO Box 88, Manchester M60 1QD.

Transformation of Integrands for Lattice Rules.
Marc Beckers and Ann Haegemans, Katholieke Universiteit Leuven, Department of Computer Science, Celestijnenlaan 200A, B–3001 Leuven, Belgium.

Finite Difference Schemes for Fourth–order Two–point Boundary–value Problems based upon Mixed Interpolation.
G. Vanden Berghe, M. Van Daele and H.E. De Meyer, University of Gent, RUG Laboratorium voor Numerieke Wiskunde en Informatica, Krijgslaan 281–S9, B–9000 Gent, Belgium.

For ODEs, the condition (stability) of the spectral Čebyšev methods is not a problem.
Jean–Paul Berrut, Universite de Fribourg, Department de Mathematiques, Perolles, CH–1700 Fribourg, Switzerland.

Fast Direct and Iterative Solvers for Orthogonal Spline Collocation Equations.
B. Bialecki, G. Fairweather and K. R. Bennett, University of Kentucky, Department of Math., Lexington, Kentucky 40506, USA.

Fitting of Variogram Functions Generated by the Hankel Transform.
J.D. Botha, University of South Africa, Department of Mathematics, P O Box 392, Pretoria 0001, South Africa.

Diagonally Implicit Multi Stage Integration Methods for Ordinary Differential Equations.
J. C. Butcher, University of Auckland, Department of Mathematics and Statistics, Private Bag, Auckland, New Zealand.

---

[1]The addresses given are those of the first named author, the presenter of each paper.

A Range of Numerical Approaches to the Nonlinear Burgers' Equation.

J. Caldwell, City Polytechnic of Hong Kong, Department of Applied Mathematics, 83 Tat Chee Avenue, Kowloon, Hong Kong.

A spectral collocation and path–following method for studying reaction–diffusion equations.

J.M.Cameron, Heriot–Watt University, Department of Mathematics, Edinburgh EH14 4AS.

Symmetrizers for symmetric Runge–Kutta methods.

R.P.K. Chan, University of Toronto, Department of Computer Science, Toronto, Canada M53 1A4.

Numerical solutions to Nekrasov's equation.

G.A. Chandler, University of Queensland, Department of Mathematics, Qld 4072, Australia.

The Numerical Treatment of Wiener–Hopf and Related Integral Equations with an Application in Outdoor Sound Propagation.

Simon Chandler–Wilde, University of Bradford, Department of Civil Engineering, Bradford BD7 1DP.

On a finite element method for 2D Semiconductor Diffusion Process Simulation.

K. Chen, M.J. Baines and P.K. Sweby, University of Reading, Department of Mathematics, Whiteknights, PO Box 220, Reading RG6 2AX.

Recent Developments in the Application of Infinite Elements to Wave Diffraction.

P.J. Clark, Newcastle University, Department of Marine Technology, Armstrong Building, Newcastle Upon Tyne NE1 7RU.

Analysis of a Family of Chebyshev Methods for $y'' = f(x, y)$.

John P. Coleman and Andrew S. Booth, University of Durham, Department of Mathematical Sciences, South Road, Durham DH1 3LE.

Cubpack: A Suite of Codes for Numerical More–dimensional Integration.

Ronald Cools, and Ann Haegemans, Katholieke Universiteit Leuven, Dept Computerwetenschappen, Celestijnenlaan 200 A , B3001 Leuven, Belgium.

Decoupled fixed point and Newton schemes for the nonlinear systems arising in semiconductor modelling.

Robert K. Coomer and Ivan G. Graham, University of Bath, Mathematics Department, Claverton Down, Bath BA2 7AY.

A Modified Numerov Integration Method for General Second Order Initial Value Problems.

M. Van Daele, H. De Meyer and G. Vanden Berghe, RUG Laboratorium voor Numerieke Wiskunde en Informatica, Krijgslaan 281–S9, B–9000 Gent, Belgium.

Numerical solution of uncertain matrix equations with correlated entries, arising from modelling steady–state flow problems through heterogeneous porous structures.

Mark Dainton, University of Reading, Department of Mathematics, Box 220, Reading RG6 2AX.

Instability in the approximation of the electric field integral equation.

P.J. Davies, University of Dundee, Department of Mathematics and Computer Science, Dundee DD1 4HN.

A New Finite–difference Approach to Convection–dominated Flows with Application to Acoustics and Thermo–convection.

Sanford Davis, NASA Ames Research Center, Moffett Field, California 94035, USA.

On resolvent conditions related to stability questions in the numerical solution of partial differential equations.

J. L. M. van Dorsselaer, University of Leiden, Niels Bohrweg 1, 2333 CA Leiden, The Netherlands.

Block Lanczos techniques for accelerating the block Cimmino method.

I.S. Duff, M. Arioli, D. Ruiz and M. Sadkane, Rutherford Appleton Laboratory, Numerical Analysis Group, Central Computing Department, Atlas Centre, Didcot, Oxon OX11 0QX

Optimal Parallel Finite Difference Algorithms.

D.B. Duncan, Heriot Watt University, Mathematics Department, Riccarton, Edinburgh, EH14 4AS.

Algorithms for L_p approximation.

Hakan Ekblom, Lulea University, Department of Mathematics, S–951 87 Lulea, Sweden.

DQAINT: An Algorithm for Adaptive Quadrature (Vector Function) over a Collection of Finite Intervals.

Terje O. Espelid, University of Bergen, Department of Informatics, Thormohlensgate 55, N–5008 Bergen, Norway.

A finite difference procedure for a class of free boundary value problems.

B. Fornberg, Exxon Research, Clinton Township, Route 22 East, Annandale, New Jersey, 08801, USA.

Parallel Optimization Techniques for Optimal Heat Integrated Separation Sequence Generation.

E. S. Fraga and K. I. M. McKinnon, University of Edinburgh, Department of Mathematics, Kings Buildings, Edinburgh EH9 3JZ.

A Boundary Integral Equation Method for the Numerical Solution of a Nonlinear Boundary Value Problem.

M. Ganesh, University of Bath, School of Mathematical Sciences, Bath BA2 7AY.

A supraconvergent finite–volume scheme for the convection–diffusion equation.

B. García–Archilla and J. MacKenzie, Universidad de Valladolid, Departmento Matemática Aplicada y Computación, Facultad de Ciencias, Valladolid, Spain.

The numerical detection of hopf bifurcations in large systems arising from discretisations of PDEs.

T.J. Garratt, A. Cliffe and A. Spence, University of Bath, School of Mathematics, Claverton Down, Bath BA7 7AY.

Numerical Computation of Multivariate Normal Probabilities.

Alan Genz, Washington State University, School of Electrical Engineering, Pullman, Washington 99164–2752, USA.

Computation of Takens–Bogdanov Points by a Direct Characterization.

W. Govaerts, E. Chu and A. Spence, Rijksuniversiteit Gent, Seminarie voor Hogere Analyse, Krijkslaan 281, B–9000 Gent, Belgium.

Analysis of numerical methods for Nekrasov's equation.
I.G. Graham, University of Bath, School of Mathematics, Claverton Down, Bath BA2 7AY.

Extrapolation of a Sequence of Functions.
P.R. Graves–Morris, University of Bradford, Bradford, West Yorkshire BD7 1DP.

Approximate Methods for Volterra–Fredholm Integral Equations.
Lechoslaw Hacia, Technical University of Poznan, Institute of Mathematics, Piotrowo 3a, 60–965 Poznan, Poland.

Applications of Preconditioned CG· to Singular or Ill–Posed Linear Equations.
Martin Hanke, Universitaet Karlsruhe, Institut fuer Praktische Mathematik, Englerstr.2, D–7500 Karlsruhe, Germany.

Making waves II: a numerical continuation method for finding periodic solutions in a class of differential difference equations.
P.S. Hansen and D.F. Griffiths, Technical University of Denmark, Department of Graphical Communication, b116, DK 2800 Lyngby, Denmark.

Backward Error and Condition of Structured Linear Systems.
Desmond J. Higham and Nicholas J. Higham, University of Dundee, Department of Mathematics and Computer Science, Dundee DD1 4HN, Scotland, UK.

Optimization by Direct Search in Matrix Computations.
Nicholas J. Higham, University of Manchester, Department of Mathematics, Manchester M13 9PL.

Introduction to CONFPACK, a Numerical Conformal Mapping Package.
David M. Hough, Coventry Polytechnic, Department of Mathematics, Coventry CV1 5FB.

Parallel Diagonal–implicit Iterated Runge–Kutta methods.
P.J. van der Houwen, CWI, Amsterdam, P O Box 4079, 1009 AB Amsterdam, The Netherlands.

Spurious Solutions of Numerical Methods for initial value problems.
A.R. Humphries, University of Bath, School of Math Sciences, Bath BA2 7AY.

Parallel Optimization Using Variants of the Conjugate Gradient Algorithm to Solve a Finite Element Problem in Non–Linear Elasticity.
P.K. Jimack, University of Leeds, School of Computer Studies, Leeds.

Iterated Defect Correction Methods for Nonlinear Differential Algebraic Equation of Index One.
Bülent Karasözen, Middle East Technical University, Department of Mathematics, 06531 Ankara, Turkey.

Some generalizations of wavelets: an algebraic approach.
Jaroslav Kautsky, Flinders University, School of Information Science, GPO Box 2100, Adelaide SA 5001, Australia.

A Parallel L–stable Splitting Method for Multidimensional Parabolic Equations.
A.Q.M. Khaliq, and D.A. Voss, Western Illinois University, Department of Mathematics, Macomb, Illinois 61455, USA.

Fast Rectangular Matrix Multiplication and QR Decomposition.
P.A. Knight, University of Manchester, Department of Mathematics, Manchester M13 9PL.

On the Numerical Solution of the Heat Equation.
J.F.B.M. Kraaijavanger, University of Leiden, Institute of Applied Mathematics, PO Box 9512, 2300 RA Leiden, The Netherlands.

Derivatives of Eigenvalues and Eigenvectors of Matrix Functions.
Peter Lancaster, Alan L. Andrew and K.-W. Eric Chu, University of Calgary, Department of Mathematics, Calgary T2N 1N4, Alberta, Canada.

Stratified sequences of nested rules for fully adaptive quadrature.
Dirk P. Laurie, Potchefstroom University, Department of Mathematics, P O Box 1174, Vanderbijlpark 1900, South Africa.

Solving Mixed Integer Nonlinear Programs by Quadratic Outer Approximation
S. Leyffer and R. Fletcher, University of Dundee, Department of Mathematics and Computer Science, Dundee DD1 4HN, Scotland, UK.

A Petrov–Galerkin Formulation for Metal Forming Processes.
Xin Kai Li, University of Leicester, Department of Engineering, Leicester LE1 7RH.

A Method for creating practical parallel algorithms for large hyperbolic problems.
Mark Lynch, Heriot–Watt University, Department of Mathematics, Riccarton, Edinburgh, EH14 4AS.

A dispersion correction method for FEM models of vibration.
R.I. Mackie, University of Dundee, Department of Civil Engineering, Dundee DD1 4HN.

Data Dependent Triangular Grid Generation.
Andrew Malcolm, University of Reading, Department of Mathematics, Whiteknights, PO Box 220, Reading RG6 2AX.

The Polar Decomposition.
Roy Mathias, College of William and Mary, Department of Mathematics, Williamsburg, VA 23185, USA.

Geometric Methods for Computing Invariant Manifolds.
G. Moore, Imperial College, Department of Mathematics, Huxley Building, Queen's Gate, London SW7 2BZ.

Order Barriers and Characterizations of Continuous Mono–Implicit Runge–Kutta Schemes.
Paul Muir and Bryn Owren, University of Toronto, Department Computer Science, Toronto, Ontario, Canada M5S 1A4.

QMR: A Quasi–Minimal Residual Method for Non–Hermitian Linear Systems.
Noel M. Nachtigal and Roland W. Freund, MIT, Department of Mathematics, Cambridge, Massachussetts 02139, USA.

Numerical computation of an analytic singular value decompostition of a matrix valued function.
N.K. Nichols, A. Bunse–Gerstner, R. Byers and V. Mehrmann, Reading University, Department of Mathematics, Whiteknights, PO Box 220, Reading, RG6 2AX.

Chebyshev pseudo–spectral methods for the KdV equation.
F.N. Nouri, University of Annaba, cité des 88 logts., Bloc 2 Appts 12 Bd d'Afrique, Annaba 23000, Algeria.

Non–negative Area Matching "Histosplines", a Local Approach.

Gerhard Opfer, University of Hamburg, Inst für Angewandte Math, Bundesstr. 55, D 2000 Hamburg 13, Germany.

Rosenbrock Methods for Partial Differential Equations.

A. Ostermann, Universitaet Innsbruck, Institut für Mathematik, Technikerstrasse 13, A–6020 Innsbruck, Austria.

Using divergent iterations in high accuracy, high speed computations, part 2.

P.W. Pedersen, Technical University of Denmark, Department of Mathematics, Bg 303, DK 2800 Lyngby, Denmark.

The numerical solution of a boundary integral formulation for the two–dimensional Helmholtz equation in a perturbed half–plane..

Andrew Peplow and Simon Chandler–Wilde, University of Bradford, Department of Civil Engineering, Bradford BD7 1DP.

Construction of Smooth Surfaces by Piecewise Tensor Product Polynomials.

A.R.M. Piah, University of Dundee, Department of Mathematics and Computer Science, Dundee DD1 4HN.

On Preconditioning for Finite Element Equations on Irregular Grids.

Alison Ramage and Andrew J. Wathen, University of Bristol, School of Mathematics, University Walk, Bristol BS8 1TW.

Parallel Power–of–Two FFTs on Hypercubes.

R. Renaut, Arizona State University, Department of Mathematics, Tempe, AZ 85287 1804, USA.

On the Numerical Inversion of the Laplace Transform in Reproducing Kernel Hilbert Spaces.

G. Rodriguez and S. Seatzu, Universita di Cagliari, Dipartimento di Matematica, Viale Merello, 92, 09123 Cagliari, Italy.

Moving Mesh Techniques for Solving One–Dimensional PDEs.

R.D. Russell, and Y. Ren, Simon Fraser University, Department of Mathematics and Statistics, Burnaby, B.C., Canada V5A 1S6.

The effect of high order nonlinear diffusion in a discrete form of Fisher's equation.

S.W. Schoombie, University of the Orange Free State, Department of Applied Mathematics, P O Box 339, Bloemfontein 9300, South Africa.

The Sticky Syrup Problem.

P.W. Sharp, Queen's University, Department of Mathematics and Stats, Kingston, Canada.

On the choice of basis functions in the finite element solution of the shallow water wave equations.

S.T. Sigurdsson, University of Iceland, Science Institute, Dunhaga 3, IS–107 Reykjavik, Iceland.

Fast Iterative Techniques for Finite Element Solution of the Stokes Problem.

David Silvester and Andy Wathen, UMIST, Department of Mathematics, PO Box 88, Manchester M60 1QD.

Computation of Inertial Manifolds.

D.M. Sloan, R.D. Russell and M.R. Trummer, University of Strathclyde, Department of Mathematics, 26 Richmond Street, Glasgow, G1 1XH.

On a Conjecture by LeVeque and Trefethen Related to the Kreiss Matrix Theorem.
M.N. Spijker, University of Leiden, Department of Mathematics and Computer Science, Niels Bohrweg 1, 2333 CA Leiden, The Netherlands.

Numerical wave propagation in an advection equation with a nonlinear source term.
Andrew Stuart, D.F. Griffiths and H.C. Yee, University of Bath, School of Mathematical Sciences, Claverton Down, Bath BA2 7AY.

Analysis of the Cell Vertex Finite Volume Method.
M. Stynes and K.W. Morton, University College Cork, Department of Mathematics, Cork, Ireland.

Extraction of Car Resistance Coefficients from Costdown Data.
Adrian Swift, Massey University, Department of Mathematics and Statistics, Palmerston North, New Zealand.

Numerical Solution of Differential/Algebraic Boundary Value Problems with Application to Detonation Modelling.
R.M. Thomas, UMIST, Department of Mathematics, PO Box 88, Manchester M60 1QD.

Estimation of the Blow–up Parameters for a Class of Ordinary Differential Equations.
Yves Tourigny and Michael Grinfeld, University of Bristol, Department of Mathematics, University Walk, Bristol, BS8 1TW.

The Structure of Jacobians in Spectral Methods for Nonlinear PDEs.
M. R. Trummer, Simon Fraser University, Department of Mathematics and Stats, Burnaby, B.C., Canada V5A 1S6.

On non–parametric constrained interpolation.
K. Unsworth and B.H. Ong, University of Dundee, Department of Mathematics and Computer Science, Dundee DD1 4HN, Scotland.

Some New Families of Explicit Runge–Kutta Methods.
J.H. Verner and P.W. Sharp, Queen's University, Department of Mathematics, Kingston, Ontario, Canada K7L 3N6.

Numerical Simulation of the Becker Döring Equations.
C.H. Walshaw, Heriot–Watt University, Department of Mathematics, Edinburgh, EH14 4AS.

The eigenvalue spectra of banded Toeplitz and quasi–Toeplitz matrices.
R.F. Warming and R.M. Beam, NASA Ames Research Center, N202A–1, Computational Fluids Branch, Moffett Field, California 94035, USA.

Variational Bounds on the Entries of the Inverse of a Matrix.
A.J. Wathen, and P.D. Robinson, University of Bristol, School of Mathematics, University Walk, Bristol BS8 1TW.

Minimizing a smooth function plus the sum of the k largest eigenvalues of a symmetric matrix.
G.A. Watson, University of Dundee, Department of Mathematics and Computer Science, Dundee DD1 4HN, Scotland.

Superconvergence of Recovered Gradients of Finite Element Approximations to Elliptic and Parabolic Problems.
J.R. Whiteman, Brunel University, Institute for Computational Mathematics, Kingston Lane, Uxbridge, Middlesex UB8 3PM.

Best Chebyshev Approximation from Families of Ordinary Differential Equations.
Jack Williams and Z. Kalogiratou, University of Manchester, Department of Mathematics, Manchester, M13 9PL.

Parallel Algorithms for QR Decomposition on a Shared Memory Multiprocessor.
K. Wright, University of Newcastle, University Computing Laboratory, Claremont Tower, Claremont Road, Newcastle upon Tyne NE1 7RU.

Tidal Barrage Optimal Control.
Z.G. Xu and N.K. Nichols, University of Reading, Department of Mathematics, Whiteknights PO Box 220, Reading RG6 2AX.

Parallel orthomin for general sparse matrices.
Zahari Zlatev, Danish Environmental Research Institute, Frederiksborgvej 399, DK–4000 Roskilde, Denmark.